MATERIALS SCIENCE AND TECHNOLOGIES

ANISOTROPY RESEARCH

NEW DEVELOPMENTS

MATERIALS SCIENCE AND TECHNOLOGIES

Additional books in this series can be found on Nova's website
under the Series tab.

Additional e-books in this series can be found on Nova's website
under the e-book tab.

MATERIALS SCIENCE AND TECHNOLOGIES

ANISOTROPY RESEARCH

NEW DEVELOPMENTS

HIRPA LEMU
EDITOR

nova publishers
New York

Library of Congress Cataloging-in-Publication Data

Anisotropy research: new developments / [edited by] Hirpa Lemu.
p. cm.
Includes bibliographical references and index.
ISBN 978-1-62081-977-7 (hardcover)
1. Anisotropy. I. Lemu, Hirpa.
QD921.A663 2011
530--dc23

2012018141

Published by Nova Science Publishers, Inc. † New York

CONTENTS

PREFACE

This book contains collection of recent research developments in study of anisotropy. Anisotropy behavior, i.e. being directionally dependent, is exhibited in diverse fields and thus the research approaches accordingly varies. This book has attempted to collect research approaches and corresponding research results from different disciplines with particular focus on material property variations and how recent developments are handled in those disciplines. The research in anisotropy behavior of materials is vital in, among others, many engineering disciplines, computer graphics, geology and geophysics, medicine and neurosciences and natural sciences.

The material in the book is organized in 11 chapters. Chapter 1 presents the research on material anisotropic behavior in metal forming process. This is today an interesting area of research not only because many mechanical forming processes such as casting, forging and rolling processes introduce some form of directional dependence of the mechanical or physical properties, but also that computer simulation based approach is one of the modern research tools. In this study, numerical investigation of anisotropic properties of sheet metal forming using finite element method in ABAQUS is executed and the simulation results are compared with experimental results. The research is intended to develop understanding of how plastic anisotropy in ductile sheet metal influences the deformation behavior of the material. Accordingly, a comparison of quadratic Hill anisotropic yield criterion and von Mises yield criterion with isotropic hardening has been done.

Composite materials play a significant role as modern engineering materials due to their attractive properties. Ferroelectrics represent one of such modern composite materials that convert mechanical energy into electric energy and vice versa. Within the last decades these materials have been studied and developed in an attempt to improve their electromechanical performance. The research reported in Chapter 2 has focused on the large anisotropy of piezoelectric coefficients and the variable anisotropy of electromechanical coupling factors that characterize the relaxor-ferroelectric single crystal/polymer and ferroelectric ceramic/polymer composites.

Chapter 3 discusses the research on pore anisotropy in porous solids. This research attempts to develop better understanding of pore structure of solid materials by describing the anisotropic behavior in terms of morphological parameters such as the surface area, the pore volume, the pore size distribution as well as pore connectivity and pore tortuosity. As the pores in a real solid are complicated involving the pore diameter and length, the researchers developed the concept of pore anisotropy and a methodology for its estimation. The geometry is a crucial parameter for the industry of catalysts and absorbents as well as in separation and crude oil recovery technologies since this can greatly affect the diffusion and mass transport processes within the porous structures.

The study of the experimental evidence on exhibition of the anisotropic behavior in Vendian and Cambrian clay formations regarding radionuclide diffusion is presented in Chapter 4. In this research, the coupled diffusion and sorption processes in the clay sediments are studied using different experimental set-ups. Experiments are executed in order to investigate if clays exhibit diffusion anisotropy when the longitudinal apparent diffusion coefficient is appreciably larger than the transverse coefficient. The research has also attempted to compare the obtained results with other published results with clay samples taken from clay formations considered as potential host rocks for the underground repository for radioactive disposal in Western Europe.

Medical and neuroscience fields are one of the major areas where anisotropy behavior is of interest. This phenomenon is observed in diffusion tensor imaging in MRI technique that involves measuring the fractional anisotropy of the random motion of water molecules in the brain. The difference in fractional anisotropy in the brain has led researchers to study the white matter anisotropy and theory of mind deficit in patients at ultra-high-risk of psychosis as reported in Chapter 5. Using a recently developed technique called tract-based spatial statistics; each subject's diffusion tensor imaging is mapped onto a common white matter tract center, i.e. the skeleton. This research focuses on the investigation of deficits in the above-mentioned patient category and the association with white matter microstructure and connectivity abnormalities.

Chapter 6 presents a research approach on study of angle and polarization anisotropy of phonons and phonon-plasmon modes in GaAs/AlAs superlattices based on the Raman spectroscopy technique. The technique has been applied to study phonons and phonon-plasmon coupled modes in undoped and Si doped set of GaAs/AlAs superlattices grown on (100), (311)A and (311)B substrates. In a related topic, Chapter 7 discusses the study of fluorescence anisotropy based on the principle of photoselective excitation of a fluorophore by a polarized light. The research is intended to provide information about rotational motions of fluorophores or fluorescently labeled molecules between photon absorption and emission. The researchers have developed a fluorescence assay that can be used to study simultaneously the DNA-binding step and the subsequent catalytic reaction in the same assay.

Research on surface magnetic anisotropy in magnetic substances induced by reversible absorption is discussed in Chapter 8. Further, the chapter reviews recent studies on the influence of molecules adsorbed through the mechanism of the formation of hydrogen bonds on magnetic properties of ferromagnetic alloys and ferrodielectrics. This study is intended to explore if significant anisotropic behavior is observed when the domain structures of amorphous magnetically soft iron-based ferromagnetic ribbons exist in the atmosphere of water vapor and methyl alcohol. In addition to identifying existence of such influence, the study has attempted to see that this influence can considerably change the magnetic properties of the substance.

Chapter 9 presents the research results on the effect of layered structures on realization of anisotropy in thermoelasticity. The field dependences of thermoelectromotive coefficient are calculated and analysed for the different scattering mechanisms in the case of both quntising magnetic field and temperature gradient parallelism to superlattice axis of crystal. In this study, models of constant lifetime and constant free path for electrons scattering by charged impurities as well as models of lifetime proportional to total and longitudinal electron velocity are used in order to see if the effects manifest themselves in reduction of thermoelectromotive coefficient value in the region of the quantum oscillations and phase retardation of quantum oscillations.

Though over twenty-five years have been elapsed since the cuprate superconductor was discovered, the anisotropic behavior of the pairing symmetry still remains a major topic of debate due to its highly anisotropic nature. This nature along with the presence of inhomogeneities has made the problem highly controversial and ambiguous. Apart from the unconventional state of high temperature superconductors, the order parameter can have various anisotropies as seen in several theoretical models and experimental measurements. Chapter 10 presents a short review on anisotropic nature of the pairing state and discusses the possible nature of the pairing and its effect on penetration depth. In addition to demonstrating using numerous experiments, the authors of the chapter have dedicated their effort to show to the readers how the structure of the cuprates along with various pairing symmetries could explain the anisotropic nature of the penetration depth from a theoretical point of view.

The last chapter, Chapter 11, reports recent research developments on magnetically active anisotropic fluids based on liquid crystals. These fluids are commonly referred to as ferronematics that are manifestation of the idea of Brochard and de Gennes who suggested that doping liquid crystals by fine magnetic particles may enhance their sensitivity to magnetic fields. The research reported in this chapter introduces the basic field-induced reorientational effect of liquid crystals that serve as a basis for most experiments on ferronematics. The experiments give emphasis on how the behavior of ferronematics depends on the host liquid crystal, type, shape, size and concentration of the magnetic particles.

In general, the book contains new and important research developments in the field anisotropy behaviour. It is, therefore, expected to give significant insight into existing research approaches and show possible future research perspectives in those disciplines where anisotropy behavior is involved.

Finally, I would like to thank all contributors whose accumulated knowledge and experience is reflected in this scientific collection and who directly or indirectly helped me to make this book a reality.

Hirpa G. Lemu
Stavanger, Norway

In: Anisotropy Research
Editor: Hirpa Lemu

ISBN 978-1-62081-977-7

Chapter 1

STUDY OF MATERIAL ANISOTROPY IN METAL FORMING USING FINITE ELEMENT METHODS

***Tomasz Trzepieciński*[1] *and Hirpa G. Lemu*[2,*]**
[1]Department of Materials Forming and Processing, Rzeszow University of Technology, Rzeszów, Poland
[2]Department of Mechanical and Structural Engineering and Materials Science, University of Stavanger, Stavanger, Norway

ABSTRACT

The chapter discusses the numerical investigation of anisotropic properties of sheet metal forming using finite element method (FEM) in ABAQUS, a computer-aided engineering tool that deals with the most influent parameters in a forming process such as simulating sheet metal forming.

Stamping of rectangular parts was simulated by the static and explicit approach in the presence of contact conditions with friction. The experimental and numerical results of rectangular cup drawing are presented. The aim of experimental study is to analyze the material behavior under deformation and further use the results to verify numerical simulation results taking into consideration friction and material anisotropy. Implicit and explicit nonlinear analyses were carried out using ABAQUS/Standard and ABAQUS/Explicit.

Frictional properties of the deep drawing quality steel sheet were determined by using the pin-on-disc tribometer. A comparison of quadratic Hill anisotropic yield criterion and von Mises yield criterion with isotropic hardening has been done. It is found that plastic anisotropy in ductile sheet metal has influence on deformation behavior of the material.

The study indicates that FEM analysis undoubtedly guarantees the most approximate numerical results to real process when material and friction anisotropy are taken into consideration.

[*] Corresponding author: Hirpa@ux.uis.no

INTRODUCTION

Sheet metal forming is an important manufacturing technology in many forming processes, and it allows the production of thin walled parts with complicated shapes. As conventional design and analysis processes for sheet metal forming are usually based on empirical approaches, computer based simulation is today particularly attractive due to high cost of materials and dies in trial-and-error based procedures. Among others, design and analysis of drawing operations requires in-depth knowledge of the deformation mechanisms. This is because metal drawing processes are characterized by *nonlinearity* both in geometry, material behavior and nature of contact or friction. Previous research works show that analytical analysis of the deformation mechanism of rectangular cup drawing is very complicated [1] due to the strong non-linear nature of the procedure. Cup drawing has been studied using both theoretical and numerical analysis approaches. The study by Daxin *et al.* [2] on blank thickness changes indicates that the maximum strain is observed near the edge of blank, while the minimum strain is at the punch shoulder. Further studies by Wen and Daxin [3] that focused on comparison of experimental and simulation results using *Finite Element Methods* (FEM) concluded that there is large discrepancy between the two approaches.

Numerical simulation using FEM is currently a preferred and widely applied approach because it gives better flexibility to control and simulate large number of variables at a lower cost. Among others, FEM simulation in sheet metal forming allows predicting the forming defects and provides knowledge of deformed shape, stress and strain distribution and punch loading. On the other hand, the accuracy of FEM simulation depends on several parameters including:

- type of finite element and its mesh size,
- type of constitutive equations used,
- description of material behavior and
- friction phenomenon.

Among the available element types in finite element analysis (FEA) tools, shell elements are often implemented in analyzing drawing problems because they have advantages of obtaining deformed shapes and contact conditions with moderate computing time. On the other hand, these element types are unable to consider the thickness reduction in the cup wall when the clearance between the punch and the die is smaller than the sheet thickness [3]. This is because shell elements cannot offer correct solutions in the compression loading state along the thickness direction.

One of the most important elements of numerical simulation is the *constitutive equation* that describes the flow stress as a function of the deformation. As briefly reviewed in [4], the correctness of the material model depends on the mathematical structure of the model and the proper experimental determination of the material properties used in the model.

Plastic anisotropy is another important material behavior that highly influences metal forming. It is the result of the distortion of the yield surface shape due to the material micro structural state and can be classified into two types [5]: normal and planar anisotropy. In *normal anisotropy* the properties differ in the thickness direction; in *planar anisotropy*, on the other hand, the properties vary with the orientation in the plane of the sheet. Whereas drawability of sheets increases with normal anisotropy, planar anisotropy leads to the formation of ears in cup drawing [6]. Flange earing is a classical defect of anisotropy in plastic deformation of sheet metals. However, after cold forming, the crystallographic directions of a metal gradually rotate towards a common axis thus creating anisotropy in the metal. Tests reveal that metals that are initially isotropic may become anisotropic after extensive plastic deformation [7].

When a metal is subjected to further forming without annealing; the hardening relationship and the associated flow rule need to be based on an *anisotropic yield criterion*. Thus, defining the anisotropic yield criteria is one central issue in analysis of sheet metal cup drawing. Various approaches have emerged for developing this criterion. Earlier proposals of anisotropic yield criteria by Hill [8, 9] and Hosford [10] could not completely represent the general state of anisotropy, even in plane stress conditions. Hill has also proposed a new and user-friendly yield criterion in which the material properties were treated as independent parameters [11]. Hill's criterion is in general simple to implement, but possesses certain anomalies.

The modified version of Hill's yield criterion is reported to be free of the above-mentioned defects but does not contain the shear stress term. Barlat and Lian [12] applied a linear transformation to modified Hosford [10] criterion to develop an anisotropic yield function for the plane stress conditions. The technique has been further extended to the three-dimensional state of stress [13, 14], though the anisotropic yield functions, as implemented on aluminum sheets, were not able to capture the full anisotropy of the material. Generally there are several quadratic and non-quadratic yield criteria that may be associated with isotropic hardening for industrial applications. These criteria can also be combined with kinematics or more advanced anisotropic hardening [15 - 17]. Other recent studies [18, 19] indicate that Hill 1948 [8] yield criterion and its modifications are the most often used for anisotropic model description of steel sheets. Zhying and Xianghuai [20] used continuum theory of rupture developed by Gurson-Tvergaard-Needleman [21] based on Hill 1948 [8] yield criterion to predict the damage evolution of anisotropic plastic voided ductile materials. Following the development in this area and based on the same criterion, other approximate yield criterion for anisotropic porous sheet metals under plane stress condition has been developed [22]. Further, the flange earing of strong anisotropic sheet metals is analyzed by implementation of Barlat-Lian anisotropic [12] yield criterion and quasi-flow corner theory in elastic-plastic large deformation finite element formulation [23]. In those studies, special emphasis was put on numerical analysis and the constraining scheme of flange earing of circular sheets in square cup drawing process. The Barlat-Lian criterion is a non-quadratic planar anisotropy yield function that describes the yield behavior of anisotropic metal sheet with strong texture property. The study by Habraken and Duchene [24] presents reviews of state of the art numerical models in sheet forming analysis dealing with mechanical anisotropy resulting from crystallographic texture.

The frictional behavior depends on several parameters including contact pressure, sliding velocity, tools and sheet surface roughness and lubricant conditions. Many experimental studies [25] show that the frictional behavior can change drastically with the sliding direction and requires an anisotropic modeling. The origin of this anisotropy can be attributed to two different sources. The first one is the material itself where the material anisotropies manifest themselves on the contact surface. The second one is technological. Currently, there are not so many publications focusing on frictional anisotropy and its implementation in numerical simulations of sheet metal forming.

Numerical analysis using FEM involves, among others, stress integration procedure that can be executed using either the explicit or implicit method. In the *explicit method,* the yield function and the plastic parameters are evaluated at known stress states, and it is not necessary to use an iterative procedure. In the *implicit method*, however, the current stresses are unknown, and the system of non-linear equations is generally solved by *iterative methods* [26]. To overcome convergence problems, an implicit model requires very small time increments. As a result, longer CPU times would be expected.

Research results from the past two decades show that explicit method is computationally efficient at solving large plastic deformations and contact portions of sheet metal forming. Comparative studies [27 -30] of the two methods indicate that the explicit method has an advantage of having greater ease with which complicated contact problems are solved. The reported works on comparison of explicit and implicit methods focused on both quasi-static non-linear problems [31] and linear dynamic problems [32]. These works indicate that analysis using the explicit method is ideally suited to high-speed dynamic events in sheet metal forming.

While the explicit method is more effective to analyze complex cases with large and more refined meshes [33], the implicit method is widely used because of its excellent accuracy and reliability. This can be one reason that earlier simulations of sheet metal forming mostly focused on the implicit method. Particularly, the implicit method is known to be more accurate and economical for 2D quasi-static problems with simple contact formulation requiring not so large computation time.

As a closer study of the literature shows, application of explicit and implicit methods to simulation of sheet metal forming has attracted many researchers in the past. As a result, different simulation codes such as DYNA-3D [34], ABAQUS [28, 30, 32] and several other tailor-made algorithms [35] that are embedded into the commercial tools have been developed and implemented. Due to its importance and the involved complexities, however, research in this area is far from being sufficient. The aim of this chapter is to contribute to this huge research challenge.

2. Material Model

The sheet metal drawing process presented in this chapter was done using a form of deep drawing quality (DDQ) cold-rolled steel sheet with a sheet thickness of 1 mm. The mechanical properties of the sheet metal were determined through tensile tests along three directions with respect to the rolling direction. The elastic behavior was specified in numerical simulations by the value of Young's modulus, E = 210000 MPa, and of Poisson's ratio ν = 0.3. The *isotropic hardening* behavior implemented in the FEM model uses the *Holomon power-type law* expressed as:

$$\sigma = C\varepsilon^{n} \tag{1}$$

where C is material constant and n is the *strain-hardening exponent*.

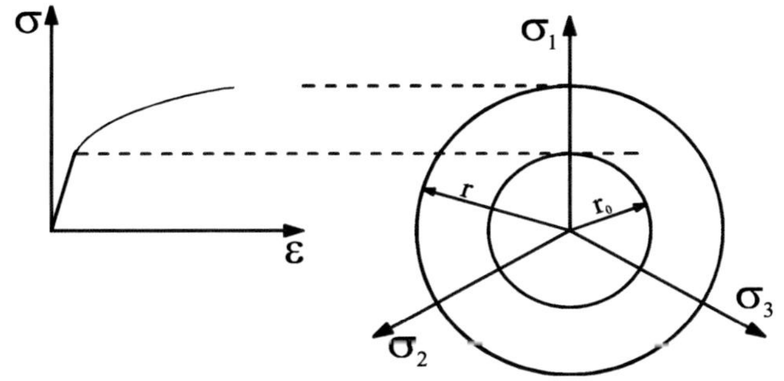

Figure 1. Isotropic hardening model.

The parameters in this Holomon equation, Eq. (1), have been fitted on stress-strain curve of the tensile test. According to continuum theory of plasticity, Bauschinger effect may be neglected in isotropic hardening models. It is also assumed that the subsequent yield surface is a uniform expansion of the initial yield surface (Figure 1) and hence the isotropic response of the material to yielding does not change during plastic deformation.

Different material properties working in different directions can have a significant effect on the degree of difficulty of the forming operation. In particular, orientations and textures of the crystal structure to the rolling direction of the sheet metal lead to anisotropic directional behavior.

The anisotropy of plastic behavior of sheet metals is characterized by the *Lankford's coefficient r* that is determined by uniaxial tensile tests [36]. The value of this coefficient is defined by the ratio:

$$r = \frac{\varepsilon_2}{\varepsilon_3} \tag{2}$$

where ε_2 and ε_3 are the true strains in width and thickness direction of the specimen respectively.

Due to the fact that measuring the longitudinal elongation is easier to perform, the condition of volume constancy $\varepsilon_1 + \varepsilon_2 + \varepsilon_3 = 0$ was assumed. A measured r-value that differs from unity shows that there is a difference between the mechanical properties measured in plane and through-thickness. This difference is usually characterized by the normal *plastic anisotropy ratio* defined as:

$$\bar{r} = \frac{r_0 + 2r_{45} + r_{90}}{4} \tag{3}$$

where r_0 is the strain ratio in the longitudinal direction, r_{45} is the strain ratio measured 45° to the rolling direction and r_{90} is the strain ratio in the transverse direction.

For isotropic materials $\bar{r}$ is equal to 1. The value $\bar{r} > 1$ is obtained when the strength in the thickness direction is greater than the average strength in the directions lying in the plane of the sheet. The higher the $\bar{r}$ value, the deeper the draw can be achieved [37]. The mechanical parameters of DDQ steel sheet are presented in Table 1 below.

Table 1. Mechanical properties of the sheet metal

β	σ_Y [MPa]	σ_U [MPa]	σ_Y/σ_U	C [MPa]	n	r
0°	162	310	0.523	554	0.21	1.55
45	163	322	0.506	542	0.20	1.27
90°	168	312	0.538	530	0.21	1.67

where β = Orientation, σ_Y = Yield stress, σ_U = Ultimate tensile stress, σ_Y/σ_U = Yield ratio.

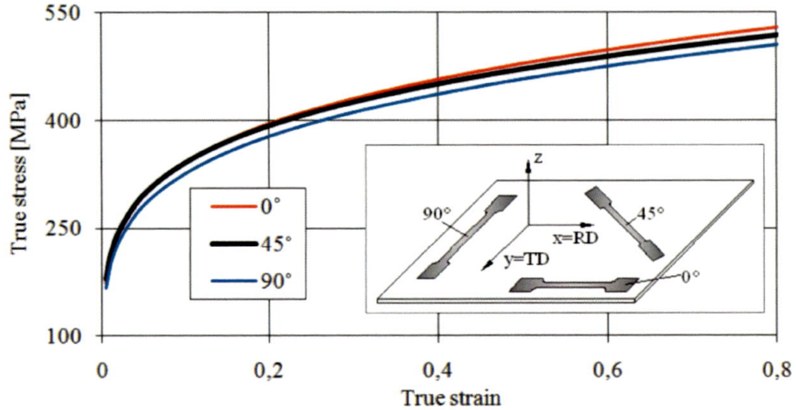

Figure 2. True stress-strain curves for DDQ steel sheet.

As the study by Yi *et al.* [37] asserts, the variation of the r value in different loading directions has a strong relationship with the texture. In general, the variation of yield stress and ultimate tensile stress shown in Table 1 are very small. Further, the values in the table show that the yield stress measured in the rolling direction is lower than those values in the transverse direction, and the higher calculated material constant C occurs for rolling direction and tends to decrease from the rolling to the transverse direction. The values in the table also indicate that the strain-hardening exponent (n) is closely related with the yield ratio that is often given as a percentage share.

The stress-strain curves of the DDQ steel sheet tested in the three selected loading directions are shown in Figure 2. The sheet metal exhibits in-plane anisotropy in the yield stress and the r value, while the hardening exponent value is not significantly affected by the sample orientation. The r value in the rolling direction is smaller than the measured value in the transverse direction because it is inversely proportional to the thickness strain.

The material behavior in the numerical simulations was described by *von Mises yield criterion* with isotropic hardening **and Hill's (1948) anisotropic yield condition.** Hill's formulation is the most frequently used yield function for steel sheet metals and it can be regarded as an extension of the isotropic von Mises function.

For ideal case of isotropic materials, von Mises yield condition is expressed as:

$$\sigma_v = \sqrt{3J_2} \tag{4}$$

where $J_2 = s_{ij}s_{ji}/2$ is the second invariant of s_{ij}. The deviatoric stress s_{ij} is obtained by removing the hydrostatic stress $\sigma_{kk}/3$ from the stress tensor.

$$s_{ij} = \sigma_{ij} \frac{1}{3}\sigma_{kk}\delta_{ij} \tag{5}$$

Written with matrix notation, the von Mises stress gets the form

$$\sigma_v = \sqrt{\frac{(\sigma_{11} - \sigma_{22})^2 + (\sigma_{22} - \sigma_{33})^2 + (\sigma_{33} - \sigma_{11})^2 + 6(\sigma_{12} + \sigma_{23}^2 + \sigma_{31})}{2}} \tag{6}$$

As mentioned previously, Hill's [8] formulation is the most frequently used yield function for steel sheet metals. It can be regarded as an extension of the isotropic von Mises function, which can be expressed in terms of rectangular Cartesian stress components as follows:

$$\bar{\sigma} = \sqrt{F(\sigma_{22} - \sigma_{33})^2 + G(\sigma_{33} - \sigma_{11})^2 + H(\sigma_{11} - \sigma_{22})^2 + 2L\sigma_{23}^2 + 2M\sigma_{31}^2 + 2N\sigma_{12}^2} \tag{7}$$

where $\bar{\sigma}$ is the equivalent stress, and indices 1, 2, 3 represent the rolling, transverse and normal direction to the sheet surface.

Constants F, G, H, L, M and N are defined anisotropy state of material and are given by:

$$F = \frac{1}{2}\left(\frac{1}{R_{22}^2} + \frac{1}{R_{33}^2} - \frac{1}{R_{11}^2}\right), \quad G = \frac{1}{2}\left(\frac{1}{R_{11}^2} + \frac{1}{R_{33}^2} - \frac{1}{R_{22}^2}\right),$$
$$H = \frac{1}{2}\left(\frac{1}{R_{11}^2} + \frac{1}{R_{22}^2} - \frac{1}{R_{33}^2}\right), \quad L = \frac{3}{2R_{23}^2}, \quad M = \frac{3}{2R_{13}^2}, \quad N = \frac{3}{2R_{12}^2} \tag{8}$$

Parameters R_{11}, R_{22}, R_{33}, R_{12}, R_{13}, R_{23} are defined in ABAQUS as user inputs consisting of ratios of yield stress in different directions with respect to a reference stress according to Eq. (9).

$$R_{11} = \frac{\sigma_{11}}{\sigma_0}, \quad R_{12} = \frac{\sigma_{12}}{\tau_0}, \quad R_{13} = \frac{\sigma_{13}}{\tau_0},$$
$$R_{22} = \frac{\sigma_{22}}{\sigma_0}, \quad R_{33} = \frac{\sigma_{33}}{\sigma_0}, \quad R_{23} = \frac{\sigma_{23}}{\tau_0} \tag{9}$$

3. Contact Condition and Friction Anisotropy

The friction properties of the steel sheets used in this experiment were determined assuming dry friction condition using the pin-on-disc tribometer T01-M [38] as shown in Figure 3. This is because pin-on-disc method allows determining variation of the value of friction coefficient in the rolling direction of the sheet metal.

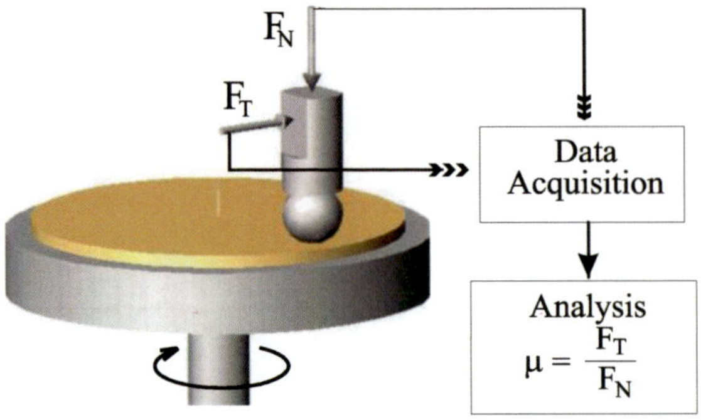

Figure 3. Scheme of the friction test.

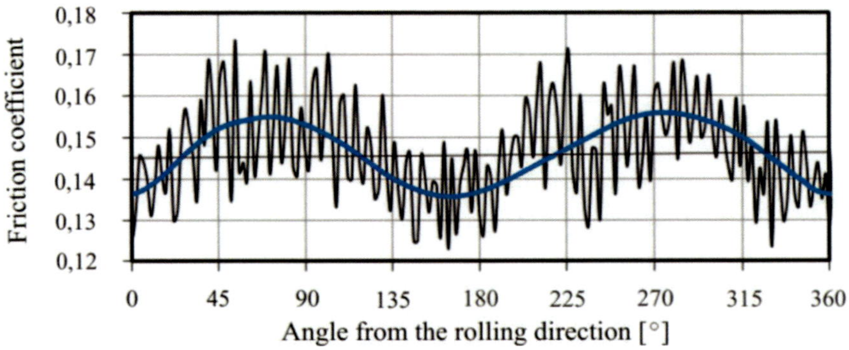

Figure 4. Friction coefficient as a function of the angle from the rolling direction.

To confirm that steel sheets are characterized by the anisotropy of tribological properties, friction anisotropy on a given surface has to be clearly distinguished from friction anisotropy for different normal orientations. In this study, the friction coefficient as a function of angular position with respect to the rolling direction was measured as shown in Figure 4. It can be observed from this plot that there are two maxima within a 360° rotation. These values correspond nearly to the transverse directions to the rolling direction.

The anisotropic friction model was implemented by specifying different friction coefficients in the two orthogonal directions on the contact surface. These orthogonal directions coincide with the defined slip directions. To use an anisotropic friction model two friction coefficients ($\mu_1 = 0.142$ and $\mu_2 = 0.157$) were specified. The first is the coefficient of friction in the first slip direction along the rolling direction and second is the coefficient of friction in the perpendicular slip direction. For simulation models with isotropic friction an average value of 0.1495 was obtained. The critical shear stress surface (Figure 5) is a piece of ellipse defined by

$$\tau_1 = \tau_1^{\max} \cos\alpha \, , \; \tau_2 = \tau_2^{\max} \sin\alpha \, , \; \alpha \in \left\{0, \frac{\pi}{2}\right\} \tag{10}$$

This elliptic surface has two extreme points given by $\tau_1^{crit} = \mu_1 P$ and $\tau_2^{crit} = \mu_2 P$. The size of the ellipse will change with the change in contact pressure between the surfaces.

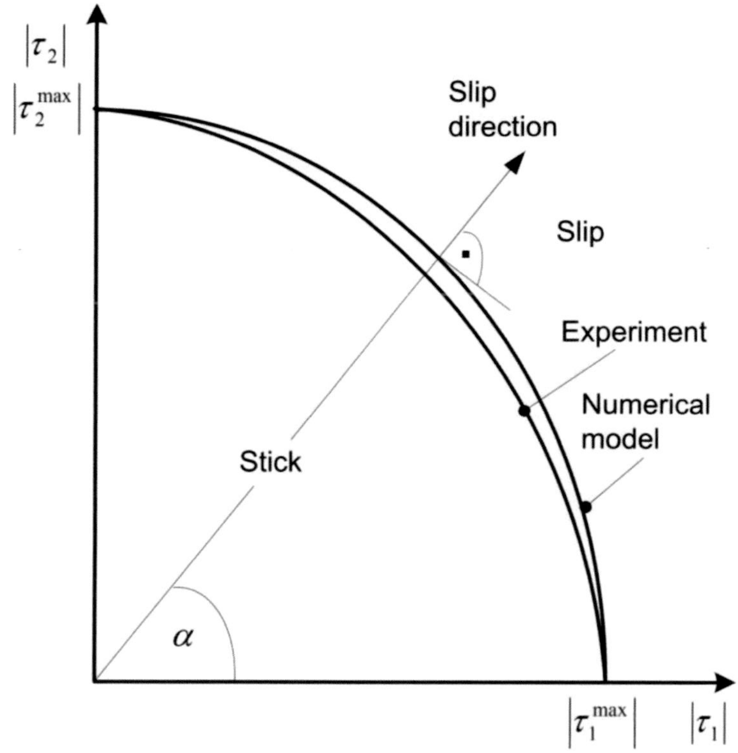

Figure 5. Anisotropic friction model.

4. Experimental Procedure

The deep drawing was done using a tool-set consisting of die, punch and a blank holder (Figure 6). The die is a flat surface with a rectangular hole of size 62.4 mm by 42.4 mm (Figure 6a), while the punch is rectangular with dimensions 60 mm by 40 mm. Both the die and the punch have rounded edges with two radii: drawing radius and the radius at the circumferential corners of the tool shape. The drawing radius for punch is 3 mm, and for die 3.5 mm. The radius in circumferential corner of the punch is 12 mm, and for die 13.2 mm.

The die set is constructed of cold-worked NC6 tool steel hardened to a minimum of 58 HRC. The drawing process was conducted using the Schenck UTS 100 hydraulic test machine with forming speed of 0.3 mm/s at a room temperature. The approximate dimensions of the blank were analytically determined (Figure 7). The radius R_c = 25 mm of the blank satisfying the strain ratio condition and the material dimensions of the flat walls are calculated for a bending profile.

Discontinuity of the shape was improved by making correction of the profile (theoretical line). The correction was done in such a way that the volume of added material to rounded corner (⊕) is equal to the volume of removed material (⊖) from the lateral side. Such theoretical contour of the blank does not take into account the material flow non-homogeneity resulting from material and friction anisotropy. As material anisotropy and friction anisotropy cause non-uniformity of flow intensity on drawpiece perimeter, the correct profile of the blank was determined experimentally.

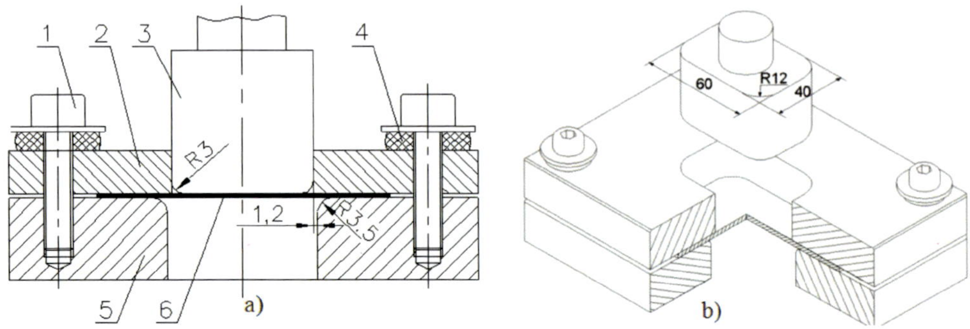

Figure 6. (a) Dimensions of the stamping tool and (b) schematic view of setup in testing machine 1 – set bolt, 2 - blank holder, 3 - punch, 4 – elastomeric washer, 5 - die, 6 – blank.

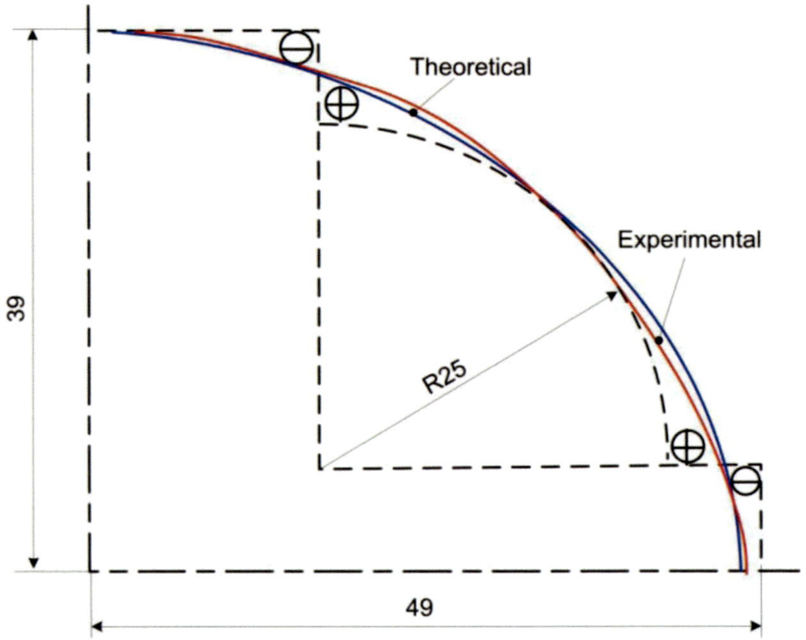

Figure 7. Shape and approximate dimensions of the blan.

Deep drawing subjects the blank to radial stresses due to the blank being pulled into the die cavity. In addition, compressive stress normal to the element arises due to the blank holder pressure. This pressure is the result of the applied load modeled as a distributed load on the contact surface between the blank and the holder. For regular shapes like axis-symmetric and rectangular shapes, there exist analytical methods that can be used to calculate the *blank holder force*. For this shape, which is almost elliptic, the force is better determined experimentally by trial-and-error method as a necessary minimal force to prevent a wrinkle formation. Further, the radial tensile stresses lead to compressive hoop stresses because of the reduction in the circumferential direction. Unless constrained by the blank holder force, the flange of the blank attempts to wrinkle because of the hoop stress. On the other hand, application of the blank holder force increases the frictional resistance in flange region and decreases the value of the limit strain of the sheet.

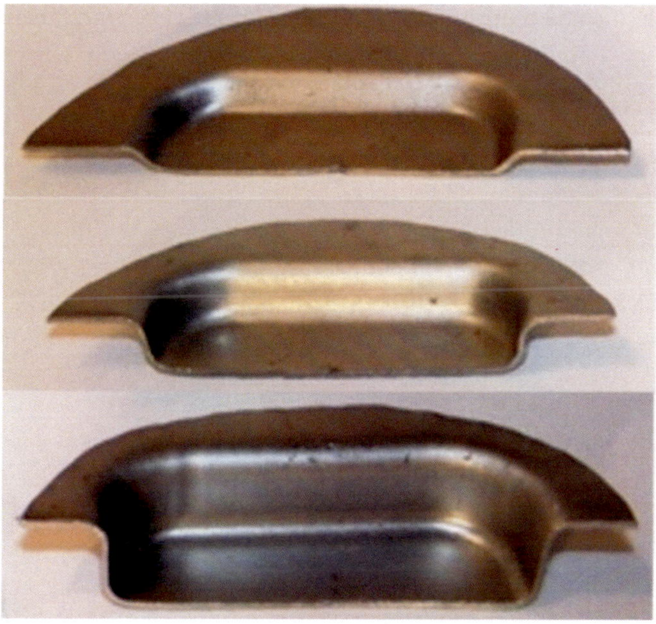

Figure 8. Cup sections corresponding to punch strokes 5 mm, 10 mm and 15 mm.

As reported in [39], varying the blank holder pressure during stamping by application of variable *blank holder force* (VBHF) is advantageous. This pressure improves formability of the sheet and can decrease the maximum tensile stress along the punch direction. The used elastomeric washers allow increasing the blank thickness in blank holder contact region and protecting against wrinkling formation simultaneously.

The force to the blank holder was applied by tightening a bolt with suitable torsion moment. The specimens were carefully positioned at the center of the die to ensure a uniform distribution of the blank holder force; and the applied blank holder force was then determined experimentally by *trial-and-error method* and kept constant at 7.85 kN throughout the drawing process. While drawing, i.e. when the punch transmits the drawing force through the walls of the cup and the holder as it is drawn into the die cavity, the walls of the cup primarily experience a longitudinal tensile stress. Additional tensile hoop stress is also caused by the cup being held tightly over the punch. Further, friction and strain resistance cause the variation in thickness of the sheet metal.

Figure 8 shows three cups with different height corresponding to punch strokes of 5, 10 and 15 mm on which full drawing was carried out. Particular cups were cut along three directions: transverse, longitudinal and at 45° angle, defined with respect to the rolling direction. The distribution of thickness was measured along vertical cross section of the cup formed using a microscope device.

5. Finite Element Modeling

ABAQUS is one of the major finite element codes used for the computer-aided analysis of sheet metal forming process in the industry. Using this tool, the predicted formability of sheet metal forming processes were compared using different *shell element* types available in the code. The blank, die and punch were modeled in accordance with the experimental set-up. Symmetry properties were utilized to reduce the CPU time, i.e., only one quarter of blank and tool with appropriate boundary conditions were modeled (Figure 9). The blank was modeled with 4-node reduced integration, doubly curved shell elements called S4R in ABAQUS terminology (ABAQUS v 6.7). S4R is a 4-node, quadrilateral, stress/displacement shell element with reduced integration and a large-strain formulation. These element types allow transverse shear deformation that becomes very small as the shell thickness decreases. They use thick shell theory as the shell thickness increases and become discrete Kirchhoff thin shell elements as the thickness decreases. The general purpose conventional shell elements are widely used because they provide robust and accurate solutions in many applications. According to ABAQUS theory manual, the continuum shell elements can be used for any thickness.

The S4R shell element type is intentionally applied for analysis of sheet metal forming processes [20, 39] because, unlike solid and plane strain elements, this element type accounts for the change of thickness in its output variables. The tools in this study are considered to be rigid, and hence no deformation is assumed in these parts. The blank model is composed of 3431 4-node elements, while the tools consisted of 10589 linear quadrilateral elements. The boundary conditions applied to the blank holder allow displacement in the normal direction to the blank surface so that the wrinkling formation is prevented and the frictional resistance in flange region is minimized.

Penalty formulation of friction contact was used to enforce the intermittent contact and sliding boundary condition between the sheet metal and tools. The contact between the blank and the tool was a standard master and slave contact interface. Both isotropic and anisotropic formulations, together with two models of material descriptions, were considered in the numerical simulations. The sheet material was treated as an elasto-plastic material with anisotropy governed by the *Hill's yield criterion* and isotropy governed by the von Mises. Further, an isotropic work hardening described by the *Holomon law* (Eq. 1) was assumed.

The calculations in the finite element code were done based on internal forces that balance with the external forces through an iterative procedure that gives the deformed state after a time increment. One of the merits of this method is that the time increment can be relatively large because of conditional stability of the time integrator and static solutions can always be obtained by natural characteristics of the method.

To solve the *Newton-Raphson's iterations*, the time step is often controlled by ABAQUS automatic increment technique. The double precision executable is used in this analysis.

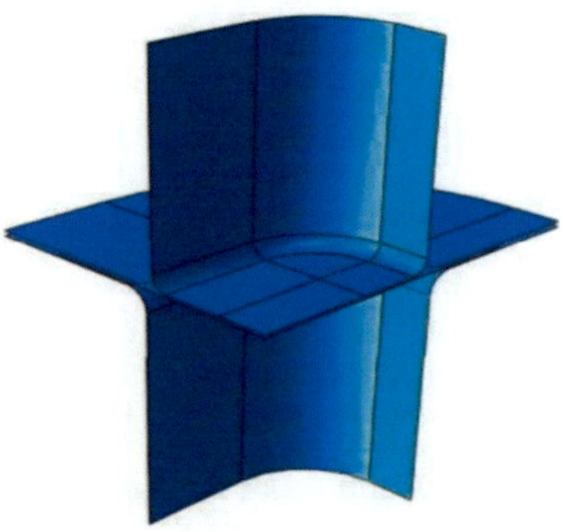

Figure 9. Schematic illustration of the tools and blank for the finite element model.

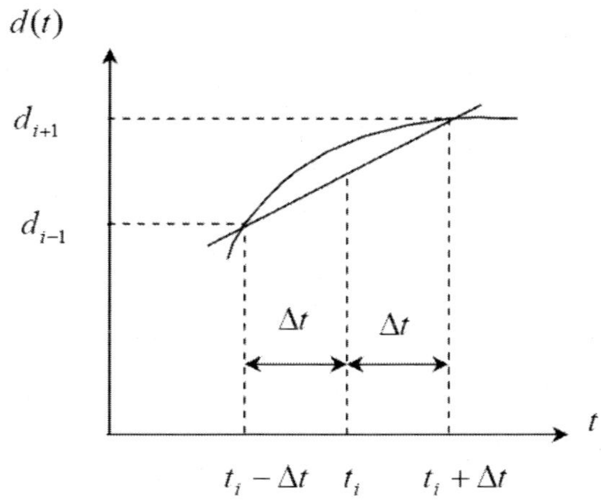

Figure 10. Numerical integration.

The global equation in FEM based models is represented as

$$\{F(t)\} = [K]\{d\} + [M]\{\ddot{d}\} \tag{11}$$

where *[K]*, *[M]* and *[F]* are the global stiffness, mass and force matrices, respectively. *d* and $\ddot{d}$ represent the nodal displacements and nodal accelerations respectively. Upon discretization of Eq. (11) with respect to time the nodal displacements at different time integration can be found using direct integration as depicted in Figure 10.

There are two classifications of direct integration: *explicit* and *implicit*. These direct integration approaches are commonly referred to as the *central difference method* [40] and *Newmark's method* respectively. Further detailed theories and formulations of these methods are given in the literature elsewhere [29, 30, 33] and only the key fundamentals are briefly presented in the following two sections.

5.1. Static Implicit Procedure

The static analysis available from ABAQUS that is used in the model handles non-linearity from large displacements, material non-linearity and boundary non-linearity such as contact, sliding and friction. Solution is obtained as series of increments with iterations to obtain equilibrium within each increment. The implicit method is based on static equilibrium and is characterized by the simultaneous solution of a set of equations. The net force acting on each node for the model to be in static equilibrium must be zero. After a full set of linear equations are assembled, this set of equilibrium equations is then solved simultaneously and the unknown nodal displacements are obtained. An important feature of the implicit solution procedure is the assembly of a global stiffness matrix given by *Newmark's direct integration* equations as

$$\dot{d}_{i+1} = \dot{d}_i + (\Delta t)\left[(1-\gamma)\ddot{d}_i + \gamma\ddot{d}_{i+1}\right] \tag{12}$$

$$d_{i+1} = d_i + (\Delta t)\dot{d} + (\Delta t)^2 \left[(0.5 - \beta)\ddot{d}_i + \beta \ddot{d}_{i+1} \right] \tag{13}$$

where β and γ are parameters chosen by the user. By multiplying Eq. (13) by mass matrix M and extracting from Eq. (11) $\ddot{d}_i$

$$\ddot{d}_i = \frac{F_i - Kd_i}{M} \tag{14}$$

then substituting Eq. (14) we obtain

$$K' d_{i+1} = F'_{i+1} \tag{15}$$

where

$$K' = K + \frac{1}{\beta (\Delta t)^2} M \tag{16}$$

$$F'_{i+1} = F_{i+1} + \frac{M}{\beta (\Delta t)^2} \left[d_i + (\Delta t)\dot{d}_1 + \left(\frac{1}{2} - \beta \right) (\Delta t)^2 \ddot{d}_i \right] \tag{17}$$

The values of d_{i+1}, $\dot{d}_{i+1}$ and $\ddot{d}_{i+1}$ are then determined by using the above equations. Due to the iterative nature of the solution procedure, a successful solution requires the satisfaction of convergence criterion at each step. Generally the convergence speed is quite problem dependent and failure to converge results in premature termination of the analysis. Implicit method enables full static solution of the deformation problem with convergence control.

Therefore, the size of the time increments can theoretically be large. In practice however, it is reduced by the contact conditions present between the sheet and die. As a result, the computer time increases almost four times with increase in number of elements [41]. Another disadvantage of an implicit approach is the singularities of the stiffness matrix at divergence points, such as instabilities at wrinkling initiation.

5.2. Dynamic Explicit Procedure

In the explicit procedure, the displacements, velocities and accelerations of each node are advanced explicitly through time. This means that the state of the model at the end of time t+Δt is solely based on the displacements, velocities and accelerations at the beginning of time t.

The nodal velocities at time t+Δt are obtained assuming that the accelerations are constant during a small time increment Δt. Similarly, the nodal displacements at the end of the increment can be computed by adding the displacements during the time increment Δt to the displacements at time t.

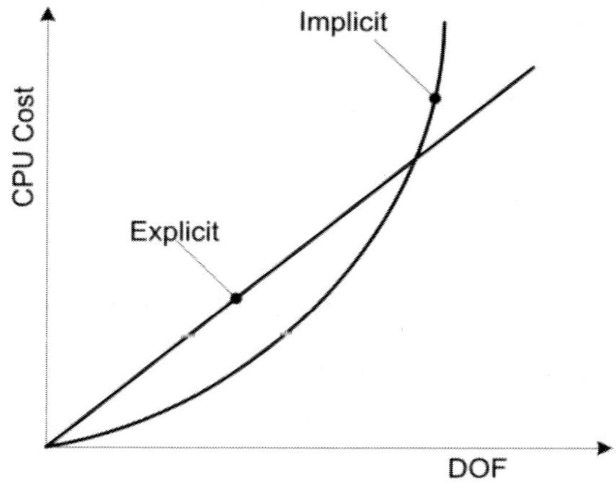

Figure 11. CPU cost versus numbers of degrees of freedom.

Dynamic solution of the FE method is used instead of static solution because the computational time using dynamic code increases linearly, whereas the time for using static code increases as a self-multiplication function of the number of degrees of freedom (DOF) as shown in Figure 11.

The explicit solver scheme is, as well as the implicit, built upon the equation

$$\mathbf{M}\ddot{\mathbf{u}} + \mathbf{C}\dot{\mathbf{u}} + \mathbf{R}(\mathbf{u}) = \mathbf{P} \tag{18}$$

where **C** is the damping matrix, **P** is the external load vector and **R** is the vector for reaction forces [42].

A single iteration obtained from the central difference method is sufficient for the explicit scheme. This iteration approximates the velocity and the acceleration in the next time step using only information from the previous step, where all state variables are known. The advantage of the explicit FE code is that there is no need to solve a system of equations when they are combined with a lumped mass matrix because all of the equations are uncoupled and, therefore, considerable smaller CPU time is required, compared with traditional implicit code. As stated earlier, the dynamic explicit method is advantageous for analysis of sheet metal forming where the real process time is just a few seconds. The method has the characteristic of less memory requirement and greater computer efficiency since there is no need for *consistent stiffness matrix.*

6. Discussion of Results

To study the variations in the wall thickness of rectangular cups, several experiments were conducted through gradual increment of punch displacement. Thickness distributions were investigated using numerical modeling and the formation process of the drawpiece was examined. Verification of numerical results was carried out on the basis of the measured wall thickness at characteristics sections. Thickness changes, after stamping experiments, were examined by cutting the formed parts along the rolling and transverse directions starting from the center of the part. Further, the cups were cut in the corner at an angle of 45° with respect to the rolling direction and the cut sides were trimmed and thickness changes were measured by using an optical microscope. Some of the obtained results in a form of comparative study are briefly discussed in the following sections.

6.1. Influence of Material Models

Plots of the strain distribution of wall thickness, measured and calculated on the symmetrical lines of long and short sides and the diagonal line passing through the corner, are shown in Figure 12. The experimental results for the three directions are compared with respect to the following proposed strategies of material models:

- Hill anisotropy model plus anisotropic friction (AF)
- Hill anisotropy model plus isotropic friction (IF)
- von Mises model plus anisotropic friction (AF) and
- von Mises model plus isotropic friction (IF)

As shown in the plots, the patterns of the thickness distribution contours for the adopted analysis strategies are nearly similar. The maximum wall thickness in all directions is observed near the corner of the blank, while the minimum thickness is at the punch shoulder.

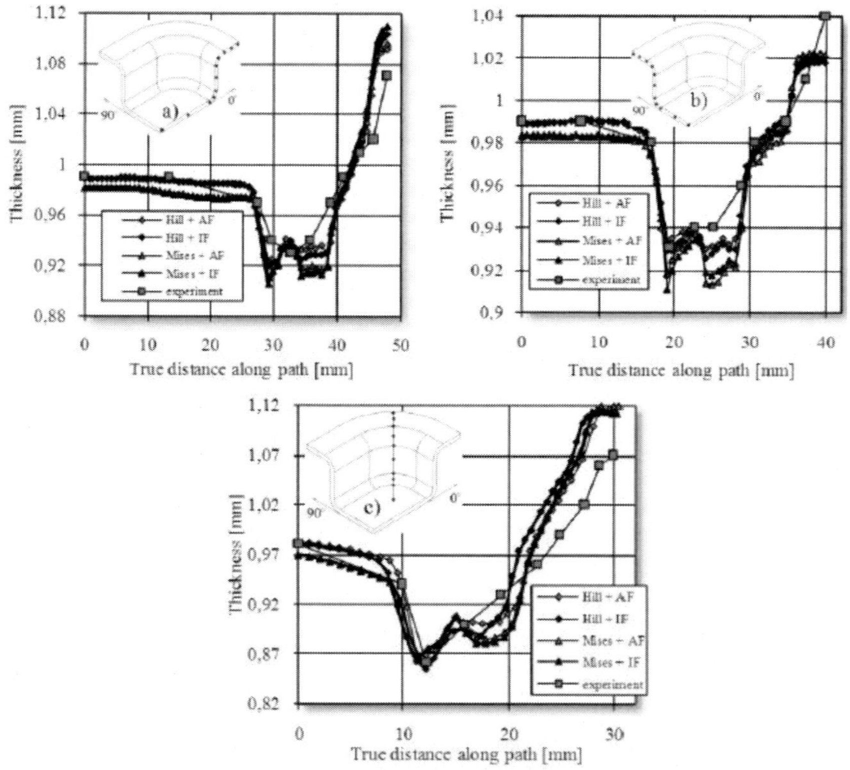

Figure 12. Plot of thickness strain under punch displacement of 15 mm measured at a) 0^0 b) 90^0 and c) 45° direction.

Compared with the deformation at the corner, the variation of thickness at the bottom of drawpiece is relatively small.

Replacing the anisotropy friction model by isotropic friction model in both models of the material resulted in slight decrease in thickness. This is clearly visible at the bottom of drawpiece where the character of the thickness variation is similar. In other parts of drawpiece, this phenomenon is more complicated. Moreover, the largest difference between the analyzed models exists at the corner.

For all FEM models, the thickness variation in the drawpiece wall measured along the diagonal line passing through the corner is almost identical. For von Mises material model with IF model, however, differences between experimental and numerical data of maximal thickening in risk section are observed. The plots also illustrate the good predictive capability of the models. The prediction error is calculated using the following expression:

$$error = \left| 1 - \frac{g_{\exp}}{g_{FEM}} \right| \cdot 100\,\% \tag{19}$$

where $g_{\exp}$ is the experimental and g_{FEM} is the simulated thickness.

The prediction error varied between 2.1% for measurement in transverse direction to 0.8% for measurement passing through the corner. The prediction errors of Hill model with anisotropy friction were 0.11% and 1.06%, respectively. Model of Hill material with isotropic friction generated error of 1.14% along the rolling direction, 0.55% in the diagonal direction and 0.47% when measured transverse to the rolling direction.

Three deformation contours of thickness distribution are plotted in Figure 13 to Figure 16 for different punch penetrations. The deformations were calculated for the isotropic and anisotropic material model as well as at different friction conditions. In the actual cup drawing, blank thickness changes in each part are not equal. As can be observed from these plots however the wall thickness varies especially at the corners. The maximal predicted flange thickness is similar for all examined strategies and varies between 0 and 0.01 mm. Considering the *anisotropic friction models*, the minimal predicted values for Hill material are larger than the von Mises model by 0.2, 0.3 and 0.58 mm for following punch penetrations of 5, 10 and 15 mm respectively. The von Mises material model with isotropic friction produces a lower minimal thickness when compared with the anisotropic friction. Similarly, the *Hill material model* with *isotropic friction* has predicted lower thickness value at the corner for all punch penetrations.

The above results show that greatest strain gradients appear in the sidewall for punch displacement of 5 mm. With the strain evolution follows the strain hardening of the material and compensation of the strain gradient in sidewall and at the bottom. Moreover, it can be seen that the thickness of sidewall along the longer edge is higher than along the shorter edge. Compared with the other parts of the drawpiece, the bottom part has lower thickness variation with punch displacement. Moreover, the thickness distribution is uniform for different models.

Thickness in flange region measured in the rolling direction is higher than in the remaining directions. Compared with the long side of the flange, there is a higher fraction of circumferential compression stress on the short side. This increases deformation resistance and as a result higher thickening.

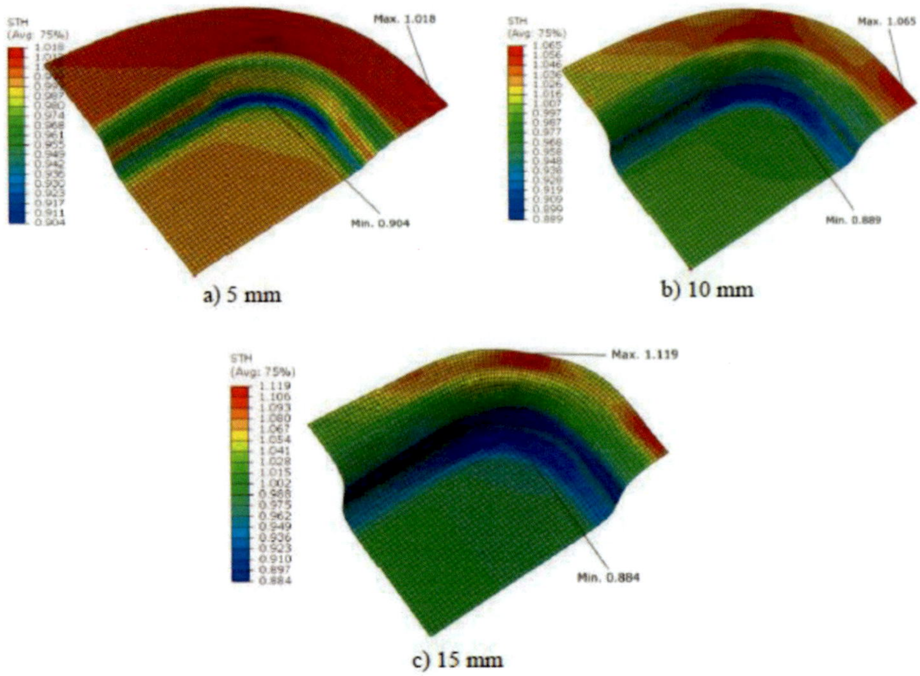

Figure 13. Contour of the sheet thickness (Hill model + AF) at punch displacement of a) 5, b) 10 and c) 15 mm.

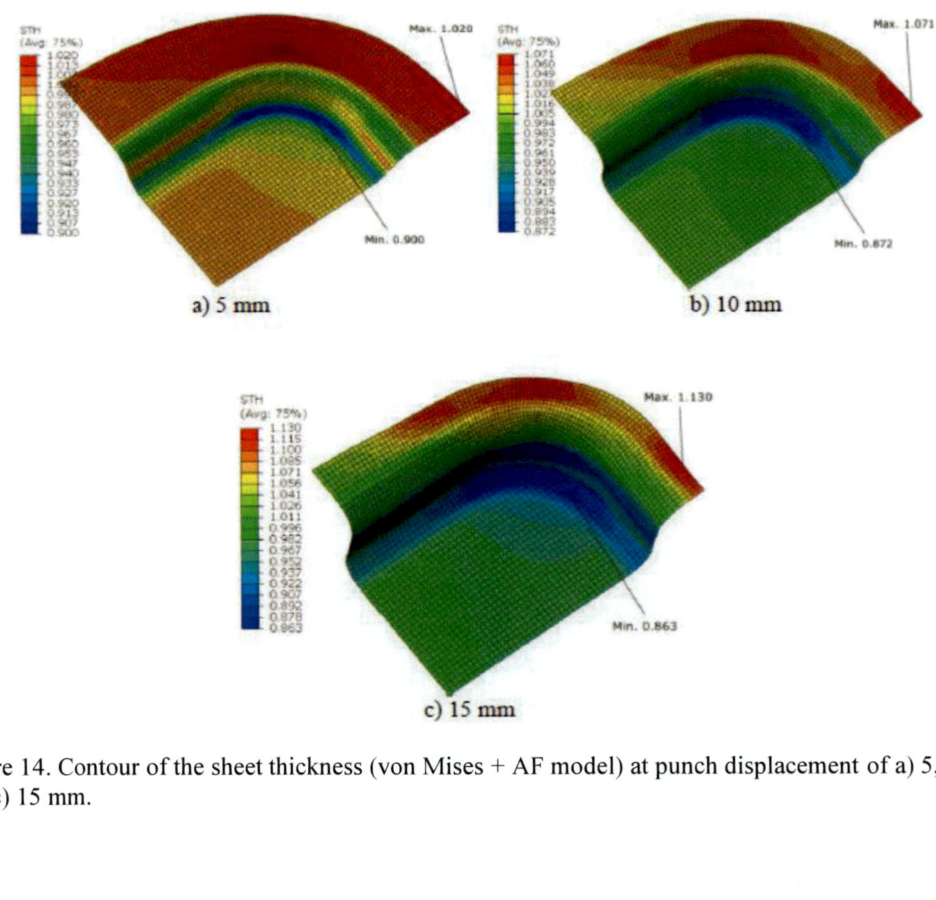

Figure 14. Contour of the sheet thickness (von Mises + AF model) at punch displacement of a) 5, b) 10 and c) 15 mm.

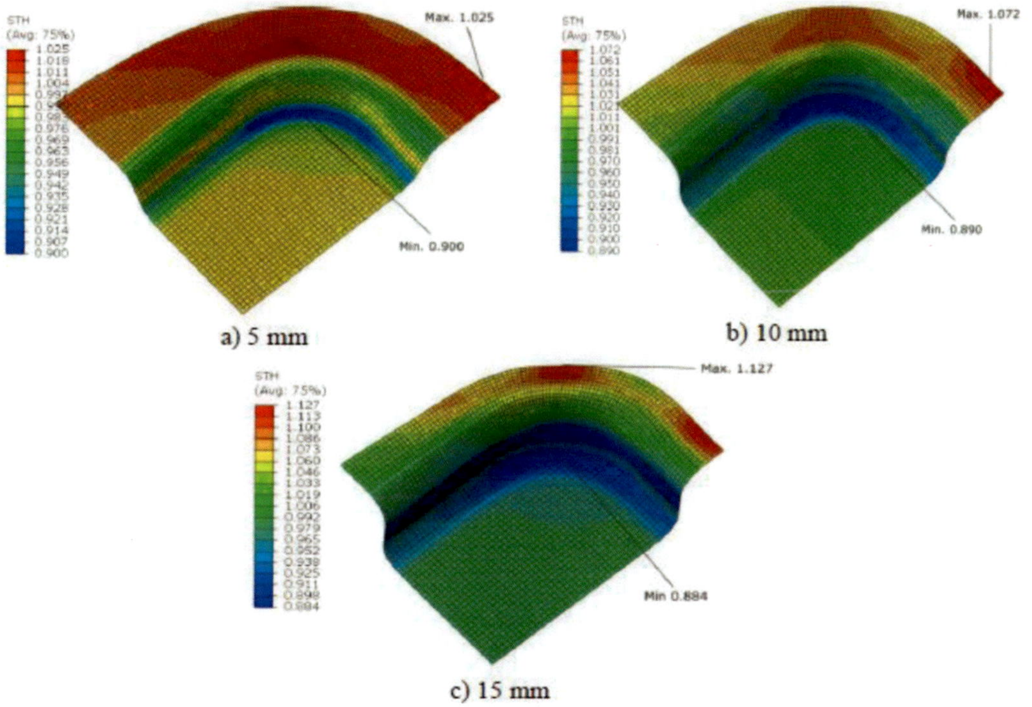

Figure 15. Contour of the sheet thickness (Hill model + IF) at punch displacement of a) 5, b) 10 and c) 15 mm.

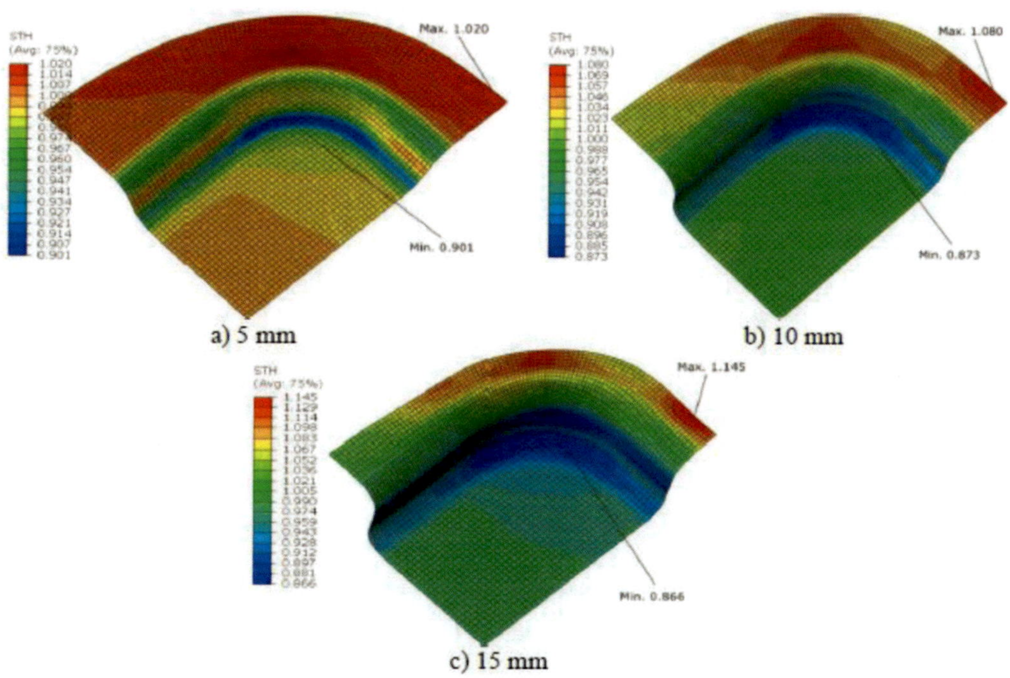

Figure 16. Contour of the sheet thickness (von Mises + IF model) at punch displacement of a) 5, b) 10 and c) 15 mm.

6.2. Strain Distribution and Punch Penetration

Figure 17a) – c) show the thickness strain distributions along the three directions: 0°, 45° and 90°. The figures illustrate the comparison of FEM analysis results with experimental measurements as a function of true distance along path. As expected, the thickness strain distributions for all punch displacements are different along these directions. The decisive impact on this character exerts the non-axisymmetrical shape of parts. The thickness distribution of the sheet under the punch is found to be more or less uniform to the initial sheet thickness. However, the thickness of the sheet, which is above the flat portion of the die, is observed to be slightly larger than the initial thickness. The thickness measured in the rolling direction (the long side) is smaller than that of the transverse direction (the short side) near the punch shoulder. The main reason for this effect is the unsymmetrical shape of the process. As a previous study [41] shows, the short sidewall of the rectangular drawpiece near the punch edge is most exposed to fracture.

The wall thickness varies especially in the corner, and the maximum magnitude of the thickness strain for all analyzed cases is observed along 45°. Apart from the fact that the neck, which develops on the radius of the punch, is more pronounced than in the experiments, the strain distributions are well described by the FEM simulations. The prediction errors of Hill model plus AF of thickness distribution along 45° for following punch displacements are 9.57%, 3.16% and 1.06%. This means that at higher deformations the FEM model prediction becomes less accurate compared with the experimental results. While the experimental values of wall thickness are measured with precision of 0.01 mm, the computation of ABAQUS program was assumed with double precision. Two different responses can clearly be seen during the lower punch penetration:

1. almost undeformed pattern at the bottom and
2. large radial stretching with thickness increase at the wall.

Once the upper punch acts, highly localized strains take place at the bottom due to the small radii of curvature existing in the lower punch that adopts the role of a die at the final stage of the process.

The trends of thickness variation for all punch penetrations are coincident. Further, the variation of the position where the minimum thickness occurred is very small for different punch penetrations. There seems a close similarity between experiment and theory, and the variation of thickness in every part is very similar. The minimum thickness occurred at the punch corner (on the arc of drawpiece) most likely due to stress concentration and the tensile stress. Peak strains are observed, particularly at the corner. The thickness drawpiece wall along 0° and 90° with respect to the rolling direction increased initially, but slightly decreased in the die corner. Despite the increase in wall thickness, the yield stress increased because of strain hardening phenomenon.

The process may be continued without risk of rapture. This has been confirmed on the basis of the analysis of minor and major strain at minimal thickness of 0.757 mm. Taking into account the forming limit diagram (FLD) for DDQ steel sheet, the work by Stachowicz and Spisak [43], based on Marciniak-Kuczyński method, has confirmed that the thickness in the punch corner is safe for continuing stamping process. Finally, the thickness increased gradually and exceeded the initial thickness in the flange under the circumferential pressure.

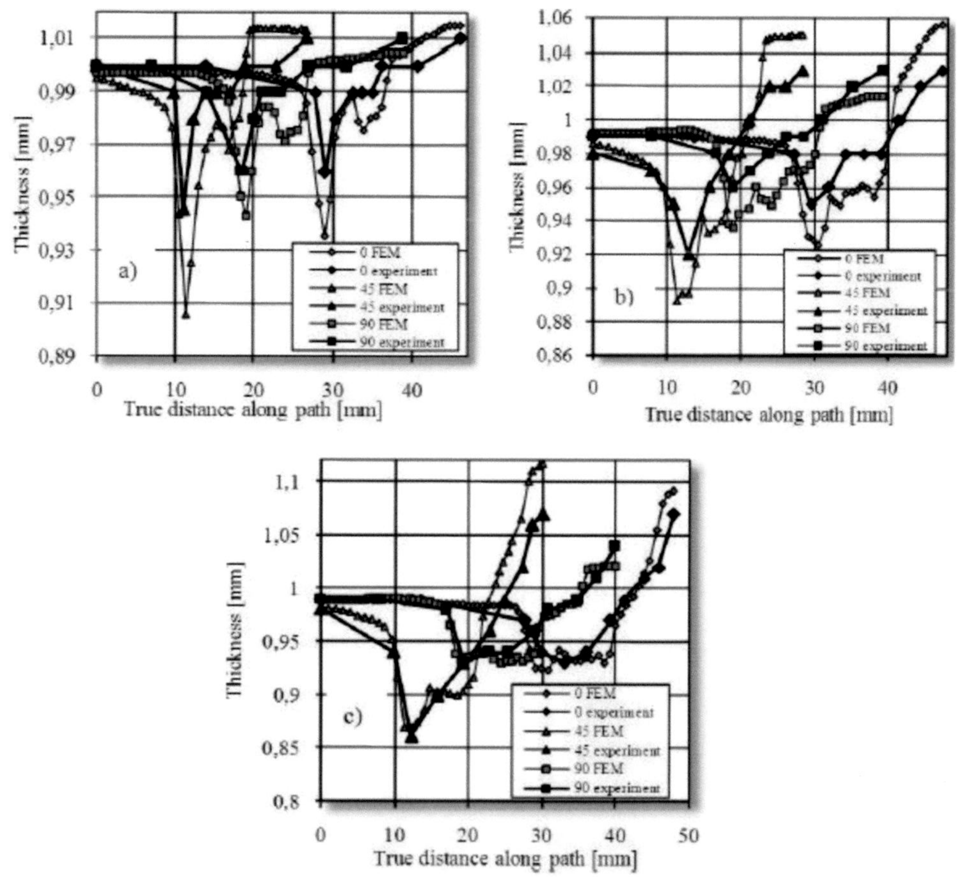

Figure 17. Plot of thickness strain (Hill model + AF) under punch displacement of a) 5, b) 10 and c) 15 mm.

6.3. Explicit and Implicit Schemes

Figure 18 shows the comparison of the thickness distribution obtained with dynamic explicit and static implicit schemes. Thickness strain results from both schemes, for *Hill material model* and anisotropic friction description, are plotted.

The strain variations are measured along the rolling direction (left) and at an angle of 45^0 to the rolling direction (right). As can be observed from the two plots, the pattern of thickness distributions for both schemes is quite similar. On the other hand, there are differences in the CPU time used to process the models. The 14020-node ABAQUS/Static model takes more than 2.7 hrs on Duo CPU time P8600 @ 2.4 GHz machine. The same computation using dynamic explicit model took over 19% less computation time. The thickness strain predicted with the explicit model in cup bottom is in general higher than the results with the static model.

In addition to the processing time, a difference in thickness reduction is observed from the results of the two methods where the explicit model gives less thickness decrease in the flange edge. The thickness strain distributions for punch displacement of 5, 10 and 15 mm are plotted in Figure 19. Comparing the results in Figure 13 and Figure 19, the thickness strain distribution and the predicted minimal and maximal values of thickness for punch displacement of 5 mm are almost identical. Both models show a clear tendency of a wrinkling flange. The predicted strain distribution in the flange along rolling direction and at an angle of 45° is different for the same punch penetrations (Figure 18). Thickness in the rolling direction increases from die corner to flange edge for all punch penetrations. Further, the thickness distribution for punch displacement of 5 mm and 10 mm is uniform along 45° direction.

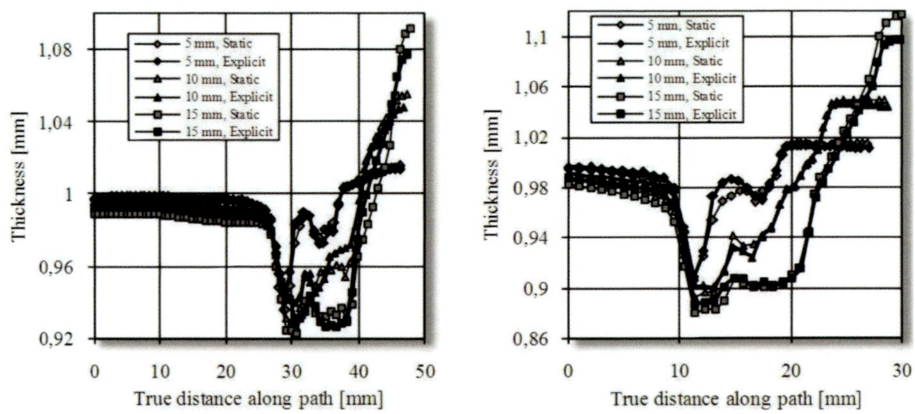

Figure 18. Plot of thickness strain for Hill model + AF measured along rolling direction (left) and in the corner at an angle of 45° (right).

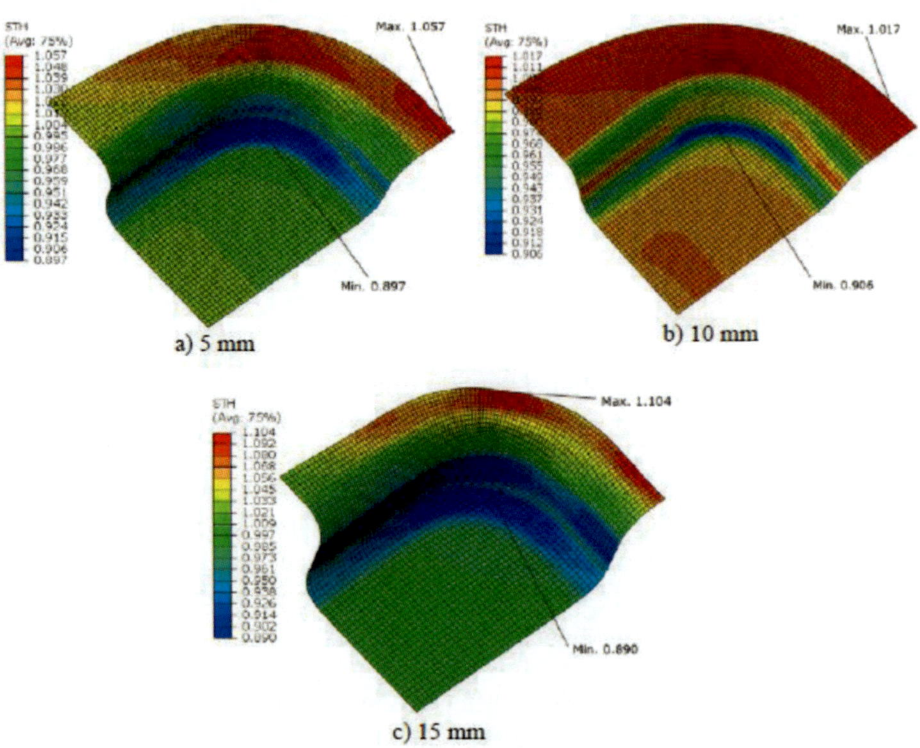

Figure 19. Plot of thickness strain (Hill model +AF) determined using explicit procedure for punch displacements of a) 5, b) 10 and c) 15 mm.

CONCLUSION

Anisotropy study of metal forming using finite element method has been presented in this chapter. Different material and friction models are also considered and dynamic explicit and static implicit procedures in modeling sheet metal forming are implemented. Though both the explicit and implicit procedures are successful in analysis of sheet metal forming, the simulations show that dynamic explicit approach is computationally more efficient. The distribution of wall thickness obtained by means of experiments and FEM, including material and friction anisotropy, generally demonstrates a good harmony, and that minor differences are due to the fact that conditions in the finite element methods are considered ideal.

Thus, the results of this study indicate that finite element methods can be successfully implemented to study the anisotropy property in complex sheet metal forming problems. In general, the process efficiency of the dynamic explicit procedure makes this approach a good method of choice in simulating large three-dimensional parts where more refined meshes are required.

ACKNOWLEDGMENT

The research reported in this chapter is financially supported by Island, Liechtenstein and Norway and is co-financed by European Economic Area and Norwegian Financial Mechanism under the Scholarship and Training Fund. The authors would like to acknowledge this financial support.

REFERENCES

[1] E. Daxin, M. Takaji and L. Zhiguo, *J. Mat. Proc. Technol.* 205 (2005).
[2] E. Daxin, P. Yuping and M. Takaji, *J. Plasticity Eng.* 11, 39 (2004).
[3] T. Wen and E. Daxin, *Modern Manuf. Eng.* 4, 40 (2004).
[4] Z. Gronostajski, *J. Mat. Proc. Technol.* 106 (2000).
[5] R. Padmanabhan, M.C. Oliveira, J.L. Alves, and L.F. Menezes, *Finite Elements in Analysis and Design*, 43, (2007).
[6] D. Banabic, H.-J. Bunge, K. Pohlandt, and A.E. Tekkaya, *Formability of Metallic Materials*, Springer-Verlag (2000).
[7] P. van Houutte, *Anisotropic plasticity*, in Hartley P., Pillingar I., Sturgess C. (eds.), Numerical modeling of material deformation process: research, development and applications, Springer-Verlag, London (1992).
[8] R. Hill, *Proceedings of the Royal Society of London* A 193, 281 (1948).
[9] R. Hill, *Mathematical Proceedings of the Cambridge Philosophical Society* 85, 179 (1979).
[10] W.F. Hosford, *Trans. ASME, J. App. Mech.* E39, 607 (1972).
[11] R. Hill, *Int. J. Mech. Sci.* 35, 19 (1993).
[12] F. Barlat and J. Lian, *Int. J. Plasticity* 5, 51 (1989).
[13] F. Barlat, D.J. Lege and J.C. Brem, *Int. J. Plasticity* 7, 693 (1991).

[14] A.P. Karafillis and M.C. Boyce, *J. Mech. and Physics Solids* 41, 1859 (1993).
[15] B. Haddag, T. Balan and F. Abed-Meraim, *Int. J. Plasticity* 23, 951 (2007).
[16] S. Bouvier, J.L Alves, M.C. Oliveira and L.F Menezes, *Comp. Mat. Sci.* 32 301 (2005).
[17] M.G. Lee, D. Kim, C. Kim, M.L Wenner, R.H. Wagoner and K. Chung, *Int. J. Plasticity* 21 883 (2005).
[18] Tikhovskiy, D. Raabe and F. Roters, *Mat. Sci. Eng.* A 488, 482 (2008).
[19] J.P. De Magalh˜aes Correia, G. Ferron and L.P. Moreira, *Int. J. Mech. Sci.* 45, 1167 (2003).
[20] C. Zhiying and D. Xianghuai, *Comp. Mat. Sci.* 44, 1013 (2009).
[21] A.L. Gurson, *J. Eng. Mater. Technol.* 99, 2 (1977).
[22] K.C. Liao, J. Pan and S.C. Tang, *Mech. Mater.* 26, 213 (1997).
[23] Y.Q. Liu, J.C. Wang and P. Hu, *J. Mat. Proc. Technol.* 120, 45 (2002).
[24] A.M. Habraken and L. Duchêne, *Int. J. Plasticity* 20, 1525 (2004).
[25] M. Hjiaj, Z-Q.Feng, G. de Saxc, and Z. Mróz, *Int. J. Eng. Sci.* 42, 1013 (2004).
[26] C. Betegon, J.J. del Coz and I. Penuelas, *Comput. Meth. Appl. Mech. Eng.* 195, 6146 (2006).
[27] J.C. Negtegaal and L.M Taylor, VDI, Berichte, 894, 705 (1991).
[28] L. Taylor, J. Cao, A.P. Karafillis and M.C. Boyce, *J. Mater. Process. Technol.* 50, 168 (1995).
[29] D.Y. Yang, D.W. Jung, I.S. Song, D.J. Yoo and J.H. Lee, *J. Mater. Process Technol.* 50, 39 (1995).
[30] F.J. Harewood and E. McHugh, *Comput. Mater. Sci.* 39, 481 (2007).
[31] N. Rebelo, J.C. Negtegaal and L.M. Taylor, *Numerical Methods in Industrial Forming Processes*. 99 (1992).
[32] J.S. Sun, K.H. Lee and H.P. Lee, *J. Mater. Process. Technol.* 105, 110 (2000).
[33] D.W. Jung, *J. Mater. Eng. Perform.* 7, 479(1998).
[34] A.G. Mamalis, D.E. Manolakos and A.K. Baldoukas, *J. Mater. Process. Technol.* 72, 48 (1997).
[35] W. Hammami, R. Padmanabhan, M.C., Oliveira, H. BelHadjSalah, J.L. Alves and L.F. Menezes, *Int. J. Mech. Mater. Des.* 5, 303 (2009).
[36] D. Banabic, H.-J Bunge, K. Pohlandt and A.E. Tekkaya, *Formability of Metallic Materials*. Springer-Verlag (2000).
[37] S. Yi, J. Bohlen, F. Heinemann and D. Letzig, *Acta Materialia* 58, 592 (2010).
[38] X. Zhang X., S. Long, Z. Ma, Y. Pan and Y. Zhou, *Trans. Nonferrous Met. Soc. China*, 17 37 (2007).
[39] A.B. da Rocha, A.D. Santos, P. Teixeira and M.C. Butuc, *J. Mat. Proc. Technol.* 209, 5097 (2009).
[40] D.L. Logan, *A First Course in Finite Element Method*, 3rd Ed., Wadsworth Group (2002).
[41] P.V. Deshmukh, *MSc Thesis*, University of Kentucky, Lexington (2003).
[42] P. Wriggers, *Computational Contact Mechanics*, Springer-Verlag (2006).
[43] F. Stachowicz and E. Spišák, *Politechniki Rzeszowskiej*, Rzeszów, (1999), in Polish.

In: Anisotropy Research
Editor: Hirpa Lemu

ISBN 978-1-62081-977-7

Chapter 2

PIEZOELECTRIC ANISOTROPY IN MODERN COMPOSITES BASED ON FERROELECTRICS

V. Yu Topolov[1,*] and P. Bisegna[2]

[1]Department of Physics, Southern Federal University, Rostov-on-Don, Russia

[2]Department of Civil Engineering, University of Rome "Tor Vergata", Rome, Italy

ABSTRACT

Modern composite materials based on ferroelectrics (either single crystals or ceramics) show an important ability to convert mechanical energy into electric energy and vice versa. In the last decades these materials have been studied and developed in an attempt to improve their performance, especially concerned with the electromechanical properties. Relaxor-ferroelectric single crystals with engineered domain structures and poled ferroelectric ceramics are of interest due to their considerable influence on the anisotropy of the effective properties of the related composites. In this chapter, physical, microgeometric and poling factors influencing the anisotropy are studied on the following directions: the large anisotropy of piezoelectric coefficients d_{33}^{*}/d_{3j}^{*}, the large anisotropy of piezoelectric coefficients e_{33}^{*}/e_{3j}^{*} and the variable anisotropy of electromechanical coupling factors k_{33}^{*}/k_{3j}^{*}. These ratios characterizing the anisotropy are analyzed for the relaxor-ferroelectric single crystal / polymer and ferroelectric ceramic / polymer composites. Hereby among the ferroelectric components, materials with the known electromechanical properties and various parameters of the anisotropy of the electromechanical properties are chosen. The polymer component is chosen among either conventional or auxetic materials. Research results are discussed for the composites with 1–3, 2–2, and 0–3 connectivity patterns. Possibilities of varying the anisotropy of the effective electromechanical properties in the composites promote their piezotechnical applications.

* Corresponding author: vutopolov@sfedu.ru

1. Introduction

As is known, *anisotropy* in physics is the quality of exhibiting properties with different values when measured along axes in different directions. Anisotropy is often observed in single crystals that are characterized by a regular arrangement of atoms, ions, or molecules, in polycrystals (ceramics) wherein the grains are not completely random in the crystallographic orientation and form so-called texture, and in biological and composite materials [1, 2]. There are many examples of *anisotropic composite materials* whose physical properties strongly depend on *microgeometry*, arrangement of components, and their properties. The typical examples of the anisotropic composite materials studied intensively in the last decades are *piezo-active composites* based on either ferroelectric ceramics (FCs) [3–5] or relaxor-ferroelectric single crystals (SCs) [4, 6, 7]. These composite materials represent an important group of *smart materials* with predictable properties [3, 4]. The widespread piezo-active composites are FC/polymer materials with the 1–3, 2–2, 0–3, and 3–3 *connectivity patterns* [3–8]. Among the SC/polymer composites developed and studied in the last decade, of interest are materials with the 1–3, 2–2, and 0–3 connectivity patterns [4, 6, 7, 9, 10]. In these composites, ratios of the effective piezoelectric coefficients d_{33}^{*}/d_{3j}^{*} and e_{33}^{*}/e_{3j}^{*} and electromechanical coupling factors k_{33}^{*}/k_{3j}^{*} (j = 1 and 2) can be varied in wide ranges [4, 9]. An increase in the *piezoelectric anisotropy*, activity, sensitivity, figures of merit, and electromechanical coupling of the aforementioned composites is often concerned with a modification of the composite structure and with improving the elastic and other characteristics of the matrix [9–14]. The aim of this chapter is to show the *microgeometric* and physical factors that promote the considerable piezoelectric anisotropy and its variation in the 0–3, 1–3, and 2–2 composites based on ferroelectrics. We focus attention on the 0–3, 1–3, and 2–2 connectivity that are related to the continuous distribution [3, 4] of the main piezo-active component (FC or SC) along zero, one, and two coordinate axes, respectively.

2. Models of Composites and Their Effective Electromechanical Properties

In this section of the chapter we consider the piezo-active composites with the regular distribution of components and with the 0–3, 1–3, and 2–2 connectivity patterns.

2.1. The 0–3 Composite

In case of the 0–3 composite, it is assumed that centers of symmetry of the spheroidal inclusions (Figure 1) occupy sites of a simple lattice with unit-cell vectors parallel to the OX_j axes. A shape of the isolated inclusion is described by the aspect ratio $\rho = a_1/a_3$ that is assumed to be fixed over the composite sample, where $a_1 = a_2$ and a_3 are semiaxes of the spheroidal inclusion. In each SC inclusion of the 0–3 composite, the main crystallographic axes are oriented as follows: $X \parallel OX_1$, $Y \parallel OX_2$ and $Z \parallel OX_3$. The SC inclusions have a spontaneous polarization $\boldsymbol{P}_s^{(1)} \parallel Z \parallel OX_3$, and the composite sample as a whole is poled along the OX_3 axis. In the 0–3 FC/polymer composite, each inclusion is characterized by the remnant polarization $\boldsymbol{P}_r^{(1)} \parallel OX_3$. Hereafter we denote the inclusion component with subscript "(1)" and the matrix component with subscript "(2)".

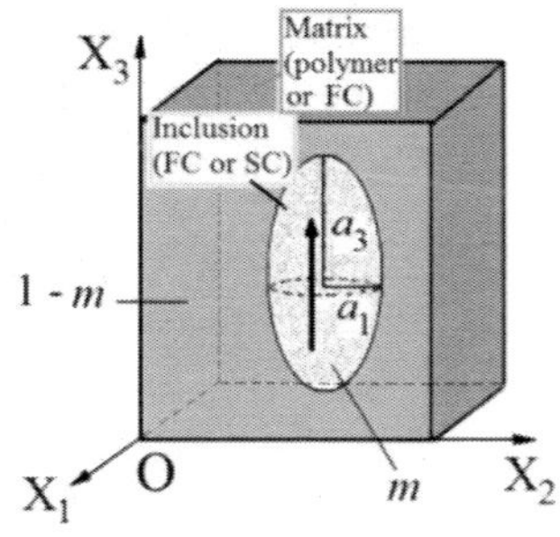

Figure 1. Schematic of the 0–3 composite.

To determine the effective electromechanical properties of the 0–3 composite with the above-described microgeometry, it is possible to apply various averaging procedures [4, 15–19] that allow for an electromechanical interaction between the piezo-active SC inclusions in the matrix. In the present chapter, the prediction of the effective properties of the 0–3 piezo-active composite is concerned with *finite element method* (FEM). Hereby the COMSOL package [20] is applied to obtain the *volume-fraction* dependence of the effective electromechanical properties of the 0–3 composite. In particular, a unit cell, containing the spheroidal inclusion (Figure 1) with radius adjusted to yield the appropriate volume fraction *m,* is discretized using tetrahedral elements. In this schematic figure, the polarization vector of the spheroidal inclusion (i.e., the *remanent polarization* $\boldsymbol{P}_r^{(1)}$ in FC or the *spontaneous polarization* $\boldsymbol{P}_s^{(1)}$ in SC) is denoted by an arrow; $(X_1X_2X_3)$ is the rectangular coordinate system; a_1 and a_3 are semiaxes of the inclusion and m and $1 - m$ are volume fractions of components.

The number of the inclusions, depending on the aspect ratio ρ of the spheroidal inclusion varies from 320,000 to 760,000. To find the volume-fraction dependence of the effective properties of the composite, the following electromechanical constants of components are taken into consideration: *elastic moduli* $c_{ab}^{(n),E}$ measured at electric field E = const., *piezoelectric stress coefficients* $e_{ij}^{(n)}$, and *dielectric permittivity* $\varepsilon_{kk}^{(n),\xi}$ measured at mechanical strain ξ = const. The aforementioned constants are involved in *piezoelectric relations* [21], given below, that link the variables of strain ξ_q and field E_f.

$$\sigma_p = c_{pq}^{E}\,\xi_q - e_{fp}\,E_f$$

and

$$D_k = e_{kl}\xi_l + \varepsilon_{kr}^{\xi}\,E_r$$

The unknown displacement and electric-potential fields are interpolated using linear Lagrangian shape functions. The corresponding number of degrees of freedom varies from

200,000 to 500,000. Periodic boundary conditions are enforced on the boundary of the unit cell, and the matrix of effective constants of the composite is computed column-wise, performing calculations for diverse average strain and electric fields imposed to the structure.

The *Geometric Multigrid* [22] iterative solver (V-cycle, successive over-relaxation pre- and post-smoother, direct coarse solver) is employed. After solving the electro-elastic equilibrium problem, the effective electromechanical constants of the composite are computed, by averaging the resulting local stress and electric-displacement fields over the composite unit cell. As a result, the set of effective elastic moduli c_{pq}^{*E}, piezoelectric coefficients e_{fp}^{*}, and dielectric permittivity $\varepsilon_{kk}^{*\xi}$ is represented for each pair of fixed values m and ρ.

The effective electromechanical properties of the 0–3 composite are determined in a long-wave approximation, i.e., in a case when the wavelength of an external acoustic field is much longer than the spacing between the neighboring inclusions. The interrelationships of the effective piezoelectric coefficients e_{fq}^{*} (stress coefficients), d_{fq}^{*} (charge coefficients), g_{fq}^{*} (voltage coefficients), and h_{fq}^{*} (strain coefficients) are expressed by the following well-known equalities [21]:

$$d_{fp}^{*} = \varepsilon_{fk}^{*\sigma} g_{kq}^{*} = e_{fq}^{*} s_{qp}^{*E}, \quad (1)$$

$$e_{fp}^{*} = \varepsilon_{fk}^{*\xi} h_{kp}^{*} = d_{fq}^{*} c_{qp}^{*E}, \quad (2)$$

$$g_{fp}^{*} = \beta_{fk}^{*\sigma} d_{kp}^{*} = h_{fq}^{*} s_{qp}^{*D}, \quad (3)$$

and

$$h_{fp}^{*} = \beta_{fk}^{*\xi} e_{kp}^{*} = g_{fq}^{*} c_{qp}^{*D}. \quad (4)$$

In the above equations, $\varepsilon_{fk}^{*\sigma}$ is *dielectric permittivity* measured at constant mechanical stress, s_{qp}^{*E} and s_{qp}^{*D} are *elastic compliances* measured at constant electric field and electric displacement, respectively, c_{qp}^{*D} is *elastic modulus* measured at constant electric displacement, and $\beta_{fk}^{*\sigma}$ and $\beta_{fk}^{*\xi}$ characterize *dielectric impermeability* at constant stress and strain, respectively. Taking into account the electromechanical properties from Eqs. (1) – (4), one can evaluate *electromechanical coupling factors* (ECFs) [21] of the composite

$$k_{3j}^{*} = d_{3j}^{*} / (\varepsilon_{33}^{*\sigma} s_{jj}^{*E}). \quad (5)$$

ECFs from Eq. (5) characterize an effectiveness of the conversion of energy from the electric form into the mechanical form and vice versa along the poling axis OX_3 (i.e., longitudinal ECF at $j = 3$) and along the perpendicular axes OX_1 and OX_2 (i.e., transversal ECFs at $j = 1$ and 2).

2.2. The 1–3 Composite

Following the *model concepts* [12], we consider the 1–3 composite (Figure 2) that has a cellular structure and a regular distribution of the cylindrical rods in the large matrix.

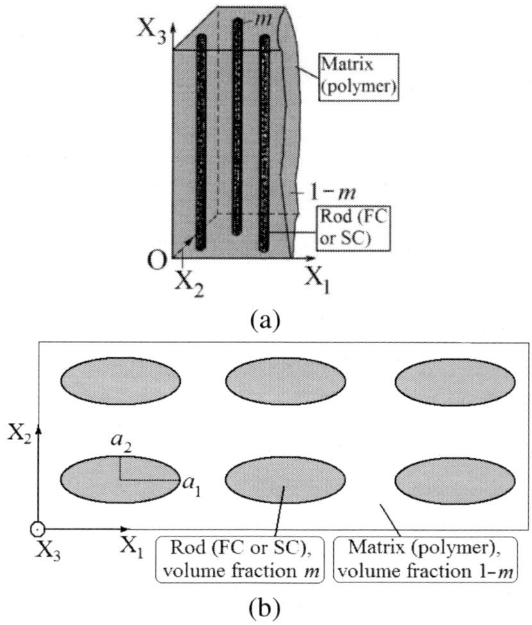

Figure 2. Schematic of (a) the 1–3 composite and (b) cross section of the 1–3 composite by the X_1OX_2 plane. ($X_1X_2X_3$) is the rectangular coordinate system, a_1 and a_2 are semiaxes of the ellipse, m and 1 – m are volume fractions of components.

Cross sections of these rods by the X_1OX_2 plane of the rectangular coordinate system ($X_1X_2X_3$) (Figure 2b) are described by an equation (Eq. (6)) relative to the axes of the coordinate system ($X_1X_2X_3$).

$$(x_1 / a_1)^2 + (x_2 / a_2)^2 = 1 \qquad (6)$$

Semiaxes of the ellipse a_f ($f = 1$ and 2) from Eq. (6) are constant in the whole composite sample. Centers of symmetry of these ellipses are arranged periodically on the OX_1 and OX_2 directions (Figure 2b). All the rods of the 1–3 composite studied are aligned along the OX_3 axis, and the height of each rod obeys the condition $h >> a_f$. The *remanent polarization* vector of each FC rod is $\boldsymbol{P_r}^{(1)} \parallel OX_3$. In each SC rod, the main crystallographic axes are oriented as follows: $X \parallel OX_1$, $Y \parallel OX_2$ and $Z \parallel OX_3$, and the spontaneous polarization is $\boldsymbol{P_s}^{(1)} \parallel Z \parallel OX_3$.

The effective electromechanical properties of the 1–3 composite are determined in a long-wave approximation. It means that sizes of FC grains are much smaller than the shorter axis $2a_2$ of the ellipse from Eq. (6). Moreover, wavelengths of waves being propagated from acoustic radiators are assumed to be much larger than the spacing between the neighboring rods of the composite sample. As in the case of the 0–3 composite, elastic moduli $c_{ab}^{(n),E}$, piezoelectric coefficients $e_{ij}^{(n)}$ and dielectric permittivity $\varepsilon_{pq}^{(n),\xi}$ of the components (n = 1 and 2) are used in calculations of the effective electromechanical properties. These calculations are carried out within the framework of either the *effective field method* (EFM) [4, 12, 15–17] or the FEM [12].

Based on the EFM concepts [4, 15, 16], one can describe an electromechanical interaction in the system "rods – matrix" (Figure 2) using a local field that acts on each rod. This effective field is determined taking into account a system of the interacting piezo-active rods and boundary conditions concerned with the rod shape. The boundary conditions involve components of electric and mechanical fields at the rod – matrix interface. The 9×9 matrix characterizing the electromechanical properties of each component (n = 1 or 2) is written in the form [4]

$$\| C^{(n)} \| = \begin{pmatrix} \| c^{(n),E} \| & \| e^{(n)} \|^t \\ \| e^{(n)} \| & \| -\varepsilon^{(n),\xi} \| \end{pmatrix}, \tag{7}$$

where the superscript t denotes the *transposition*. The effective electromechanical properties of the 1–3 composite are described by the 9×9 matrix

$$\begin{aligned} \| C^* \| = \| C^{(2)} \| + m\,(\| C^{(1)} \| - \| C^{(2)} \|)\ [\| I \| \\ + (1-m)\, \| S \|\, \| C^{(2)} \|^{-1} (\| C^{(1)} \| - \| C^{(2)} \|)]^{-1} \end{aligned} \tag{8}$$

In Eq. (8), the matrices of electromechanical constants of components $\|C^{(1)}\|$ and $\|C^{(2)}\|$ have a form that is shown in Eq. (7), m is the volume fraction of FC, $\| I \|$ is the identity matrix, and $\|S\|$ is the matrix that contains the *Eshelby tensor* components [19, 23] depending on the elements of $\|C^{(2)}\|$ and the rod shape. Hereafter, the ratio $\eta = a_2/a_1$ of semiaxes of the ellipse in the rod cross section (see Figure 2b and Eq. (6)) and the volume fraction m of FC are regarded as two independent parameters to be varied in ranges [12] $0 < \eta \le 1$ and $0 < m < \pi / 4$, respectively.

The limiting case of $\eta = 0$ corresponds to the 2–2 parallel-connected composite, and the limiting case of $\eta = 1$ is relevant to the 1–3 connectivity pattern with circular rod cross section. The upper limit $m = \pi/4$ corresponds to the ratio of the area of the ellipse to the area of the rectangular on condition that this ellipse has been inscribed into the rectangular. This rectangular is regarded as a base of a representative unit cell [12, 24] of the 1–3 composite.

The prediction of the effective electromechanical properties of the 1–3 composite is also carried out using the FEM. It should be mentioned that in the last decades, the FEM has been applied to various 1–3 FC/polymer composites with a periodic arrangement of the rods (see, e.g., papers [24–26]). Following the FEM, the unit-cell model of the 1–3 composite with the regular arrangement of the rods (Figure 2) is put forward, and periodic boundary conditions for the representative unit cell are considered. In the present work, the COMSOL package [20] is applied to obtain the volume fraction dependence of the effective electromechanical properties. The rectangular unit cell, containing the FC rod in the form of the elliptic cylinder with semiaxes a_f (Figure 2b) adjusted to yield the appropriate volume fraction of FC, is discretized using triangular elements. The unknown displacement field is interpolated using second-order Lagrange shape functions, leading to a problem with approximately 120,000 to 200,000 degrees of freedom. The number of the degrees of freedom depends on the ratio of semiaxes η.

Periodicity is enforced at the boundary of the rectangular unit cell of the composite studied. The $||C^*||$ matrix of the effective electromechanical properties of the composite is computed column-wise, performing calculations for diverse average strain and electric fields imposed to the 1–3 structure with elliptic cylinders.

The $||C^*||$ matrix has the form coinciding with that shown in Eq. (7). After solving the equilibrium problem, the *effective elastic moduli* c_{ab}^{*E} (m, η), *piezoelectric coefficients* e_{ij}^{*} (m, η) and *dielectric permittivity* $\varepsilon_{pq}^{*\xi}$ (m, η) of the composite studied are computed [12] by means of averaging the resulting local stress and electric-displacement fields over the unit cell.

2.3. The 2–2 Composite

The next composite pattern to be considered in the present chapter is 2–2 (Figure 3). It is assumed that the layers of the components are parallel-connected and separated by planar interfaces x_1 = const. The effective electromechanical properties of the 2–2 composite are determined using the *matrix method* [4, 10, 27]. Electromechanical constants of the components are given by the 9×9 matrices

$$\| C^{(n)} \| = \begin{pmatrix} \| s^{(n),E} \| & \| d^{(n)} \|^t \\ \| d^{(n)} \| & \| \varepsilon^{(n),\sigma} \| \end{pmatrix}, \tag{9}$$

where $||s^{(n),E}||$ is the matrix of elastic compliances, $||d^{(n)}||$ is the matrix of piezoelectric coefficients and $||\varepsilon^{(n),\sigma}||$ is the matrix of dielectric permittivity of the components, and superscript t denotes the transposition, $n = 1$ is related to the ferroelectric (either FC or SC) component in layers of the 1st type (Figure 3), and $n = 2$ is related to polymer in layers of the 2nd type.

An important advantage of the matrices $||C^{(n)}||$ from Eq. (9) over those from Eq. (7) is concerned with orders-of-magnitude of electromechanical constants therein: in Eq. (9) all constants are simply expressed in 10^{-12} SI units, while constants in Eq. (7) are expressed in SI units with multipliers 10^{-12}, 10^{12}, and 10^{0}.

The matrix of the effective electromechanical properties of the 2–2 composite shown in Figure 3 is given by [4, 27]

$$\| C^* \| = [\| C^{(1)} \| \cdot \| M \| \, m + \| C^{(2)} \| \, (1 - m)] \, [\| M \| \, m + \| I \| \, (1 - m)]^{-1} \tag{10}$$

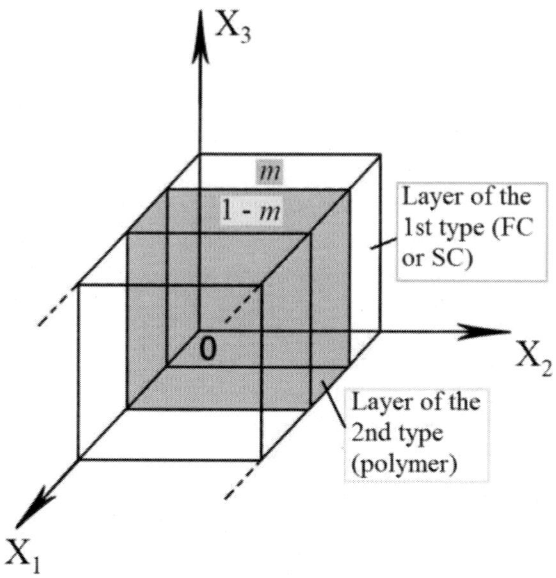

Figure 3. Schematic of the 2–2 parallel-connected composite. ($X_1X_2X_3$) is the rectangular coordinate system, m and 1 – m are volume fractions of components.

The structure of the matrix $\| C^* \|$ from Eq. (10) is similar to that shown in Eq. (9). In Eq. (10), $\| M \|$ is the matrix related to the boundary conditions at interfaces x_1 = const, and $\| I \|$ is the identity matrix.

The boundary conditions at x_1 = const (Figure 3) imply a continuity of components of mechanical stress σ_{11}, σ_{12} and σ_{13}, strain ξ_{22}, ξ_{23} and ξ_{33}, electric displacement D_1, and electric field E_2 and E_3. Taking into account the matrix elements C^*_{ab} from Eq. (10) and using Eqs. (1) – (5), we find the effective piezoelectric coefficients $d^*_{kl}(m)$, $e^*_{kl}(m)$, $g^*_{kl}(m)$, and $h^*_{kl}(m)$ and ECFs $k^*_{3j}(m)$ of the 2–2 composite, and the volume fraction m can be varied from 0 to 1.

To describe the anisotropy of effective parameters in the studied composites, we introduce the following *anisotropy factors*:

$$\zeta_{d31} = d^*_{33}/d^*_{31} \text{ and } \zeta_{d32} = d^*_{33}/d^*_{32} \tag{11}$$

(for the piezoelectric coefficients d^*_{3j} or $g^*_{3j} = d^*_{3j}/\varepsilon^{*\sigma}_{33}$),

$$\zeta_{e31} = e^*_{33} / e^*_{31} \text{ and } \zeta_{e32} = e^*_{33} / e^*_{32} \tag{12}$$

(for the piezoelectric coefficients e^*_{3j} or $h^*_{3j} = e^*_{3j}/\varepsilon^{*\xi}_{33}$), and

$$\zeta_{k31} = k^*_{33}/k^*_{31} \text{ and } \zeta_{k32} = k^*_{33}/k^*_{32} \tag{13}$$

(for ECFs k^*_{3j}).

Among the components (Table 1) of the *anisotropic composites* to be considered, we choose the PZT-type FC (PCR-7M, the abbreviation means "piezoelectric ceramics from Rostov-on-Don") with the moderate *piezoelectric anisotropy* [28], the $PbTiO_3$-type FC with the large piezoelectric anisotropy [29], the relaxor-ferroelectric SC of solid solutions of $(1 - x)Pb(Mg_{1/3}Nb_{2/3})O_3 - xPbTiO_3$ (PMN–xPT) [30], and piezo-passive polymers [31–33] whose elastic properties are varied in a wide range.

It should be noted that in contrast to the overwhelming majority of FCs and related SCs, the $PbTiO_3$-type FC is characterized by the piezoelectric coefficients $e_{3j} > 0$ [4, 29], and this feature is to be taken into consideration in piezotechnical applications. We add that the anisotropy factors from Eqs. (11) and (13) are of value for high-anisotropy piezo-active elements in medical diagnostic devices with ultrasonic antennae based on pulse-echo principles and control of the preferred direction, in hydrophones, and other acoustic devices [34].

Transducer applications of high-anisotropic piezoelectric elements are often based on effective deleting the interaction between the thickness and lateral vibration modes in the vicinity of the resonance of the thickness mode [34].

Table 1. Elastic moduli (in 10^{10} Pa), piezoelectric coefficients and dielectric permittivity measured at room temperature

Component	c_{11}^E	c_{12}^E	c_{13}^E	c_{33}^E	c_{44}^E	e_{31}	e_{33}	e_{15}	$\varepsilon_{11}^{\xi}/\varepsilon_0$	$\varepsilon_{33}^{\xi}/\varepsilon_0$
PCR-7M FC	13.3	9.2	9.1	12.5	2.28	–9.5	31.1	20.0	1980	1810
Modified $PbTiO_3$ FC	14.33	3.221	2.417	13.16	5.59	0.459	6.50	2.96	212	133
PMN–0.33PT *	11.5	10.3	10.2	10.3	6.9	–3.9	20.3	10.1	1434	680
Araldite	0.78	0.44	0.44	0.78	0.17	0	0	0	4.0	4.0
Elastomer	0.11	0.09	0.09	0.11	0.01	0	0	0	5.0	5.0
Polyurethane	0.44	0.26	0.26	0.44	0.09	0	0	0	3.5	3.5

* PMN–0.33PT domain-engineered SC (4*mm* class, (001) cut).

Table 1 shows values of elastic moduli c_{ab}^E, piezoelectric coefficient e_{ij} and dielectric permittivity ε_{kk}^{ξ} that were measured at room temperature (data from [28–33]). For FCs and polymers, the equation $c_{66} = (c_{11}^E - c_{11}^E)/2$ is valid [21]. In PMN–0.33PT domain-engineered SC, $c_{66}^E = 6.6 \cdot 10^{10}$ Pa [30].

Anisotropy factors ζ_{d31} equal –2.17 (PCR-7M FC), –11.6 (modified $PbTiO_3$ FC), and –2.11 (PMN–0.33PT SC), and for the aforementioned piezo-active components, the equalities $\zeta_{d31} = \zeta_{d32}$, $\zeta_{e31} = \zeta_{e32}$, and $\zeta_{k31} = \zeta_{k32}$ are valid.

3. Anisotropy of Piezoelectric Coefficients and Electromechanical Coupling Factors in Composites

In the present section we give a few examples of behavior of the anisotropy factors from Eqs. (11) – (13). These parameters related to the two-component composites based on either FC or SC show that the *piezoelectric anisotropy* as a function of the volume fraction of the ferroelectric component can vary in wide ranges.

3.1. Characteristics of 0–3 Composites

Based on the *microgeometry* of the studied 0–3 composite (Figure 1) and in the presence of ferroelectric components with ∞mm or $4mm$ symmetry, the anisotropy factors from Eqs. (11) – (13) obey the following conditions:

$$\zeta_{d31} = \zeta_{d32},\ \zeta_{e31} = \zeta_{e32},\ \text{and}\ \zeta_{k31} = \zeta_{k32}. \tag{14}$$

Taking into account Eqs. (2) and (5), we represent ζ_{d31} from Eq. (11) and ζ_{k31} from Eq. (13) as

$$\zeta_{d31} = (2e^*_{31}\, s^{*E}_{13} + e^*_{33}\, s^{*E}_{33}) / [e^*_{31}(s^{*E}_{11} + s^{*E}_{12}) + e^*_{33}\, s^{*E}_{13}]$$

$$= (2s^{*E}_{13} + \zeta_{e31}\, s^{*E}_{33}) / (s^{*E}_{11} + s^{*E}_{12} + \zeta_{e31}\, s^{*E}_{13}) \tag{15}$$

and

$$\zeta_{k31} = \zeta_{d31}(s^{*E}_{11} / s^{*E}_{33})^{1/2}, \tag{16}$$

respectively. The anisotropy factors from Eqs. (15) and (16) are functions of the volume fraction of the ferroelectric component m and the aspect ratio of the inclusion ρ (Figure 1). Some trends in changing the anisotropy factors ζ_{e31} and ζ_{d31} of the SC/polymer and SC/FC composites with 0–3 connectivity are shown in Table 2.

The larger value of $|\zeta_{e31}|$ is achieved at the smaller value of ρ (if we assume that condition m = const. holds), and the smaller value of m at ρ = const. is also preferable to achieve the large values of $|\zeta_{e31}|$.

The small m and ρ values are related to the composite with the highly prolate inclusions (Figure 1) that take a small part of the volume of the composite sample as a whole. Such microgeometry promotes both the good polarizability of the composite in the electric field $\boldsymbol{E} \parallel OX_3$ and the considerable anisotropy of elastic properties of the composite on the OX_3 and OX_1 directions. In the same ranges of m at ρ = const., the anisotropy factor ζ_{d31} remains comparable to that of the piezo-active components (Table 2) and is often one order-of-magnitude lower than ζ_{e31}.

Table 2. Anisotropy factors ζ_{e31} and ζ_{d31} which were calculated for 0–3 composites based on PMN–0.33PT SCs (FEM data)

Matrix component	m = 0.05	m = 0.20	m = 0.50						
	ζ_{e31} at ρ= 0.1	ζ_{e31} at ρ= 0.3	ζ_{e31} at ρ= 0.5	ζ_{e31} at ρ= 0.1	ζ_{e31} at ρ= 0.3	ζ_{e31} at ρ= 0.5	ζ_{e31} at ρ= 0.1	ζ_{e31} at ρ= 0.3	ζ_{e31} at ρ= 0.5
Araldite	–27.5	–7.71	–4.56	–22.6	–5.76	–3.53	–12.3	–4.20	–3.31
Poly-urethane	–10.7	–6.94	–4.84	–6.60	–4.47	–3.56	–6.24	–3.94	–3.43
Elastomer	–10.9	–7.94	–5.80	–6.10	–4.49	–3.75	–4.26	–2.83	–2.74
Modified $PbTiO_3$ FC	115	88.2	68.3	–19.5	–21.4	–26.3	–8.13	–8.40	–8.86
Matrix component	m = 0.05	m = 0.20	m = 0.50						
	ζ_{d31} at ρ=0.1	ζ_{d31} at ρ= 0.3	ζ_{d31} at ρ= 0.5	ζ_{d31} at ρ=0.1	ζ_{d31} at ρ= 0.3	ζ_{d31} at ρ= 0.5	ζ_{d31} at ρ= 0.1	ζ_{d31} at ρ= 0.3	ζ_{d31} at ρ= 0.5
Araldite	–2.63	–2.43	–2.30	–2.48	–2.30	–2.19	–2.20	–2.12	–2.09
Poly-urethane	–2.47	–2.38	–2.29	–2.38	–2.25	–2.18	–2.60	–2.41	–2.35
Elastomer	–2.16	–2.14	–2.12	–2.16	–2.11	–2.09	–2.47	–2.32	–2.29
Modified $PbTiO_3$ FC	–6.34	–6.40	–6.48	–4.10	–4.13	–4.20	–2.80	–2.80	–2.81

This distinction is inseparably connected with the role of the elastic properties of the composite in accordance with Eq. (15) and with the presence of the isolated piezo-active inclusions. Only at small values of m, in the PMN–0.33PT SC/modified $PbTiO_3$ FC composite, the large values of ζ_{e31} and $|\zeta_{d31}|$ are achieved due to the strong influence of the high-anisotropic FC. In the studied 0–3 composites, the anisotropy factor ζ_{k31} also considerably depends on the elastic properties [see Eq. (16)], however ECF k_{33}^{*} remains relatively small in the wide range of m even at small values of ρ. For example, in the PMN–0.33PT SC/araldite composite, we have k_{33}^{*} = 0.678 at m = 0.50 and ρ = 0.1, while $k_{33}^{(1)}$ = 0.962 for PMN–0.33PT SC. The decrease in k_{33}^{*} is associated with the relatively small piezoelectric coefficient d_{33}^{*} of the composite that contains the isolated inclusions with the high piezoelectric activity.

3.2. Characteristics of 1–3 Composites

The 1–3 composite structure enables us to avoid the aforementioned decrease in the effective parameters peculiar to the 0–3 composite. In the 1–3 composite, the ferroelectric component (either FC or SC) is distributed continuously along the poling axis OX_3 (Figure 2a), and this circumstance improves the piezoelectric activity concerned with d_{33}^{*} and promotes the anisotropy of electromechanical properties measured along OX_3 and OX_1. Below we consider examples of the piezoelectric anisotropy in some 1–3 composites (Figure 4).

At the first stage, we consider the anisotropy factors of the 1–3 composites with circular cross sections of rods (i.e., at η = 1), and the polymer matrix surrounding the rods remains unchanged. Again in the presence of ferroelectric components with ∞mm or 4 mm symmetry, the anisotropy factors obey conditions in Eq. (14).

Simple comparison of data (Figure 4) suggests that conditions

$$|\zeta_{e31}| >> |\zeta_{d31}| \text{ and } |\zeta_{e31}| >> |\zeta_{k31}| \qquad (17)$$

are valid in the wide range of m. Moreover, the mutual arrangement of curves in graphs (cf. Figure 4a, b, and c) enables us to surmise that differences between the elastic properties of components would influence the anisotropy factors from Eqs. (15) and (16) and gives rise to considerable changes in the dependence of $\zeta_{k31}(m)$.

The presence of the high-anisotropic FC leads to the fulfillment of conditions $|\zeta_{k31}| >> 1$ and $|\zeta_{d31}| > 5$ in wide ranges of m (see curves 2 and 3 in Figure 4c).

At the second stage, we compare data on the 1–3 composites with high-anisotropic FC components.

Among these materials of interest are solid solutions of $(Pb_{1-x}Ca_x)TiO_3$ with the considerable anisotropy of the piezoelectric coefficients $e_{3j}^{(1)}$ and $d_{3j}^{(1)}$ [35].

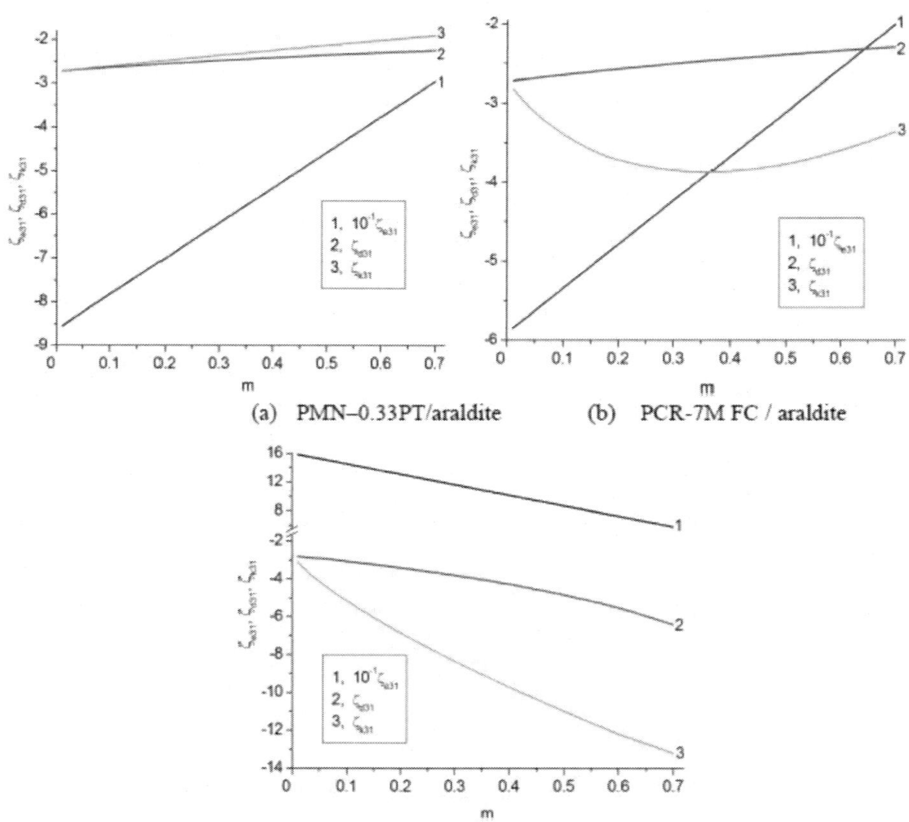

(a) PMN–0.33PT/araldite (b) PCR-7M FC / araldite

(c) Modified $PbTiO_3$ FC/araldite

Figure 4. Anisotropy factors calculated for 1–3 composites with the circular cross section of piezo-active rods ($\eta = 1$). In the whole range of m, differences between the FEM and EFM data are less than 2%.

As in the modified $PbTiO_3$FC [29], the positive values of $e_{3j}^{(1)}$ are observed in the wide x range (Table 3). The values in this table were calculated for $(Pb_{1-x}Ca_x)TiO_3$ FCs by means of the effective medium method [35]. For FCs, the equation $c_{66} = (c_{11}^E - c_{11}^E)/2$ is valid [21].

Table 3. Elastic moduli c_{ab}^{E} (in 10^{10} Pa), piezoelectric coefficients e_{ij} [C/m^2], and dielectric permittivity ε_{kk}^{ξ} calculated for $(Pb_{1-x}Ca_x)TiO_3$ FCs

x	c_{11}^{E}	c_{12}^{E}	c_{13}^{E}	c_{33}^{E}	c_{44}^{E}	e_{31}	e_{33}	e_{15}	$\varepsilon_{11}^{\xi}/\varepsilon_0$	$\varepsilon_{33}^{\xi}/\varepsilon_0$
0.10	19.1	5.22	5.15	18.6	6.85	0.676	3.90	1.62	111	109
0.15	18.8	5.09	5.01	18.3	6.76	0.751	4.15	1.73	118	115
0.20	18.5	4.76	4.68	18.0	6.79	0.844	4.31	1.77	126	123
0.24	18.4	4.72	4.61	17.9	6.76	1.07	4.69	1.87	145	141
0.25	18.9	5.19	5.06	18.3	6.76	1.33	5.09	1.95	152	147
0.30	18.6	4.89	4.80	18.1	6.77	1.08	5.48	2.27	202	197

Table 4. Anisotropy factors calculated for the 1–3 $(Pb_{1-x}Ca_x)TiO_3$FC/araldite composite with the circular cross section of piezo-active rods (η = 1, EFM data)

m	0.01	0.05	0.10	0.30	0.50	0.70
x = 0.10						
ζ_{e31}	84.4	81.2	77.2	61.3	45.5	29.6
ζ_{d31}	–2.84	–2.95	–3.10	–3.83	–4.88	–6.49
ζ_{k31}	–3.21	–4.49	–5.75	–9.51	–12.6	–15.0
x = 0.15						
ζ_{e31}	79.2	76.3	32.5	57.6	42.8	27.9
ζ_{d31}	–2.84	–2.95	–3.11	–3.88	–4.99	–6.77
ζ_{k31}	–3.21	–4.48	–5.74	–9.57	–12.8	–15.6
x = 0.20						
ζ_{e31}	71.2	68.5	65.2	51.8	38.5	25.1
ζ_{d31}	–2.85	–2.97	–3.14	–4.01	–5.34	–7.67
ζ_{k31}	–3.21	–4.50	–5.79	–9.87	–13.7	–17.6
x = 0.24						
ζ_{e31}	60.0	57.8	55.0	43.7	32.5	21.2
ζ_{d31}	–2.86	–3.00	–3.19	–4.19	–5.86	–9.21
ζ_{k31}	–3.22	–4.53	–5.86	–10.3	–15.0	–21.1
x = 0.25						
ζ_{e31}	53.5	51.5	49.0	39.0	28.9	18.9
ζ_{d31}	–2.86	–3.01	–3.21	–4.27	–6.12	–10.1
ζ_{k31}	–3.24	–4.57	–5.92	–10.6	–15.7	–23.2
x = 0.30						
ζ_{e31}	71.3	68.6	65.2	51.9	38.5	25.1
ζ_{d31}	–2.84	–2.97	–3.14	–3.99	–5.29	–7.54
ζ_{k31}	–3.21	–4.50	–5.78	–9.83	–13.5	–17.3

The presence of the high-anisotropic FC promotes the fulfillment of conditions (17) in the wide range of m. The anisotropy factors ζ_{d31} and ζ_{k31} (Table 3) correlate with those of the FC component: maximum values of $|\zeta_{d31}|$ and $|\zeta_{k31}|$ in the composite are achieved at molar concentrations $x \approx 0.24 \ldots 0.25$ and these molar concentrations are related to the FC compositions [35] with the largest values of $d_{33}^{(1)}/|d_{31}^{(1)}|$ and $k_{33}^{(1)}/|k_{31}^{(1)}|$. It should be noted that condition (16) promotes the larger anisotropy of ECFs k_{3j}^{*}, especially at volume fractions of FC $m \geq 0.30$ (Table 4). In this case the elastic anisotropy of the composite becomes considerable due to both the microgeometry and the presence of the anisotropic FC.

At the third stage, we vary the aspect ratio η and the volume fraction m in the 1–3 composite based on FC. In the recent study [12], examples of the piezoelectric response and anisotropy were analyzed for the 1–3 composites based on either "soft" or "hard" FC and having either piezo-active or piezo-passive matrix. Combinations of the FC and polymer components were first analyzed that provide different volume fraction dependences of the piezoelectric coefficients d_{3j}^* and g_{3j}^* at η = const. Now we choose the matrix material among so-called auxetic polymers with negative Poisson's ratios [36–38].

The auxetic material expands laterally when stretched [37], in contrast to ordinary materials with the positive Poisson's ratios. Hereby it should be noted that all isotropic materials (including polymers) obey the law of conservation of energy, show a positive definite matrix of elastic compliances and, as a consequence, can have the Poisson's ratio in the range from –1 to 0.5 [36, 37]. Some results on the 1–3 FC/auxetic polymer composites with the improved hydrostatic piezoelectric response were discussed by Topolov and Bowen [39]. As follows from literature data on the piezo-active composites, work to date has not attempted to analyze the influence of the composite *microgeometry* and the auxetic polymer matrix on the piezoelectric anisotropy of the 1–3 composite based on FC.

The auxetic polymer matrix of the 1–3 composite is assumed to be made of polymethacrylimide (PMI) which is a closed-cell foam piezo-passive material with Young's modulus $E^{(2)}$ = 259.93 MPa, shear modulus $G^{(2)}$ = 149.96 MPa, and Poisson's ratio $\nu^{(2)}$ = –0.12 at room temperature [40]. As in paper [39], we assume that the value of dielectric permittivity is $\varepsilon_{pp}^{(2)}/\varepsilon_0 = 3.5$, since it is a typical value of similar dense polymers (polyurethane etc. [32, 33]).

As follows from our FEM calculations, the studied 1–3 PCR-7M FC / auxetic PMI composite is characterized by the large anisotropy of both e_{3j}^* and d_{3j}^*. Conditions $|\zeta_{e31}| >> 1$ and $|\zeta_{e32}| >> 1$ are valid in the whole studied volume-fraction and aspect-ration ranges (i.e., at $0 < m \leq 0.7$ and $0.01 \leq \eta \leq 1$). Graphs of ζ_{d31} and ζ_{d32} (Figure 5) suggest that the anisotropy factors ζ_{d31} and ζ_{d32} can be infinitely large due to $d_{31}^* = 0$ and $d_{32}^* = 0$ at $0 < m < 0.2$ and $0.01 \leq \eta \leq 1$. Values of m, at which conditions

$$\zeta_{d31} \to \pm\infty \text{ and } \zeta_{d32} \to \pm\infty \tag{18}$$

hold, depend on a balance of elastic constants of the components and on the aspect ratio η. As $\eta \to 1$, a difference between the anisotropy factors from conditions in Eq. (18) reduces.

Changes in $\operatorname{sgn} d_{31}^*$ and $\operatorname{sgn} d_{32}^*$ are caused by the presence of the *auxetic polymer matrix* that would expand laterally in the stretched state. Taking into account Eq. (16), we note that the volume-fraction dependence of the anisotropy factors ζ_{k31} and ζ_{k32} is similar to that shown in Figure 5: $\zeta_{k31} \to \pm\infty$ as condition $d_{31}^* = 0$ holds and $\zeta_{k32} \to \pm\infty$ as equality $d_{32}^* = 0$ is valid. Contrary to the sign-variable behavior of d_{31}^* and d_{32}^*, the piezoelectric coefficients $e_{31}^* < 0$ and $e_{32}^* < 0$ at $0 < m \leq 0.7$ and $0.01 \leq \eta \leq 1$. Such a distinction appears due to the remarkable effect of the elastic properties of the auxetic polymer matrix on the piezoelectric properties of the composite and has no analogs among the composites analyzed earlier.

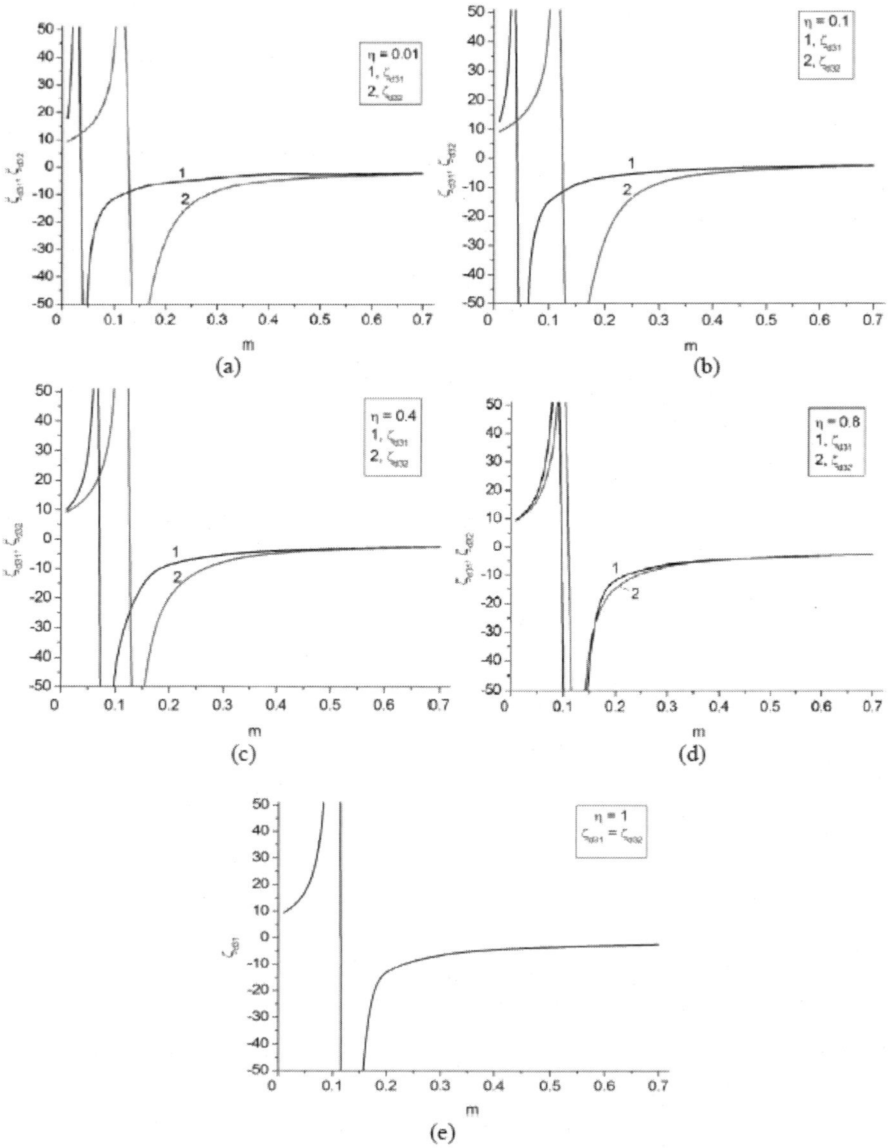

Figure 5. Anisotropy factors calculated for the 1–3 PCR-7M FC/auxetic PMI composite at η = const. (FEM data).

3.3. Characteristics of 2–2 Composites

The 2–2 composite with the parallel connection of the layers (Figure 3) is regarded as a limiting case of the studied 1–3 composite (Figure 2b) as $\eta \to \infty$, i.e., as the condition $a_2 >> a_1$ holds. In the 2–2 composite, the volume fraction m of the ferroelectric component (layers of the 1st type, see Figure 3) can be varied from 0 to 1 without microgeometric restrictions.

Combining the components, we attain large absolute values of anisotropy factors from Eqs. (11) – (13). Data from Figure 6 suggest that conditions

$$|\zeta_{e31}| >> |\zeta_{e32}| >> 1 \text{ and } \zeta_{d31} \approx \zeta_{d32} \tag{19}$$

are valid in the presence of various FC and SC components. The volume-fraction ranges, wherein conditions in Eq. (19) are valid, depend on the anisotropy of the piezoelectric and elastic properties of the ferroelectric component in the layers of the 1st type.

It is also seen that anisotropy factors ζ_{d31} and ζ_{d32} remain comparable to the $d_{33}^{(1)} / d_{31}^{(1)}$ ratio related to the ferroelectric component. Such behavior is a consequence of the laminar structure of the composite (Figure 3) with the continuous distribution of the ferroelectric component along the poling direction and along one of the perpendicular directions.

Drastic changes in anisotropy factors from Eqs. (11) – (13) are observed after replacing the *araldite layers* (polymer with positive Poisson's ratio) with the auxetic PMI layers: the latter component is characterized [40] by negative Poisson's ratio. Absolute values of anisotropy factors ζ_{e31} and ζ_{e32} become larger (Figure 7a) in comparison to those shown in Figure 6.

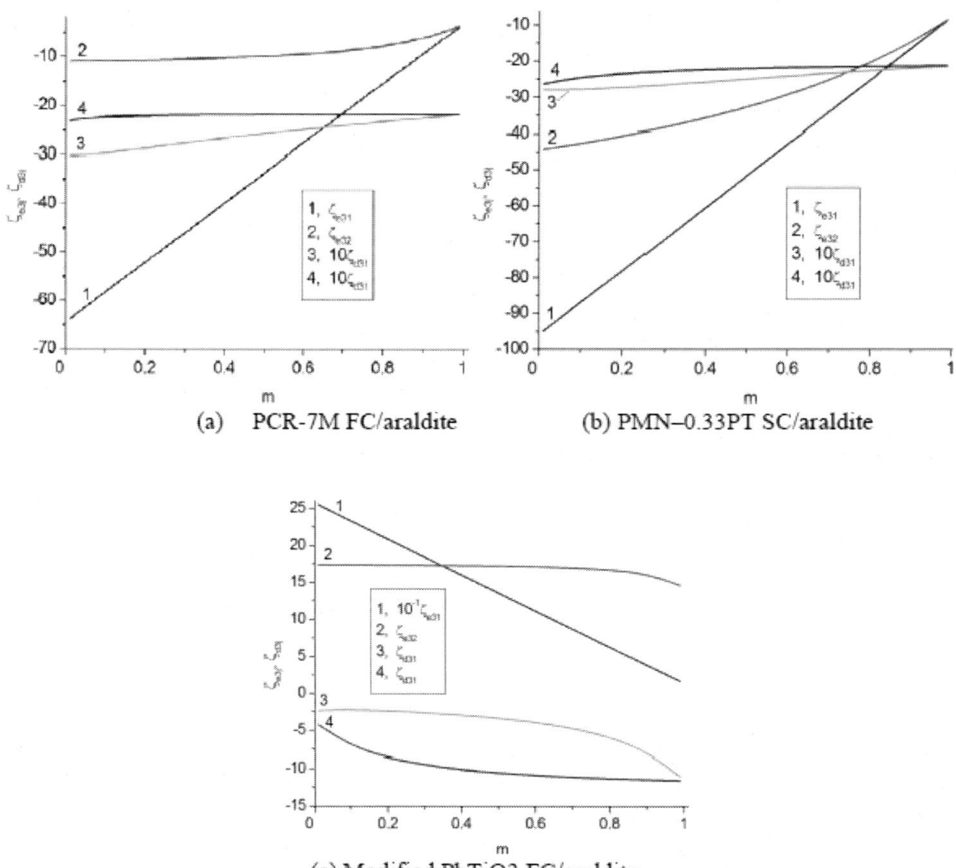

Figure 6. Anisotropy factors calculated for 2–2 composites with the following components. Calculations were performed using the matrix method.

Conditions (18) are achieved (Figure 7b) due to the presence of the auxetic polymer component that strongly re-distributes the mechanical stress field and, therefore, influences the balance of the piezoelectric coefficients d_{3j}^{*}.

As a consequence, ECFs satisfy conditions $\zeta_{k31} \to \pm\infty$ and $\zeta_{k32} \to \pm\infty$ (Figure 7c). Figure 7c represents a "deformed" Figure 7b because of the effect of elastic properties on ECFs [see, for instance, Eq. (16)]. The relatively small volume fraction m_1, at which conditions $\zeta_{d31} \to \pm\infty$ and $\zeta_{k31} \to \pm\infty$ take place (Figure 7b and c), may be shifted to the larger values at the proper combination of the FC and polymer components. To the best of our knowledge, the position of the m_1 point considerably depends on ratios of elastic compliances $s_{ab}^{(1),E} / s_{ab}^{(2)}$ of components. Hereby it is also important to reduce a difference $m_2 - m_1$, where m_2 is the volume fraction, at which conditions $\zeta_{d32} \to \pm\infty$ and $\zeta_{k32} \to \pm\infty$ hold simultaneously.

The related problem of the optimization of the volume fractions m_i is of an independent interest and can be considered in further studies on piezo-active composites with auxetic components.

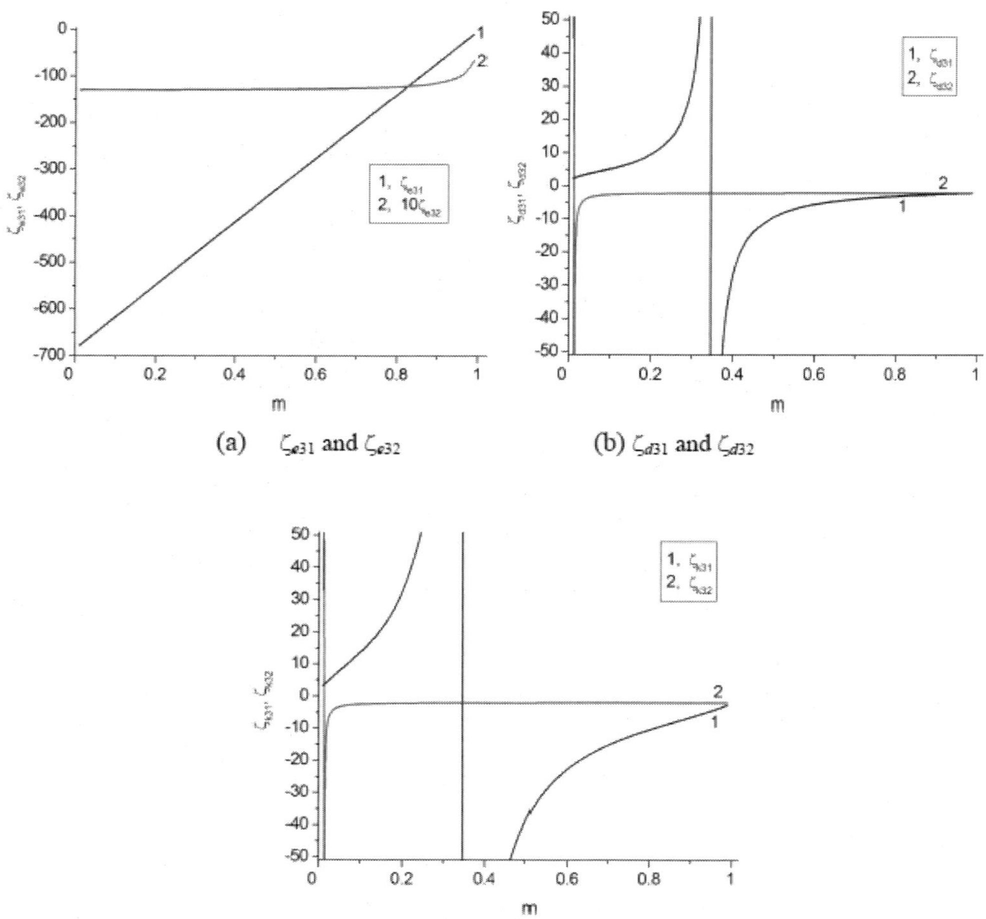

(a) ζ_{e31} and ζ_{e32} (b) ζ_{d31} and ζ_{d32}

(c) ζ_{k31} and ζ_{k32}

Figure 7. Anisotropy factors calculated for the 2–2 PCR-7M FC/auxetic PMI composite. Calculations were performed using the matrix method.

Finally, a limiting case of the 1–3 composite (Figure 2b) with $\eta \to 0$ (i.e., at $a_1 >> a_2$) leads to the 2–2 composite with the layers separated by interfaces x_2 = const. Such an orientation of the interfaces requires a denomination of subscripts of anisotropy factors in Figure 6 and Figure 7.

CONCLUSION

In this chapter, we have predicted and analyzed the effective electromechanical properties of the 0–3, 1–3, and 2–2 composites based on ferroelectrics (either FCs or SCs with the perovskite-type structure). The studied two-component composites are characterized by the certain anisotropy of the piezoelectric coefficients and ECFs [see Eqs. (11) – (13)]. Factors of the anisotropy of the composite are concerned with its electromechanical properties that depend on

(i) the electromechanical properties of components (additionally taking into account domain and heterophase structures, chemical compositions of ferroelectric components, etc.),
(ii) connectivity of the composite, and
(iii) features of the arrangement of components in the composite.

In the context of item (i), it should be noted that the auxetic polymer component promotes the large piezoelectric anisotropy of the composite even in the presence of ferroelectric components with the low or moderate anisotropy.

Concerning item (ii), one can emphasize that increasing the number of axes of the continuous distribution of the ferroelectric components in the composite favors its piezoelectric anisotropy. In the present chapter, we considered simple cases of the regular distribution of the components, and deviations from the regular distribution even for a separate connectivity might become a subject of an independent study concerned, for example, with the proper FEM modeling. Examples of the large (and even infinitely large) anisotropy are described in accordance with conditions given in Eqs. (17) – (19).

These results are of value for many piezotechnical applications of modern composite materials (transducers, sensors, actuators, hydrophones, etc.). Data calculated using different methods correlate and can be used for a prediction of the performance of the anisotropic piezo-active composites with the widespread connectivity patterns.

ACKNOWLEDGMENTS

The authors would like to thank Prof. Dr. I. A. Parinov, Prof. Dr. A. E. Panich, and Dr. V. V. Eremkin (Southern Federal University, Rostov-on-Don, Russia), Prof. Dr. C. R. Bowen (University of Bath, Bath, UK), and Prof. Dr. M. Lethiecq, Dr. F. Levassort, and Dr. D. Certon (François Rabelais University, Tours, France) for their interest in the research problems.

REFERENCES

[1] *Encyclopædia Britannica 2006*, Ultimate Reference Suite DVD, www.britannica.com, London (2006).

[2] *Physics Encyclopædia,* V. 1, Editor-in-Chief A. M. Prokhorov, Sovetskaya Entsiklopediya, Moscow (1988, in Russian).

[3] R. E. Newnham, D. P. Skinner, and L.E. Cross, *Mater. Res. Bull.* 13, 525 (1978).

[4] V. Yu. Topolov and C. R. Bowen, *Electromechanical Properties in Composites Based on Ferroelectrics*, Springer, London (2009).

[5] E. K. Akdogan, M. Allahverdi, and A. Safari, *IEEE Trans. Ultrason., Ferroelec., a. Freq. Contr.* 52, 746 (2005).

[6] T. Ritter, X. Geng, K. K. Shung, P. D. Lopath, S.-E. Park, and T. R. Shrout, *IEEE Trans. Ultrason., Ferroelec., a. Freq. Contr.* 47, 792 (2000).

[7] F. Wang, C. He, Y. Tang, X. Zhao, and H. Luo, *Mater. Chem. Phys.* 105, 273 (2007).

[8] C. R. Bowen, A. Perry, R. Stevens, and S. Mahon S, *Integr. Ferroelectrics,* 32, 333 (2001).

[9] V. Yu. Topolov, A. V. Krivoruchko, and C.R. Bowen, *Ferroelectrics,* 351, 145 (2007).

[10] V. Yu. Topolov and A. V. Krivoruchko, *Smart Mater. Struct.* 18, 065011 (2009).

[11] V. Yu. Topolov and A. E. Panich, *Ferroelectrics,* 392, 107 (2009).

[12] V. Yu. Topolov and P. Bisegna, *J. Electroceram.* 25, 26 (2010).

[13] V. Yu. Topolov, P. Bisegna, and C. R. Bowen, *Ferroelectrics,* 413, 176 (2011).

[14] V. Yu. Topolov, A. V. Krivoruchko, and P. Bisegna, *Compos. Sci. a. Technol.* 71, 1082 (2011).

[15] M. L. Dunn, *J. Appl. Phys.* 73, 5131 (1993).

[16] V. M. Levin, M. I. Rakovskaja, and W. S. Kreher, *Internat. J. Solids a. Struct.* 36, 2683 (1999).

[17] N. Fakri, L. Azrar, and L. El Bakkali, *Internat. J. Solids a. Struct.* 40, 361 (2003).

[18] C. Poizat and M. Sester, *Comput. Mater. Sci.* 16, 89 (1999).

[19] J. H. Huang and W.-S. Kuo, *Acta Mater.* 44, 4889 (1996).

[20] COMSOL, Inc. *COMSOL Multiphysics™ User's Guide* (version 3.5a, 2008), http://www.comsol.com.

[21] T. Ikeda, *Fundamentals of Piezoelectricity*, Oxford University Press, Oxford, New York, and Toronto (1990).

[22] W. Hackbusch, *Multi-grid Methods and Applications*, Springer, Berlin (1985).

[23] J. H. Huang and S. Yu, *Compos. Engin.* 4, 1169 (1994).

[24] H. Berger, S. Kari, U. Gabbert, R. Rodríguez-Ramos, J. Bravo-Castillero, and R. Guinovart-Díaz, *Mater. Sci. a. Engin. A* 412, 53 (2005).

[25] R. Kar-Gupta, C. Marcheselli, and T. A. Venkatesh, *J. Appl. Phys.* 104, 024105 (2008).

[26] V. Yu. Topolov, P. Bisegna, and A. V. Krivoruchko, *J. Phys. D: Appl. Phys.* 41, 035406 (2008).

[27] E. Akcakaya and G. W. Farnell, *J. Appl. Phys.* 64, 4469 (1988).

[28] Ya. Dantsiger, O. N. Razumovskaya, L. A. Reznitchenko, L. D. Grineva, R. U. Devlikanova, S. I. Dudkina, S. V. Gavrilyatchenko, N. V. Dergunova, and A. N. Klevtsov, *Highly Effective Piezoceramic Materials (Handbook)*, Kniga, Rostov-on-Don (1994, in Russian).

[29] S. Ikegami, I. Ueda, T. Nagata, *J. Acoust. Soc. Am.* 50, 1060 (1971).
[30] R. Zhang, B. Jiang, W. Jiang, and W. Cao, *J. Appl. Phys.* 90, 3471 (2001).
[31] F. Levassort, M. Lethiecq, D. Certon, and F. Patat, *IEEE Ultrason., Ferroelect., a. Freq. Contr.* 44, 445 (1997).
[32] A. Grekov, S. O. Kramarov, and A. A. Kuprienko, *Mech. Compos. Mater.* 25, 54 (1989).
[33] L. V. Gibiansky and S. Torquato, *J. Phys. a. Mech. Solids*, 45, 689 (1997).
[34] V. Turik and V. Yu. Topolov, *J. Phys. D: Appl. Phys.* 30, 1541 (1997).
[35] S. V. Glushanin, V. Yu. Topolov, and A. V. Krivoruchko, *Mater. Chem. a. Phys.* 97, 357 (2006).
[36] W. A. Smith, *Proc. IEEE Ultrasonics Symp. Dec. 8 – 11, 1991, Lake Buena Vista, FL, USA*. V. 1, IEEE, New York, NY, p. 661–666.
[37] R. Lakes, *Science*, 235, 1038 (1987).
[38] Y. C. Wang and R. S. Lakes, *J. Appl. Phys.* 90, 6458 (2001).
[39] V. Yu. Topolov and C. R. Bowen, *Modelling a. Simulat. Mater. Sci. a. Engin.* 16, 015007 (2008).
[40] E. O. Martz, T. Lee, R. S. Lakes, V. K. Goel, and J. B. Park, *Cellular Polymers*, 15, 229 (1996).

In: Anisotropy Research
Editor: Hirpa Lemu

ISBN 978-1-62081-977-7

Chapter 3

THE PORE ANISOTROPY IN POROUS SOLIDS

Philippos J. Pomonis[1,*] and Gerasimos S. Armatas[2]
[1]Department of Chemistry, University of Ioannina, Ioannina, Greece
[2]Department of Materials Science and Technology,
University of Crete, Heraklion, Greece

ABSTRACT

The pore structures of solid materials can be better understood when described in terms of their morphological parameters. These include the surface area, pore volume, pore size distribution as well as more elusive parameters of pore connectivity and pore tortuosity. In practice, those textural and morphological parameters are routinely obtained from gas (e.g. nitrogen) physisorption measurements using appropriate algorithms (usually the BJH method) based on pore models with cylindrical or slit-like pore geometry.

However, the main assumption in these considerations is the isotropic pore size (i.e. when the diameter is equal to the pore length) that, in many cases, may be inaccurate. The pores in a real solid are complicated. They, in addition to their diameter (D), possess an additional dimension that is the pore length (L). This leads to the concept of pore anisotropy (b), which is the ratio between the pore length and diameter ($b = L/D$), and to the development of a methodology for its estimation using standard N_2 adsorption data.

The geometry of pores is a crucial parameter for the industry of catalysts and absorbents as well as for the separation and crude oil recovery technologies. This is because the pore size and geometry can greatly affect the diffusion and mass transport processes within the porous structures.

This review chapter highlights recent developments in the determination of pore geometry and especially of pore anisotropy of various structured porous solids. Porous materials both with ordered and disordered pore structure are considered. A summary of the efforts to correlate the pore anisotropy with pore connectivity and microporosity are also discussed.

---

[*] E-mail: ppomonis@cc.uoi.gr.

1. INTRODUCTION

Anisotropy, as opposed to *isotropy*, is a term indicating that a physical, physicochemical, geometrical or biological property of a system develops in a different way along different spatial axes. The term originates from the Greek words *"αν-ισος + τροπος"*, meaning *"un-equal + way"*.

Anisotropic behavior, phenomena and applications are quite common in many branches of science and technology. In *chemistry,* for instance, one finds relevant phenomena such as the anisotropic chemical shift in NMR (nuclear magnetic resonance) spectroscopy, the fluorescence anisotropy and the anisotropic filtering of particles and liquid crystals. There are also plenty of systems exhibiting anisotropy in *physics* like the magnetic anisotropy and the "filamentation" observed in plasmas, the celebrated optical anisotropy of refractive index of light in Iceland crystal and the anisotropy of cosmic microwave background radiation. The seismic anisotropy and the anisotropy of hydraulic conductivity of aquifers are also very important *geological and geophysical* parameters in oil exploitation and the rational use of water respectively. In *medical acoustics*, anisotropy is a property of medical ultrasound imaging describing different echogenicity from various tissues. Further, diffusion tensor imaging in n*euroscience* is an MRI (magnetic resonance imaging) technique that measures the *fractional anisotropy* of the random motion-Brownian-motion of water molecules in the brain. The other important area of anisotropy behavior is in *material science and engineering*. Many properties such as the Young's modulus, i.e. a measure of the materials stiffness, depend on the anisotropy of material. For example, wood and fiber-reinforced composites are very anisotropic and exihibiting much stronger physical properties along the grain/fiber direction, while metals and alloys tend to be more isotropic.

The most common example of anisotropy in nature is certainly the development of plants, which is favored along the vertical axis than along the horizontal direction. This *directional development* becomes apparent in wood, which is easy to split along its grain (Figure 1). The details at cellular and molecular level causing the development of such structures is a very important subject of *biology* in general and plant biology in particular [1 - 3], a point to be discussed at the end of this chapter.

In contrast to the anisotropic development of plants, stemming from their need to grasp from their environment as much energy (light) and mass (CO_2) as possible, dead nature, e.g. inorganic materials, usually do not have any particular reason to develop anisotropically. There exist also few natural inorganic materials exhibiting anisotropic structural development like clays and the more recent ordered mesoporous MCM (mobile composition of matter) and SBA *(Santa Barbara amorphous)* type materials.

Perhaps the most common classes of inorganic natural material exhibiting structural anisotropy are *clays and pillared clays* [5 - 7]. Their structure is made up of aluminosilicate ($Al_2O_3{\cdot}2SiO_3{\cdot}2H_2O$) sheets forming two dimensional galleries as shown in Figure 2. The mass transfer in a large collection of such sheets is favored along the long axis parallel to sheets and inhibited along the short axis, which is perpendicular to the sheets. As a result, such materials are used for the construction of natural barriers of polluting agent (e.g. radionuclides) in soil. Between the aluminosilicate layers of thickness ~10Å and width ~1000Å, various removable molecules (e.g. H_2O) or stable particles (e.g. Keggin-type $[Al_{13}O_4(OH)_{24}(H_2O)_{12}]^{7+}$ polycations) can be intercalated. In the last case, the materials are known as *pillared clays* (PILCs) and have permanent two-dimensional (2D) anisotropic pore structure. In addition to wood and clays, which are natural anisotropic materials, there are also some very interesting anisotropic structural materials made in the laboratory. For instance, at the beginning of the 1990s, a group of scientist from Mobil Corporation developed a novel class of ordered mesoporous materials that exhibit long range pore anisotropy [8 - 11]. These solids, termed as M41S and include the MCM-41 and MCM-48 type materials that exhibit ordered pore structure in hexagonal and cubic morphology respectively, are shown in Figure 3a. The wall of the M41S materials is usually of silicate nature, but other metals like Al, Zr, Ge, etc. can be used as well. In Figure 3b [9], a typical formation mechanism of MCM-41 is illustrated schematically where surfactant molecules, such as alkyl trimethylammonium bromide (C_nTAB, n=12-18), are self-organized in micelles that, consequently, ***pack*** together ***into*** hexagonal arrays through electrostatic or *van der Waals* interactions. In other words, the molecules are arranged vertically to the long axis of rod-like micelles, developing a strongly anisotropic structure.

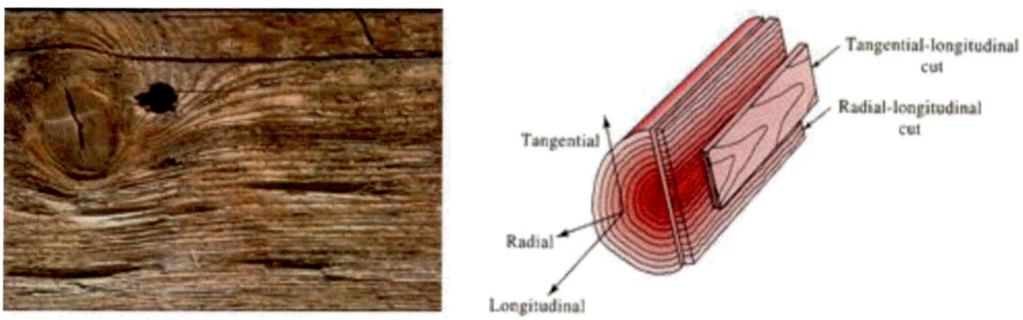

Figure 1. The most common example of anisotropy in nature is the directional development of plants exemplified in the structure of wood [4].

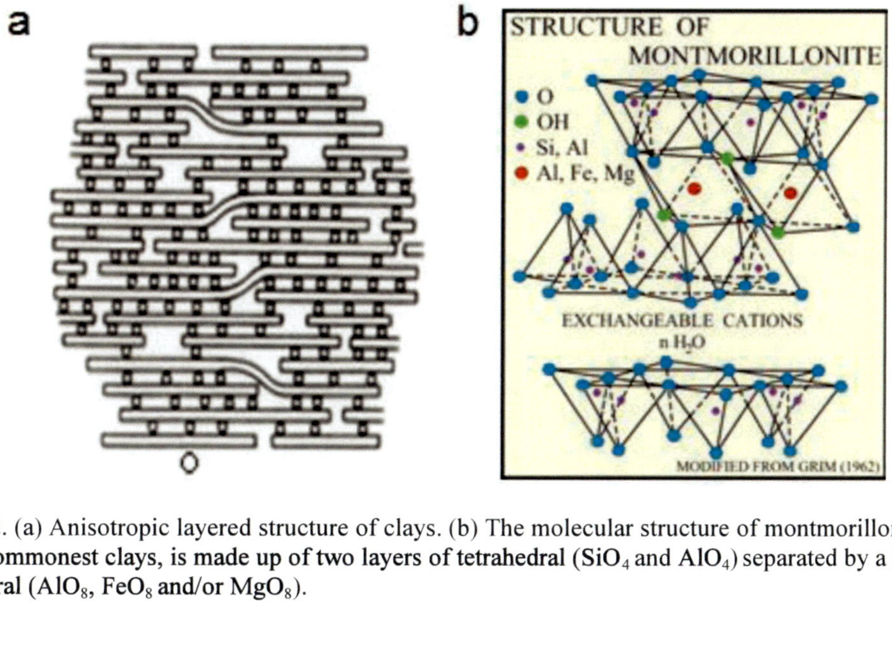

Figure 2. (a) Anisotropic layered structure of clays. (b) The molecular structure of montmorillonite, one of the commonest clays, is made up of two layers of tetrahedral (SiO_4 and AlO_4) separated by a layer of octahedral (AlO_8, FeO_8 and/or MgO_8).

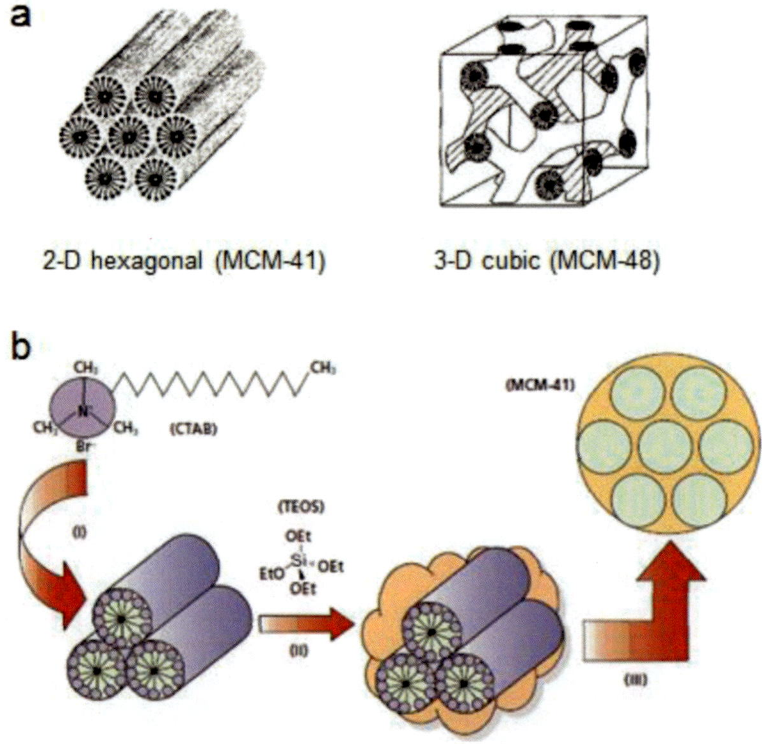

Figure 3. (a) The pores of MCM-41 and MCM-48 materials are orderly arranged in 2D hexagonal and 3D cubic morphology, respectively. (b) The mechanism of development of MCM-41[1].

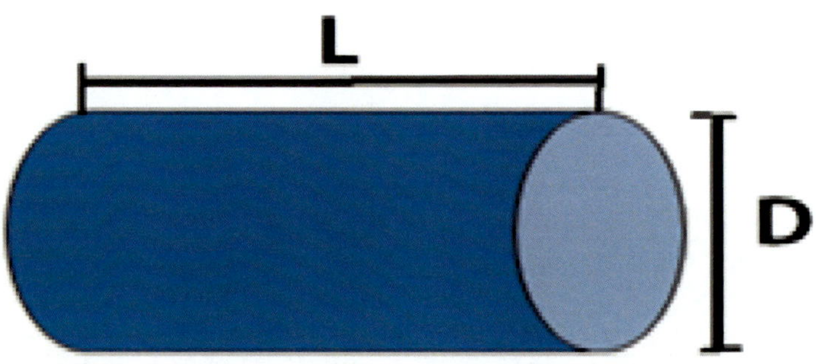

Figure 4. The anisotropy b of a cylindrical pore is defined by the ratio between the pore length L and the pore diameter D.

As depicted in Figure 3b, a layer of silica is deposited on the surface of those micelles from hydrolysis and condensation of a suitable source, often *tetraethyl orthosilicate* (TEOS). Finally, careful firing and removal of organic part leaves behind a porous, honeycomb-like, mesostructure. This chapter focuses on *pore anisotropy* (hereafter *b*) of porous solids, especially *mesoporous materials* with ordered and disordered pore structures. MCM and SBA-type silicas represent ordered pore structures while aluminas, silicas and aluminophosphates are examples of disordered porous stuructures.

[1] Aldrich Materials Science, Synthesis of Mesoporous Materials, *Materials Matters* 2008,3(1),17.

The pore anisotropy b is defined as the ratio of the pore length L to its diameter D as shown in Figure 4.

$$b = L/D = L/2r \quad (1)$$

This chapter outlines the following issues.

1. It will show how one can estimate b in porous materials with organized or random mesoporosity using, as the only input, the experimental N_2 adsorption-desorption porosimetry data.
2. Since *mesopores* in real materials are not ideally arranged in parallel patterns but connected to and intercepted by other pores, forming a complex pore network, it will present how the pore anisotropy b of these nanostructures is affected by the mean *pore connectivity (c)*.
3. Since the length of *mesopores* is usually intercepted by various micropores, it will show how the pore anisotropy b is influenced by the extent of microporosity (%micro) of porous domains.
4. It will show that the methodology for the estimation of b in (I) is directly related to the *fractal dimensionality* (D_s) of adsorption – desorption phenomena, a case with its own scientific legacy.
5. It attempts to make a comparison between the phenomenology describing the development of micelle precursors in organized mesoporous materials like MCM and SBA and the mechanism affecting the most common example of anisotropic development in nature, which is the plant cell.

It is reminded that according to the IUPAC classification, the *micropores* bear openings from very low up to 2 nm. The diameter of *mesopores* is in the range of 2 to 50 nm, while pores with diameters above 50 nm are termed *macropores*.

2. The Estimation of Pore Anisotropy in Porous Solids via N2 Porosimetry

One of the most important parameters characterizing the pore structure of porous solids is the *pore size distribution* (PSD) [12 - 14]. The PSD affects the mass transfer in adsorptive and catalytic processes taking place in the internal surface of porous materials [15, 16]. In nanoporous (micro- and mesoporous) materials, the estimation of PSD is based on gas (usually N_2) physisorption experiments carried out on routine basis in commercial or homemade porosimeters.

The data obtained from these measurements are the adsorbed and desorbed volume V of probe gas given as a function of partial pressure (p/p_o) as:

$$V(N_2)_{ads} = f\,(p/p_o) \quad (2a)$$

$$V(N_2)_{des} = f\,(p/p_o) \quad (2b)$$

The adsorption takes place in ascending pressure mode, from *p/po* = *0* to *p/po* = *1*, while the desorption sequence is the other way around. It is often observed that the desorption curve does not follow the adsorption curve, but it gives a hysteresis loop: the desorbed amount of N_2 is lower than the corresponding absorbed amount at the same partial pressure ($V(N_2)_{des} < V(N_2)_{ads}$). Such hysteresis phenomena are usually attributed to *cavitation,* i.e. thermodynamically unstable liquid-vapor phase, or *percolation* effects, i.e. trapping of adsorbed species into large cavities of solid bearing small exits to the external environment.

The next step is the treatment of the primary (V, p/p_o) data using the *Kelvin equation.* This equation relates the partial pressure (p/p_o) to the pore radius r filled during the multilayer adsorption as given in Eq. (3).

$$\ln(p/p_o) = -2K/r \quad (3)$$

where $K = \gamma V_m/RT$, γ = the surface tension, V_m = the molar volume, R = the universal gas constant and T = the temperature.

After transformation of the relative pressure (p/p_o) to pore radius, the PSD is usually determined by using the *Barret-Joyner-Halenda* (BJH) *algorithm* [17]. In this way, the pore volume distribution, dV/dr, and the surface area distribution, dS/dr, are obtained as a function of pore radius r filled at each particular pressure.

$$dV/dr = f(r) \quad (4a)$$

$$dS/dr = f(r) \quad (4b)$$

A key assumption behind these calculations is that the pores possess cylindrical shape. As a cylinder is characterized by two geometrical parameters, i.e. the diameter D and length L, the question that naturally occurs is: what is the ratio $b = L/D$ of those pores or capillaries? Perhaps up to 1992, before the invention of MCM materials that clearly possess long cylindrical pores, this question was somehow irrelevant.

The ordinary porous solids, such as silicas, aluminas, aluminosilicates, aluminophosphates, etc., possess a more or less random porous network and it was difficult to talk about distinct geometrical characteristics such as diameter and length of openings. Besides, various observations made by electron microscopy have shown that many conventional porous materials appear as a random collection of sticks, packed spherical or semispherical particles. Obviousely, it is not easy to attribute discreet Euclidean geometrical parameters, such as D and L, to these kinds of cavities. Nevertheless, Dullien and co-workers [18] showed, using various microphotographs of solids with random porosity, that in many cases the length L of pores is very close to their diameter D, i.e., $L \sim D$. But with the discovery of MCMs the distinction between L and D obtained a new and clear meaning.

Recently, Pomonis and Armatas proposed [19], for the first time, a method for the estimation of *pore anisotropy b* of nanoporous materials. This work focused on two types of materials: (i) ordered mesoporous MCM-41 silicas whose porosity gradually deteriorates by the introduction of Al heteroatoms and (ii) disordered porous silicas with tunable micro- and mesoporosity. In order to block a variable amount of pores, the internal surface of the parent material SiO_2 was modified with organosilicate groups of various lengths. The method was subsequently applied to a large variety of other nanoporous solids including:

- mesoporous MCM-48, SBA-15 and SBA-16 silicas and TiO_2 anatases [20],
- silicas with different random porosities using the adsorption isotherms of two different probe adsorbents, nitrogen and methanol [21],
- ordered mesoporous CoO_x/MCM-41 materials exhibiting long-range self-organized nanostructured morphology [22],
- organized mesoporous silicates (OMSi-CeCo) containing cobalt and cerium [23],
- organized mesoporous silico-nickelates (OMSiNi) and silico-lanthano-nickelates (OMSiLaNi) exhibiting a competitive relationship between crystallogenesis and morphogenesis on one hand and microporosity and pore anisotropy on the other [24], and
- organized mesoporous silicas (OSi) modified with small addition of various transition metal ions, i.e. Pt, Cu, Mn, Fe, and Ni in pore walls [25].

3. Method of Calculation

The method of estimating *b*, based always on *nitrogen physisorption*, is as follows: at each particular pressure P_i (= p_i/p_o), values of partial specific surface area S_i ($m^2\ g^{-1}$) and partial specific pore volume V_i ($cm^3\ g^{-1}$) are estimated by using a standard method like BJH algorithm.

Then, at each P_i, it is assumed that a number N_i of cylindrical pores with equal diameter $D_i = 2r_i$ are filled-up.

Furthermore, the length L_i of the pores is related to the radius r_i through the scaling relationship $L_i = r_i^{\alpha i}$, where α_i is a *scaling parameter* whose determination, in combination with pore radius r_i, enables the estimation of anisotropy. As a result, the *pore anisotropy* is given as:

$$b_i = L_i \,/\, D_i = r_i^{\alpha i} \,/\, 2r_i = 0.5r_i^{\alpha i\text{-}1} \qquad (5)$$

Then, the differential specific surface area S_i and the differential specific pore volume V_i corresponding to each relative pressure P_i may be related to the dimension (a_i) and the number N_i of cylindrical voids via the following relationships:

$$S_i = N_i(2\pi r_i)L_i = N_i(2\pi r_i)r_i^{\alpha i} = 2\pi N_i r_i^{\alpha i+1} \qquad (6)$$

$$V_i = N_i(\pi r_i^2)L_i = N_i(\pi r_i^2)r_i^{\alpha i} = \pi N_i r_i^{\alpha i+2} \qquad (7)$$

Relations (6) and (7) can now be used to estimate *b* either in combination or independently. These cases will be separately discussed and the results are compared.

Estimation of the Pore Anisotropy from the Scaling of the Dimensionless Ratio λ_i with the Pore Radius

From Eqs (6) and (7), the following dimensionless ratio λ_i can be easily obtained:

$$S_i^3/V_i^2 = \lambda_i = 8\pi N_i r_i^{\alpha i\text{-}1} \quad (8)$$

and after taking logarithms

$$\log(S_i^3/V_i^2) = \log(\lambda_i) = \log(8\pi N_i) + (\alpha_i\text{-}1)\log(r_i). \quad (9)$$

This means that plot of $\log(S_i^3/V_i^2) = \log(\lambda_i)$ *vs.* $\log(r_i)$ provides curves with slope $s_{i(\mathrm{SV})} = (\alpha_i\text{-}1)$ at each point *i*.

Therefore, according to Eq. (5), the anisotropy b_i for each group *i* of pores can be obtained as:

$$b_{i(\mathrm{SV})} = 0.5 r_i^{\alpha i\text{-}1} = 0.5 r_i^{si(\mathrm{SV})} \quad (10)$$

where the subscript (SV) means that the estimation took place based on both S_i and V_i parameters.

Estimation of the Pore Anisotropy from the Scaling of the Differential Surface Area with the Pore Radius

Upon taking logarithms of Eq. (6) yields:

$$\log(S_i) = \log(2\pi N_i) + (\alpha_i\text{+}1)\log(r_i). \quad (11)$$

Then, plot of $\log(S_i)$ *vs.* $\log(r_i)$ should provide curves with slope $s_{i(\mathrm{S})} = \alpha_i\text{+}1$ at each point *i*.

Consequently, according to Eq. (5), the anisotropy for each group *i* of pores can be obtained from the relationship in Eq. (12):

$$b_{i(\mathrm{S})} = 0.5 r_i^{\alpha i\text{-}1} = 0.5 r_i^{si(\mathrm{S})\text{-}2} \quad (12)$$

where the subscript (S) indicates that the estimation took place based only on the specific surface area S_i.

Estimation of the Pore Anisotropy from the Scaling of the Differential Pore Volume with the Pore Radius

In this case, Eq. (7) is used for the estimation of *b*, which in logarithmic form yields:

$$\log(V_i) = \log(\pi N_i) + (\alpha_i\text{+}2)\log(r_i) \quad (13)$$

The plot of $\log(V_i)$ *vs.* $\log(r_i)$ provides curves with slope $s_{i(\mathrm{V})} = (\alpha_i\text{+}2)$ at each point i. Then, according to eq. (5), the anisotropy b_i for each group *i* of pores can be obtained from:

$$b_{i(\mathrm{V})} = 0.5 r_i^{\alpha i\text{-}1} = 0.5 r_i^{si(\mathrm{V})\text{-}3} \quad (14)$$

where the subscript (V) means that the estimation took place based only on the specific pore volume V_i.

Note that pore anisotropies $b_{i(SV)}$, $b_{i(S)}$ and $b_{i(V)}$ are identical and their estimation should not be affected by the applied method.

$$b_i = 0.5r_i^{ai-1} = 0.5r_i^{si(SV)} = 0.5r_i^{si(S)-2} = 0.5r_i^{si(V)-3} = b_{i(SV)} = b_{i(S)} = b_{i(V)} \quad (15)$$

Figure 5 shows indeed that this is the case. This figure compares, in a bar-charts, the three anisotropies $b_{i(SV)}$, $b_{i(S)}$ and $b_{i(V)}$ for different organized or quasi-organized mesoporous silicate (OSi) materials containing 1-2% of various metal ions, namely 2% Pt (2Pt/OSi), 1% Ni (Ni/OSi), 1% Fe (1Fe/OSi), 2% Cu (2Cu/OSi), and 1% Mn (1Mn/OSi). The pore anisotropies $b_{i(SV)}$, $b_{i(S)}$ and $b_{i(V)}$ were found to be ~23.8 for 2Pt/OSi; ~20.6 for 1Ni/OSi; ~11.5 for 1Fe/OSi; ~8.4 for 2Cu/OSi; ~5.5 for OSi, and ~3.4 for 1Mn/OSi. It is as though the Pt and Ni, and to a less degree Fe and Cu, lead to better organized porosity compared to the parent material OSi, while Mn leads to worse porosity. Figure 6 shows typical results from N_2 physisorption as well as the relevant calculations that have been made to determine b for mesoporous OSi sample.

The following are depicted in Figure 6:

i) The N_2 adsorption – desorption isotherms,
ii) the pore size distribution $dV/dD = f(D)$ estimated according to BJH algorithm,
iii) the plot $\log(\lambda_i) = \log(S_i^3/V_i^2)$ *vs.* $\log(r_i)$ according to Eq. (9),
iv) the anisotropy $b_{(SV)}$ as a function of pore radius r_i according to Eq. (10),
v) the plot $\log(S_i)$ *vs.* $\log(r_i)$ according to Eq. (11),
vi) the anisotropy $b_{(S)}$ as a function of pore radius r_i according to Eq. (12),
vii) the plot $\log(V_i)$ *vs.* $\log(r_i)$ according to Eq. (13), and
viii) the anisotropy $b_{(V)}$ as a function of pore radius r_i according to Eq. (14).

The following *important points* should be kept in mind when using the above methodology:

(i) The method is strictly applied only to pores with cylindrical geometry. If the
(ii) pores possess different shapes (e.g. slit-like) then appropriate corrections should be made.
(iii) The estimation of b is based on the local slopes s_i, either by $d[\ln(\lambda)]/d[\ln(r)]$ or $d[\ln(S)]/d[\ln(r)]$ or $d[\ln(V)]/d[\ln(r)]$ plot, which in turn are directly affected by the number of points obtained in the adsorption – desorption isotherms: the more the experimental points, the greater the precision of calculations. Our experience has shown that the use of the standard software MS Origin v6.0 without insertion of additional virtual points between the real ones provides reasonable results.
(iv) The pore anisotropy b exhibits some kind of distribution as a function of pore radius r. This is quite reasonable since not all the *mesopores* are expected to bear the same length.
(v) The maximum pore anisotropy distribution appears at the same maximum pore size distribution as expected.

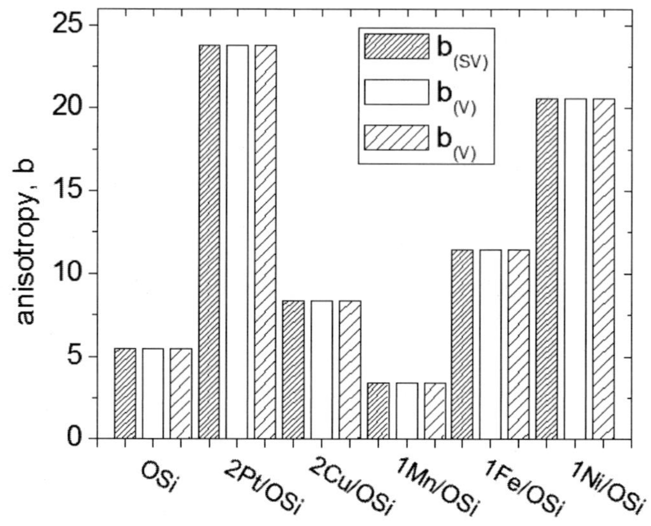

Figure 5. Comparison of pore anisotropies $b_{(SV)}$, $b_{(S)}$ and $b_{(V)}$.

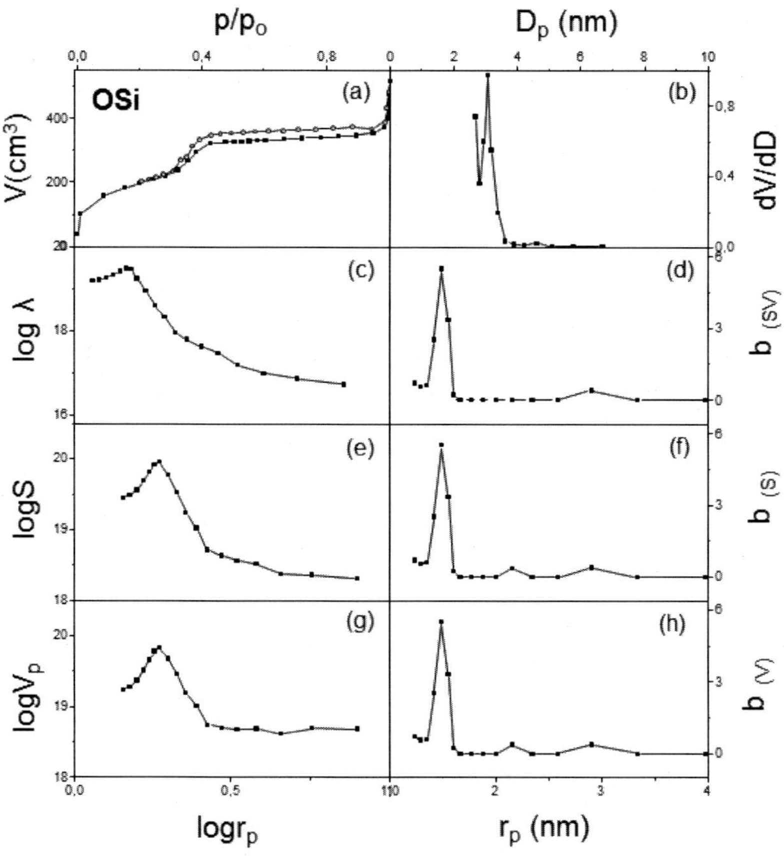

Figure 6. Plots of typical results from N_2 physisorption.

Table 1 shows the surface area, micropore volume and maximum pore anisotropy *b* values determined for various porous materials [19-25].

Table 1. The pore anisotropy *b* of various materials examined in refs [19-25]

Sample †	Surface area, S_p [m^2g^{-1}]	Micropore Volume, (%micro) [cm^3g^{-1}]	Pore anisotropy, b	Ref.
Al_0/MCM-41	1304	0.01 (1)	250	[19]
Al_5/MCM-41	1383	0.09 (10)	11	[19]
Al_{10}/MCM-41	1387	0.02 (2)	8	[19]
Al_{15}/MCM-41	1237	0.11 (16)	6	[19]
Al_{20}/MCM-41	1180	0.12 (17)	0.13	[19]
Al_{50}/MCM-41	785	0.15 (32)	0.03	[19]
SiO_2	897	0.02 (1)	0.5	[19]
SiO_2-SiH	790	0.03 (2)	0.9	[19]
SiO_2-$Si(CH_2)_2OH$	776	0.02 (1)	0.25	[19]
SiO_2-$Si(CH_2)_3OH$	675	0.01 (1)	0.15	[19]
MCM-48	1538	0.068 (8)	608	[20]
MCM-48(100)	1451	0.079 (7)	1562	[20]
Cu/MCM-48	1529	0.125 (13)	607	[20]
PyPzH-Cu/MCM-48	1143	0.049 (8)	100	[20]
PyPzH/MCM-48	1402	0.070 (9)	102	[20]
Cu-PyPzH/MCM-48	1303	0.074 (10)	13.2	[20]
SBA-15(40)	419	0.034 (9)	211	[20]
SBA-15(100)	675	0.057 (5)	1801	[20]
SBA-16/TMB(0)	692	0.063 (13)	118	[20]
SBA-16/TMB(5.6)	725	0.063 (11)	41	[20]
SBA-16/TMB(10)	747	0.059 (11)	10.5	[20]
SBA-16/TMB(20)	634	0.051 (10)	232	[20]
TiO_2-SESA	262	0.006 (2)	(4.5, 1.9)	[20]
Cu/TiO_2-SESA	197	0.006 (2)	(0.5, 0.06)	[20]
PyPzH-Cu/TiO2–SESA	148	0.016 (4)	(0.05, 0.02)	[20]
Fr1428	295	0.064 (8)	0.3	[21]
Fr474	294	0.042 (6)	0.6-0.3	[21]
Fr1386	230	0.053 (11)	0.3, 0.2	[21]
MM1164	98	0.013 (6)	0.6	[21]
CoO_x-A/MCM-41	1247	-	15	[22]
CoO_x-B/MCM-41	1030	-	382	[22]
CoO_x-C/MCM-41	977	-	244	[22]
CoO_x-D/MCM-41	911	-	63	[22]
CoO_x-E/MCM-41	967	-	120	[22]
OMSi-CeCo/1:9/5.5	1216	0.24 (30)	250	[23]
OMSi-CeCo/1:9/7.5	989	0.20 (26)	300	[23]
OMSi-CeCo/1:9/9.5	782	0.16 (27)	8	[23]
OMSi-CeCo/2.5:7.5/5.5	980	0.20 (29)	10	[23]
OMSi-CeCo/2.5:7.5/7.5	896	0.19 (26)	200	[23]
OMSi-CeCo/2.5:7.5/9.5	869	0.19 (25)	55	[23]
OMSi-CeCo/5:5/5.5	767	0.17 (17)	350	[23]
OMSi-CeCo/5:5/7.5	725	0.16 (16)	730	[23]
OMSi-CeCo/5:5/9.5	682	0.15 (17)	3000	[23]
OMSiNi/RT/10/5.5	1125	0.25 (27)	4202	[24]
OMSiNi/RT/10/7.5	920	0.15 (15)	7	[24]
OMSiNi/RT/10/9.5	800	0.21 (24)	1454	[24]
OMSiLaNi/RT/2.5:7.5/5.5	973	0.24 (22)	1335	[24]
OMSiLaNi/RT/2.5:7.5/7.5	783	0.22 (20)	11996	[24]
OMSiLaNi/RT/2.5:7.5/9.5	843	0.19 (19)	2118	[24]

Table 1. (Continued)]

Sample †	Surface area, S_p [m^2g^{-1}]	Micropore Volume, (%micro) [cm^3g^{-1}]	Pore anisotropy, b	Ref.
OMSiLaNi/RT/5.5/5.5	638	0.15 (16)	20698	[24]
OMSiLaNi/RT/5.5/7.5	600	0.15 (16)	56495	[24]
OMSiLaNi/RT/5.5/9.5	655	0.16 (16)	24403	[24]
OMSiNi/50/10/5.5	773	0.13 (13)	85	[24]
OMSiNi/50/10/7.5	659	0.10 (9)	24	[24]
OMSiNi/50/10/9.5	680	0.17 (16)	24136	[24]
OMSiLaNi/50/2.5:7.5/5.5	643	0.15 (13)	655	[24]
OMSiLaNi/50/2.5:7.5/7.5	603	0.12 (9)	224	[24]
OMSiLaNi/50/2.5:7.5/9.5	600	0.13 (13)	89	[24]
OMSiLaNi/50/5.5/5.5	596	0.13 (12)	249	[24]
OMSiLaNi/50/5.5/7.5	476	0.10 (9)	562	[24]
OMSiLaNi/50/5.5/9.5	610	0.14 (14)	4969	[24]
OSi	716	-	5.5	[25]
2Pt/OSi	682	-	23.8	[25]
2Cu/OSi	737	-	8.4	[25]
1Mn/OSi	726	-	3.4	[25]
1Fe/OSi	729	-	11.5	[25]
1Ni/OSi	775	-	20.6	[25]

† OMSi–Ce–Co/x:z/X, where x:z is the nominal ratio of Ce:Co ions in the synthesis bath and X corresponds to the pH value (A: 5.5, B: 7.5 and C: 9.5) at which the particular sample was isolated. Fr1428 is LiChrospher Si60 25 μm native silica. Fr474 is the same sample but with bonded electroneutral diol groups on the outer surface. Fr1386 is a Reduced Access Materials (RAM) with bonded electroneutral diol groups on the outer surface and sulfonic groups on the inner surface. MM1164 is a RAM hydrophobic sample with external electroneutral diol groups and internal *n*-octadecyl groups. CoOx–X/MCM-41, where X corresponds to the pH value (A: 5.5, B: 6.5, C: 7.5, D: 8.5 and E: 9.5) at which the particular sample was isolated. PyPzH/MCM-48, PyPzH-Cu/MCM-48, and PyPzH-Cu/TiO2-SESA, where PyPzH is 3(5)-(2-pyridinyl) pyrazole and SESA denote solvent evaporation self-assembly. SBA-16/TMB(N), where N refers to the 1,3,5-trimethylbenzene (TMB)/F127 surfactant ratio used in the preparation. OMSiLaNi/T/X/pH, where T is the temperature where the solution was aged, X is the La:Ni molar ratio in the reaction bath (or Ni/(Ni+Si) = 10 for OMSiNi/T/10/pH) and pH corresponds to the values 5.5, 7.5 and 9.5 where each particular sample was isolated. XM/OSi denotes organized silicates with X% (1 or 2%) of metal M (Pt, Cu, Mn, Fe, and Ni).

4. The Relation Between Pore Anisotropy and Dimensionality of Adsorption

In this section, it is attempted to search for the relationship between the pore anisotropy and the *fractal dimension* D_s of adsorption and show that the scaling dimension of pore anisotropy *b,* the last measured in units of radius *r* corresponding to the fractal dimension D_s of the adsorbing space.

The link between parameters *b* and D_s is actually the *Frenkel-Halsey-Hill* (FHH) theory [26], put forward in 1940s and 1950s and used subsequently in 1980s and 1990s by various researchers to study the fractal dimension of porous surfaces.

In an attempt to describe the multilayer adsorption on flat solid surfaces, Frenkel, Halsey and Hill proposed the following equation:

$$V/V_m \approx [\log(p_o/p)]^{-1/s} \quad (16)$$

where V/V_m represents the *surface fractional coverage*, V is the *adsorbed volume*, V_m is the volume necessary for the formation of monolayer, and s is a parameter characterizing the shape of the isotherm. The theoretical value of $s = 3$, stemming from non-retarded *van de Waals* interactions between adsorbent and adsorbate, is rarely observed, and for real solids the value of s usually varies in the range 2.5 – 2.7.

Equation (16) was revisited in the late 80s and early 90s by Pfeifer [27] and Neimark [28] who suggested that it can be used to describe the fractal dimensionality D_s of solids surface, a hot subject at that time, after the ground breaking work of Mandelbrot [29] and, later, of Avnir and co-workers [30]. This re-examination of the FHH equation led Pfeiffer to suggest the following two forms:

(i) If the van der Waals attraction between the solid and adsorbed thin film (in liquid form) is the dominating factor, then the FHH isotherm is written as:

$$V/Vm \approx [\mathrm{RT}\log(p_o/p)]^{-(3-Ds)/3} \quad (17)$$

which, for Ds = 2, is reduced to Eq. (16).

(ii) If the liquid/gas surface tension (capillary force), which tends to reduce the interface area is more important, then the FHH isotherm obtains the form:

$$V/Vm \approx [\mathrm{RT}\log(p_o/p)]^{-(3-Ds)} \quad (18)$$

Equation (18) is actually reduced to Eq. (16) for $D_s = 2.67$. Since the capillary forces dominate at higher coverage, Eq. (18) should be used at high relative pressures, while Eq. (17) can be used at low relative pressure range. Nevertheless, the exact threshold of switching between Eqs (17) and (18) is not always clear and depends on the nature of the adsorbent-adsorbate system.

At about the same time, Neimark proposed the so-called *thermodynamic fractal isotherm equation*, Eq. (19). The basis for this equation is a simple relationship between the area of the solid/fluid interface S and the mean radius r of the curvature of this interface.

$$\log S = \text{constant} - (D_s - 2) \log r \quad (19)$$

It has been shown theoretically by Jaroniec [31] and, later, experimentally by Sahouli [32] that the same results are obtained in the capillary condensation regime when either the Neimark's equation (19) or the fractal FHH equation (18) with a (D_s-3) exponent is used. Eq. (19) as well as Eqs (17) and (18) in their logarithmic forms (Eqs (20) and (21), respectively) have been used by researchers to estimate the *fractal dimension* of a large number of porous materials, including silicas [33], carbon fibers, carbon blacks and carbon films [34], titania [35], aluminosilicates [36], simple or pillared clays [37, 38], zirconia [39], hydroxyapatites [40] and aluminophosphates [41].

Table 2. Comparison of equations (19) - (21) and equations (11) and (13)

Equations to estimate fractal dimension D_s	Equations to estimate pore anisotropy b
$\log S = C - (D_s-2)\log(r)$ (19)	$\log(S_i) = \log(2\pi N_i) + (\alpha_i+1)\log(r_i)$ (11)
$\log V \approx C - ((D_s-3)/3)[\log(\log(p_o/p))]$ (20)	$\log(V_i) = \log(\pi N_i) + (\alpha_i+2)\log(r_i)$ (13)
$\log V \approx C - (D_s-3)[\log(\log(p_o/p))]$ (21)	$\log(V_i) = \log(\pi N_i) + (\alpha_i+2)\log(r_i)$ (13)

C = constant.

$$\log V \approx \text{constant} - ((D_S-3)/3)\,[\log(\log(p_o/p))] \qquad (20)$$

$$\log V \approx \text{constant} - (D_S-3)\,[\log(\log(p_o/p))] \qquad (21)$$

A careful observation of Eqs (19) - (21) in this section and Eqs (11) and (13) in the previous section shows that the equations are similar and this is shown in Table 2. The table depicts comparison of two classes of equations, Eqs (19) - (21) that are used to estimate fractal dimension D_s (left column), and Eqs (11) and (13) used to estimate pore anisotropy (right column).

The similarities between those two classes of equations are obvious. Nevertheless, before going into their detailed discussion, it is very important to draw attention to the fact that the plots $\log(V)$ *vs.* $\log(\log(p_o/p))$ and $\log(S)$ *vs.* $\log(r)$ do not exhibit a unique slope. They usually exhibit a linear part at low pressures, or small radii, from which a value of the fractal dimension D_s can be estimated.

At higher pressures or radii, however, the slope is altered. Many authors of related articles ignore this part, or approach it very cautiously because this is an unmapped territory. Nevertheless, there is a very penetrating observation by Lee and Tsay [42] drawing attention to the variation of local slope of plots $d[\log(V)]$ *vs.* $\log(\log(p_o/p))$ and $\log(S)$ *vs.* $\log(r)$ in the whole range of pressure.Therefore, care is needed as to the range of pressure (p_o/p) or the pore radius r, where such a comparison between those two sets of equations is meaningful, or otherwise the meaning of local slopes in the various ranges of P_i or r_i. The physical meaning of the results obtained in the different pressure ranges of the FHH equation have been discussed extensively by various groups of scientists [43].

4. The Relationship Between the Pore Anisotropy and the Fractal Dimension of Adsorption According to the Niemark's Relation

Comparison between Eqs (19) and (11), in Table 2, reveals that $-(D_s-2) = (\alpha_i+1) = s_{i(S)}$, where $s_{i(S)}$ is the local slope of $d[\log(S_i)]$ *vs.* $d[\log(r_i)]$ plot. Therefore, $-D_s = (s_{i(S)}-2)$ and Eq. (15) can be rewritten in the form:

$$b_i = b_{i(S)} = 0.5 r_i^{si(S)-2} = 0.5 r_i^{-Ds} \qquad (15) = (22)$$

Now, Eq. (22) can be used for the estimation of the corresponding $D_{si(S)}$ values by the following relationship:

$$D_{si(S)} = \log(b_{i(S)})/\log(0.5r_i) \quad (23)$$

where the subscript (S) indicates that D_s values have been estimated on the basis of specific surface area S in *Niemark's equation* (19). Eq. (22) is similar to the standard relation of the fractal geometry.

$$D_s = \log(M)/\log(L) \quad (24)$$

where M is the quantity to be measured (for example the perimeter of an inland, the surface area of a solid, the volume of a porous material, etc) and L is the stick or probe size (of length, area, or volume) used to measure M. Details are given the standard text of Mandelbrot "*The Fractal Geometry of Nature*" [29].

Therefore, Eq. (23) expresses the following fact: *"The dimension of pore anisotropy b, the last scaled in units of radius r, corresponds to the fractal dimension D of the adsorbing space or operation"*.

This conjunction is depicted graphically in Figure 7 where it is shown that if the pore anisotropy b is scaled in units of pore radius r, then its dimensionality D corresponds to the fractal dimension of the adsorbing space or operation. The shaded area at the beginning of the isotherm (left) is the so-called BET range where adsorption takes place in a single monolayer and where the specific surface area is determined.

The shaded area in the middle corresponds to the pressure range where the pores are filled-up by the adsorbed/condensed species. A *single* imaginary pore is shown in horizontal fashion filled with spheres of diameter equal to the pore radius r. The scaling/measuring of pore length in this way leads to b. The dimensionality of the space according to the principles of fractal geometry is $D \approx \log b/\log r$.

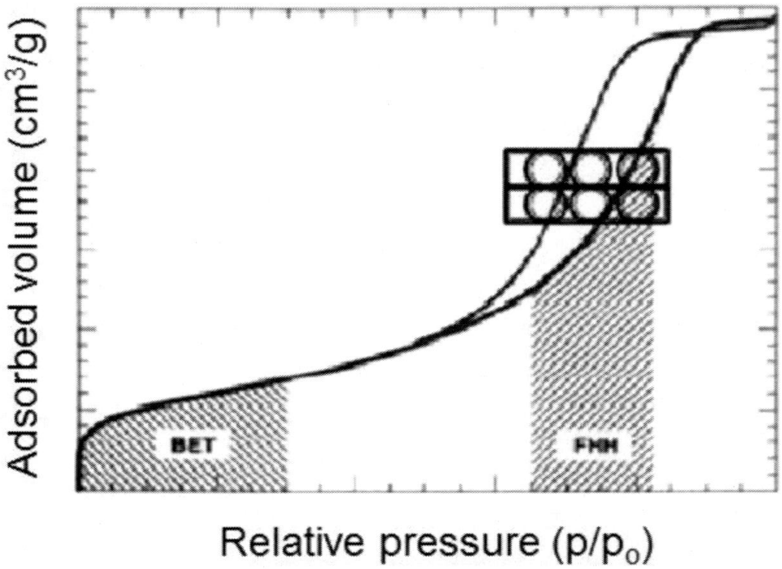

Figure 7. The correspondence of the dimensionality D of pore anisotropy b to the fractal dimension of the adsorbing space or operation.

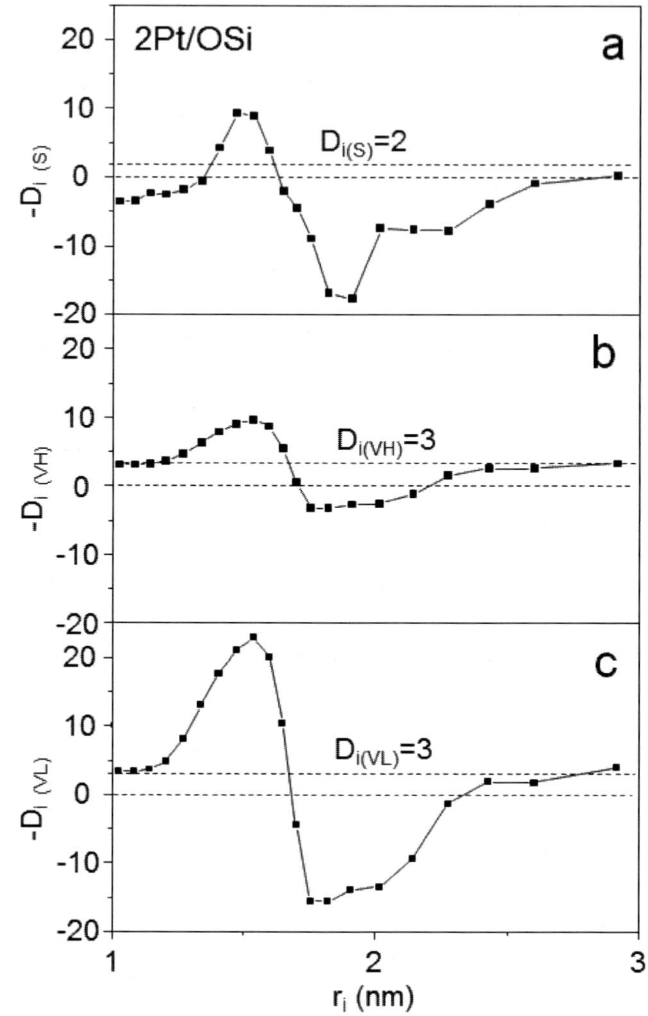

Figure 8. Variation of dimensionalities $-D_{si(S)}$ $-D_{si(VH)}$ and $-D_{si(VL)}$ as a function of pore radius r_i.

Figure 8 shows the variation of dimensionalities $-D_{si(S)} = (s_{i(S)}-2)$, $-D_{si(VH)} = (s_{i(V)}-3)$ and $-D_{si(VL)} = 3(si_{(V)}-1)$ as a function of pore radius r_i for a typical mesoporous 2Pt/OSi sample. The values are estimated according to Eq. (22), (26) and (28) respectively. The values of $D_{si} = 2$ (Figure 8a) and $D_{si} = 3$ (Figure 8b and c) are also indicated for the sake of comparison. Attention should be drawn to the fact that the values of $D_{si(S)}$ are not similar at various stages of adsorption but altered with its progress. This implies that the different physicochemical processes taking place in the pores, like surface adsorption and capillary condensation, are characterized by different dimensionality.

In order to show the relationship between pore anisotropy and fractal dimension of adsorption, the comparison of *Pfeifer's relations* (20) and (21) with Eq. (13) are given in Table 2. The comparison in this case is not straightforward since Eq. (13) is based on the pore radii r while Eqs (20) and (21) are based on the corresponding relative pressures (p/p_o). Nevertheless, it has been shown by Jaroniec and co-workers [31] that these two quantities, i.e. (p/p_o) and r, are related through the well-known *Kelvin equation* $r = 2\gamma Vm/\text{RT}\cdot\log(p_o/p)$. Therefore, $\log(r) = \text{A}-\log(\log(p_o/p))$, where $\text{A} = \log(2\gamma Vm/\text{RT})$.

Then, Eq. (13) may be rewritten in the form:

$$\log(V_i) = \log(\pi N_i)+(\alpha_i+2)\log(r_i) = \log(\pi N_i)+(\alpha_i+2)[\text{A}-\log(\log(p_o/p))]$$
$$= \log(\pi N_i)+(\alpha_i+2)\text{A}-(\alpha_i+2)[\log(\log(p_o/p))] = \text{C}-$$
$$(\alpha_i+2)[\log(\log(p_o/p))] \quad (13) = \quad (25)$$

where, $\text{C} = \log(\pi N_i)+(\alpha_i+2)\text{A} = \log(\pi N_i)+(\alpha_i+2)\log(2\gamma V_m/\text{RT})$. Now comparison of Eq. (21) with (13) reveals that $-(D_s-3) = (\alpha_i+2) = s_i(\text{V})$, where $s_i(\text{V})$ is the slope of $\log(V_i) = f(\log(r_i))$ or $\log(V_i) = f(\log(\log(p_o/p))$ plot at each point i. Therefore $-D_s = (s_{i(\text{V})}-3)$ and Eq. (15) can be written in the form:

$$b_i = b_{i(\text{V})} = 0.5\text{r}_\text{i}^{si(\text{V})-3} = 0.5r_i^{-Ds} \quad (26)$$

The last relation, i.e. Eq. (26), is similar to (22) and conveys the same inter-relation between anisotropy and dimensionality described above. Equation (26) in its logarithmic form (Eq. (27)) can be used for the estimation of the corresponding $D_{si(\text{V})}$ values as:

$$D_{si(\text{VH})} = \log(b_{i(\text{V})})/\log(0.5r_i) \quad (27)$$

The D_s values obtained by Eq. (27) will be designated as $D_{si(\text{VH})}$, meaning that they are estimated on the basis of pore volume (V) (Eq. 21) which are valid at high (H) relative pressures. Typical $-D_{i(\text{VH})}$ values as a function of r_i are shown in Figure 8b. Finally, comparison of Eq. (20) with Eq. (13) reveals that $-(D_s-3)/3 = (\alpha_i+2) = s_{i(\text{V})}$. Therefore, $-D_s = (3s_{i(\text{V})}-3)$ and Eq. (15) can be written in the form:

$$b_i = b_{i(\text{V})} = 0.5r_i^{si(\text{V})-3} = 0.5r_i^{-Ds/3-2} \quad (28)$$

The D_s values obtained by Eq. (28) will be designated as $D_{i(\text{VL})}$ meaning that they are estimated on the basis of pore volume (V) in eq. (20) and is valid at low (L) relative pressures. Typical $-D_{i(\text{VL})}$ values as a function of r_i are shown in Figure 8c.

It is observed in this figure that the dimensionalities $-D_{i(\text{S})} = (s_{i(\text{S})}-2)$, $-D_{i(\text{VH})} = (s_{i(\text{V})}+3)$ and $-D_{i(\text{VL})} = (3s_{i(\text{V})}-3)$ obtain high positive values in the shaded range where the filling of cylindrical pores takes place.

In this range the pore anisotropy b gets relatively high values (*pore anisotropy regime*). Those values of anisotropy are related to the values of dimensionality of pore filling by Eqs (22), (26) and (28). Outside of the shaded parts in Figure 7, but in the vicinity of their boundaries, the values of $-D_{si(\text{S})} = (s_{i(\text{S})}-2)$, $-D_{si(\text{VH})} = (s_{i(\text{V})}+3)$ and $-D_{si(\text{VL})} = 3(s_{i(\text{V})}-1)$ become very small or even negative and the corresponding values of b get small or very small. Those boundaries correspond actually to some kind of transitions. While the left-hand boundary corresponds to the transition from a 2D adsorbing regime, or *BET regime*, to the pore filling or the *pore anisotropy regime*, the right-hand boundary relates to the transition from the pore filling, or the *pore anisotropy regime*, to the 3D multi-slab adsorption, corresponding to the traditional *FFH regime* [43].

From the above discussions, the following broad picture emerges. During the adsorption process on a rough surface, and analysis of the experimental data with the FHH equation (16) or its more sophisticated variants (19) and (20) or (21), the scaling/probing of the surface by the adsorbed molecules leads to the estimation of the ordinary dimensionality of adsorption $D_s \approx 2$ and detection of the roughness and fractal dimension of the surface. This is true within a certain, usually small, range at low pressure.

If the surface is gradually curved towards some kind of cylindrical geometry, and the increasing adsorbed molecules reach eventually the point of condensation into that cylindrical capillary space, then the estimated local dimensionality of adsorption (i.e. the fractal dimension D_s) is directly related to the anisotropy b of the filled space. This will be true within a certain *intermediate local range* of pressure. Beyond that point, the applied FHH-based method provides again results akin to the 3D condensation of the adsorbed species onto the non-curved space since the deep pore cavities have been now flooded and no-anomalies are sensed by the adsorbing species.

It is finally mentioned that the accuracy of the results for the calculation of pore anisotropy and dimensionality of the adsorption by the above methodologies may be affected by the number of available experimental data, which in turn affect the local slopes that are the key points in such calculations.

6. The Dependence of Pore Anisotropy on Microporosity

Microporosity, i.e. the volume of pores having diameter less than 2 nm, is an important *morphometric parameter* which strongly affects the permeability of porous solids.

As a result, a large number of methods have been developed to estimate micropore volume, including non-local density function theory (NLDFT), *Dubinin-Radushkevich (DR) theory, Horvath-Kawazoe* (HK) *method, grand canonical Monte Carlo* (GCMC) *simulation* and a_s-plot and t-plot analysis [44]. This section discusses the possible relationship between the extent of microporosity (%micro) and pore anisotropy b of nanoporous solids. To address this question, we review recent works on textural properties of different types of porous solids with tunable microporosity.

The percentage of micropore volume and pore anisotropy of various porous materials has been given in Table 1. A comparison between b and %micro for the series of well-ordered mesoporous Cu/MCM-48 and SBA-16/TMB as well as disordered mesoporous TiO_2-SESA materials is also shown in Figure 9 [20].

The lower three points in this figure correspond to the TiO_2-SESA materials with random micro- and mesoporosity. The upper group corresponds to the MCM-48, Cu/MCM-48, PyPzH/MCM-48, PyPzH-Cu/MCM-48, SBA-15 and SBA-16/TMB materials with more or less ordered nanoporosity.

Interestingly, the b *vs.* %micro relationship, which seems to be logarithmic rather than linear, is not general but differentiated for different types of solids. The TiO_2-SESA samples which possess random porosity follow a different trend than other materials with more ordered pore structure.

However, all the materials obey the same trend: the pore anisotropy decreases as the microporosity increases. In general, a porous material with ordered pore structure is expected to have large anisotropy and minimal microporosity.

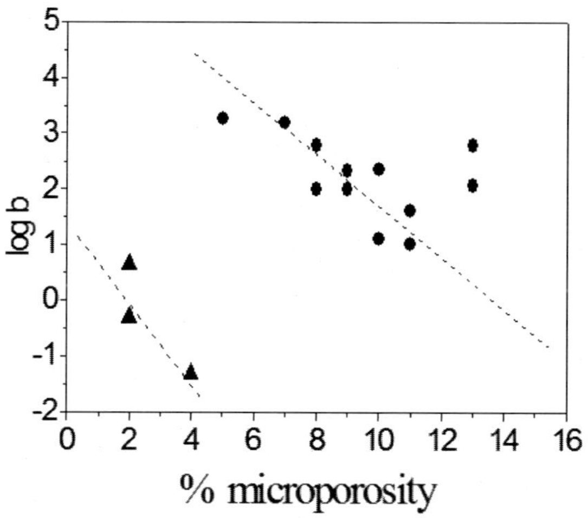

Figure 9. Interdependence between log(b) and %micro.

A similar trend was also observed for the organized mesoporous silico-nickelate (OMSiNi), silico-lanthano-nickelate (OMSiLaNi) and silico-cerium-cobaltite (OMSiCeCo) materials (Figure 10) [24]. The general relationship between b and %micro can be expressed in the following form:

$$\log(b) = \mathrm{B} - \mathrm{A}(\%\mathrm{micro}) \qquad (29)$$

where B and A are the intercept and slope of the linear regression, respectively.

For the combined collection of all examined samples in the upper part of Figure 10 (left), the trend described by Eq. (29) is blurred because the samples tend to be clustered in different groups. This differentiated dependence observed between log(b) and %micro for different groups of porous solids should be as a result of differences in the way the micropores are arranged relative to mesopores, and this is eventually expressed by the slope A in Eq. (29) (see below). In general, materials having similar ordered pore structure follow the same regression line log(b) *vs.* %micro and, therefore, characterized by the same slope A. The clustering of the (log(b), %micro) data – and the exceptions observed – triggers the question if there is perhaps an underlying parameter influencing these groupings and its use could eventually clarify the situation.

This question can be tackled if one considers the fact that the introduction of micropores within the pore walls of a porous solid decreases the anisotropy of the porous network by interrupting the length of mesochannels. Indeed, trivial calculations show that a small addition (1-2%) of microporosity can result in a drop of pore anisotropy by around one order of magnitude [24]. Quantification of such an argument can be as follows: The introduction of N_{micro} micropores into a porous system, containing N_{meso} number of mesopores, reduces the original pore anisotropy b_o by a factor of (N_{micro}/N_{meso}) according to the following relationship:

$$b = (L/[N_{micro}/N_{meso}])/D = (L/D)/[N_{micro}/N_{meso}) = b_o/[N_{micro}/N_{meso}] \qquad (30)$$

and

$$\log(b) = \log(b_o) - \log[\mathrm{N_{micro}/N_{meso}}] \quad (31)$$

Combining equations (29) and (31) reveals the following:

(i) the parameter $b_o = (L/D)$ is the possible upper limit of pore anisotropy and can be as high as 10^3-10^6 (see Figure 9) in porous solids with limited microporosity [24]. This practically happens when $\mathrm{N_{micro}} = \mathrm{N_{meso}}$ or when just one micropore interrupts each mesopore.

(ii) Eq. (32) is obtained and it defines the (%microporosity) = (V_{micro}/V_p) in terms of number of micropores N_{micro} and number of *mesopores* N_{meso}.

$$\mathrm{A}(V_{micro}/V_p) = \log[N_{micro}/N_{meso}] \quad (32)$$

where V_p and V_{micro} are the total pore volume and micropore volume respectively. The relation in Eq. (32) conveys the reasonable fact that the micropore volume (V_{micro}) depends on the number of micropores (N_{micro}) while the total pore *volume* (V_p) corresponds, more or less, to the volume of mesopores and should be proportional to their number (N_{meso}).

The differentiated dependence between log(b) and %microporosity for different groups of porous solids, observed in Figure 10 and expressed by different slopes A (see Eqs (29) and (32)), should be the result of differences in the way the micropores are arranged relative to the mesopores in those distinct groups.

In searching for an underlying parameter, which results from this scattering/ dispersion of the (log(b), % microporosity) data into different groups and the differentiation of A, we have reached the conclusion that this is the width of the distribution of micropores. This parameter tallies roughly to the range of pressure ($\Delta P_{\mathrm{micro}}$) extended from $(p/p_o) = 0$ up to that value of (p/p_o) corresponding to the upper limit of the micropore volume. This conclusion stems from the experimental observation that for the various groups of mesoporous materials containing some microporosity, a relation of the form $A \sim C/(\Delta P_{micro})$ *is observed between the* (ΔP_{micro}) *and the slope A while C is a proportionality parameter*. Therefore, Eq. (29) can be expressed in the following form:

$$\log(\mathrm{b}) = \mathrm{B} - \mathrm{C}[(\%\ \mathrm{microporosity})\ (\Delta \mathrm{P_{micro}})]. \quad (29a)$$

The data shown in Figure 10 (right) show indeed that relationship (29a) has a physical meaning since the trends become apparent for each particular group of porous materials.

As stated, the explanation for the dependence of log(b) *vs*. %micro for different types of *porous solids* can be inferred from the width of the size distribution of micropores. This parameter tallies to the range of pressure $\Delta \mathrm{P_{micro}}$ where the volume of micropores is filled-up with adsorbate (e.g. nitrogen). The $\Delta \mathrm{P_{micro}}$ is strongly related to the size of micropores in the way that very small micropores, usually termed as untra-micropores, have a narrow distribution of pore sizes that is completely filled with liquid at the lowest pressures (p/p_o). On the other hand, as the width of $\Delta \mathrm{P_{micro}}$ moves to higher values of relative pressure (p/p_o), the width of the micropore distribution is extended to wider micropores in the upper limit of the range 1-2 nm. In this case, *some of the narrower micropores might be connected not only to mesopores but also to the wider micropores.*

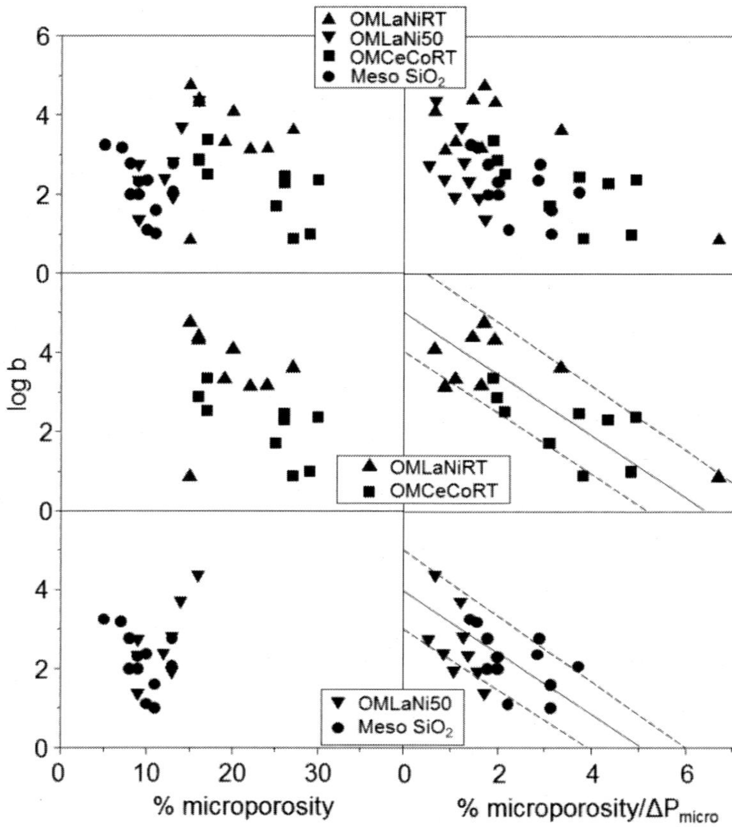

Figure 10. Variation of the pore anisotropy log b as a function of % microporosity. Note: The upper left part of the figure shows the variation of the logarithm of the pore anisotropy log b as a function of the % microporosity for thirty-nine (39) samples (*circles*: twelve MCM-48, SBA-15 and SBA-16 materials [20]; *squares*: nine organized mesoporous OMSiCeCo [23]; *triangles*: nine organized mesoporous OMSiLaNi/RT [24]; *inversed triangles*: nine organized mesoporous OMSiLaNi/50 [24]). The middle part of the left hand side figures the same variations for as above for the nine OMSiCeCo and nine OMSiLaNi/RT materials, while the lower part is for the twelve MCM-48, SBA-15 and SBA-16 materials and nine OMSiLaNi/50 materials. The plots on the right hand side show similar variation as the left part but the plots are now in the form $\log(b) = f(\%\mathbf{micro}/\Delta P_{micro})$. The parallel dashed lines are a guide for the eye and they correspond to the uncertainty according to Eq. (30).

So, some micropores are wasted not in interrupting the long mesopores and reducing their anisotropy but simply reducing the already small anisotropy of wider micropores. This effect is explained schematically in Figure 11. The figure shows the influence of the micropore range distribution on the pore connectivity. The upper figure shows the introduction of N_{mi} = 5 similar micropores and their even distribution along the long axis L_{me} of a mesopore that reduces its initial anisotropy b_0 to $b_0/6$.

In the lower figure, the same number of micropores is introduced but one of them is now larger/wider compared to the rest four. Then, one of the smaller/narrower micropores might be connected not to the original mesopore but to the large/wide micropore. As a result, the initial anisotropy b_0 is reduced to $b_0/5$. The narrow range of micropore distribution in the upper figure corresponds to a small value of ΔP and in a large decrease of anisotropy. In the lower part, however, the wider range of micropore distribution corresponds to a larger value of ΔP_{micro} and the reduction of pore anisotropy is smaller.

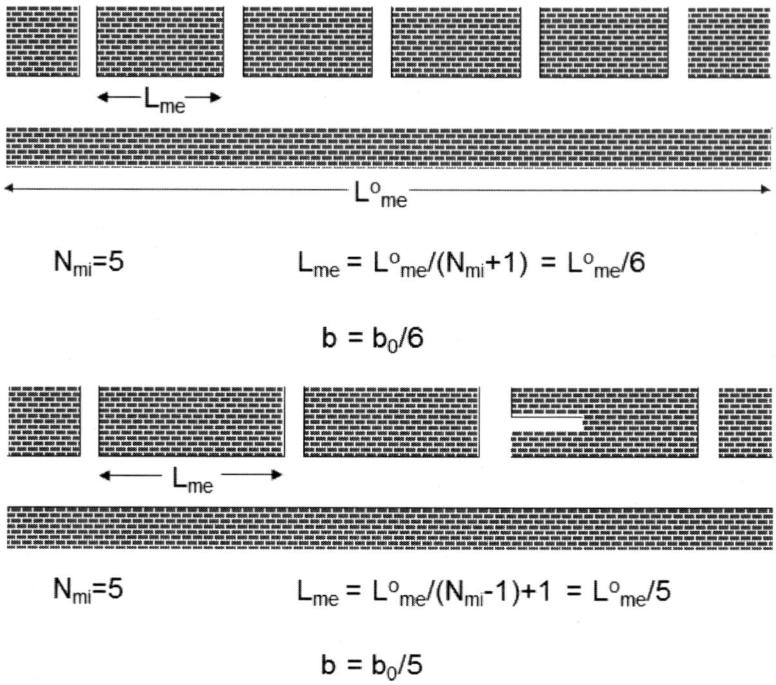

Figure 11. Influence of the micropore range distribution on the pore connectivity.

7. The Dependence of Pore Anisotropy on Pore Connectivity

The objective of this section is to probe the relationship between the pore anisotropy *b* and the *mean pore connectivity c*. Such a relationship can be taken eventually into account during preparation of porous solids for achieving enhanced performance in technological applications such as *adsorption* and *separation*. Furthermore, knowledge of this relationship may be used to improve prediction of internal diffusion effects occurring in heterogeneous catalysis. Nevertheless, the topological parameter *c* of pores is meaningful and can be estimated in porous solids with *random porosity* where the N_2 adsorption – desorption isotherms show hysteresis loop. On the contrary, it is demanding to estimate connectivity in ordered MCM-type materials with parallel cylindrical pores. This is because the hysteresis loop between the adsorption and desorption branch in these materials can be explained in terms of thermodynamic stability of liquid-vapor phase and it is associated with the *cavitations* effects, rather than *percolation* phenomena (that means it is a property of adsorbent and not of solid).

So, in order to probe the relationship $b = f(c)$ we employed in this study sixteen (16) mesoporous vanado-phosphoro-aluminate samples ($Al_{100}P_XV_Y$ where the values assigned to X and Y are 0, 5, 10 and 20) possessing random porous network. Their preparation and characterization has been discussed previously [45, 46]. The key measurements for the estimation of parameters *b* and *c* were the N_2 adsorption – desorption isotherms from which the corresponding pore size distributions were calculated according to the BJH method. The results are shown in Table 3 and Figure 12 [45, 46].

Table 3. Textural parameters of mesoporous $Al_{100}P_XV_Y$ samples

Sample	S_p(BET) [m^2g^{-1}]	V_p [cm^3g^{-1})	D_{max} [nm]	$(fwhm)_p$ $\approx 2\sigma_p$, [nm]	Connec-tivity, c	Aniso-tropy, b	$(fwhm)_b$ $\approx 2\sigma_b$ [nm]
$Al_{100}P_0V_0$	201.3	0.430	6.54	2.5	6.55	69.4	0.23
$Al_{100}P_5V_0$	245.3	0.518	6.34	2.4	12.28	23.5	0.26
$Al_{100}P_{10}V_0$	319.8	1.192	9.14	4.4	11.21	96.7	0.33
$Al_{100}P_{20}V_0$	239.5	0.928	10.74	6.3	13.95	42.1	0.40
$Al_{100}P_0V_5$	270.5	0.446	5.44	2.1	5.85	13.0	0.24
$Al_{100}P_5V_5$	352.6	0.703	6.24	2.9	6.99	27.3	0.34
$Al_{100}P_{10}V_5$	386.1	1.146	9.24	4.6	9.06	127.9	0.5
$Al_{100}P_{20}V_5$	336.0	0.475	10.34	5.9	9.56	116.5	0.57
$Al_{100}P_0V_{10}$	301.9	0.497	5.24	2.1	5.85	42.0	0.19
$Al_{100}P_5V_{10}$	359.0	0.833	6.94	3.4	7.80	158.0	0.36
$Al_{100}P_{10}V_{10}$	321.5	1.109	10.24	5.4	10.55	133.5	0.57
$Al_{100}P_{20}V_{10}$	257.9	0.358	12.44	10.2	14.97	7.0	0.76
$Al_{100}P_0V_{20}$	176.7	0.456	8.14	5.2	10.63	4.5	0.15
$Al_{100}P_5V_{20}$	286.3	0.743	7.34	4.1	8.94	19.9	0.46
$Al_{100}P_{10}V_{20}$	305.2	1.020	9.24	6.0	10.86	10.0	0.64
$Al_{100}P_{20}V_{20}$	199.9	0.245	12.64	10.3	15.17	42.1	0.67

where S_p(BET) (m^2 g^1) is surface areas, V_p (cm^3 g^{-1}) pore volume, D_{max} and $(fwhm)_p$ are maximum and full width at half maximum of PSD respectively, c is pore connectivity, b is pore anisotropy and $(fwhm)_b$ is full width at half maximum of the pore anisotropy distribution.

The calculation of pore connectivity: The pore connectivity can be calculated from N_2 adsorption – desorption isotherms and the corresponding PSD based on the percolation analysis of the porous network. Taking the *desorption isotherm* as a bond percolation phenomenon and using the assumption that pore network effects dominate, Seaton and co-workers [47 - 51] have proposed a method to determine the mean coordination number of pores, *c*, and the characteristic size of the particles, *L*, expressed as the number of the pore lengths. The procedure for the determination of *c* and *L* from the experimental adsorption and desorption data can be summarized as follows [47 - 51]: First, the *pore size distribution* is obtained using the BJH method assuming cylindrical pore model. Then, the *bond occupation probability f* is obtained as a function of *percolation probability F* from the adsorption and desorption isotherms according to the following equations.

$$f = \frac{\int_{r^*}^{r=\infty} n_r dr}{\int_0^{r=\infty} n_r dr} \qquad (33)$$

$$\frac{f}{F} = \frac{V_{flat-max} - V_{des}}{V_{flat-max} - V_{ads}} \qquad (34)$$

where n_r is the corresponding PSD for cylindrical pores, $V_{flat-max}$ is the part of desorption curve before the start of desorption, and V_{des} and V_{ads} are the corresponding volumes in the desorption and adsorption curves.

Finally, the optimum c and L values are obtained by fitting the calculated data (F, f) to a generalized scaling relation between F and f proposed by Kirkpatrick [52].

$$L^{\beta/\nu} cF = G((cf - 3/2)L^{1/\nu}) \tag{35}$$

The parameters β and ν are the so-called *critical exponents* and the obtained values are $\beta = 0.41$ and $\nu = 0.88$ [52]. The values of connectivity c for mesoporous $Al_{100}P_XV_Y$ materials are cited in Table 3.

The calculation of pore anisotropy: As discussed early in this chapter, the pore anisotropy b can be estimated using either the dimensionless ratio $\lambda_i = S_i^3/V_i^2$ or the differential specific surface area S_i or differential specific pore volume V_i. As the calculations in Figure 13 show, the results of b obtained in those three cases are identical, i.e. $b_{(SV)} = b_{(S)} = b_{(V)}$. Now the PSD, estimated via the BJH algorithm as mentioned above (see Figure 12), is characterized by the full width at its half maximum $(fwhm)_p$, which corresponds roughly to the variance $2\sigma_p$ of each distribution $(fwhm)_p \approx 2\sigma_p$, where σ_p is the standard deviation of distribution of the pore sizes (i.e. the pore radii).

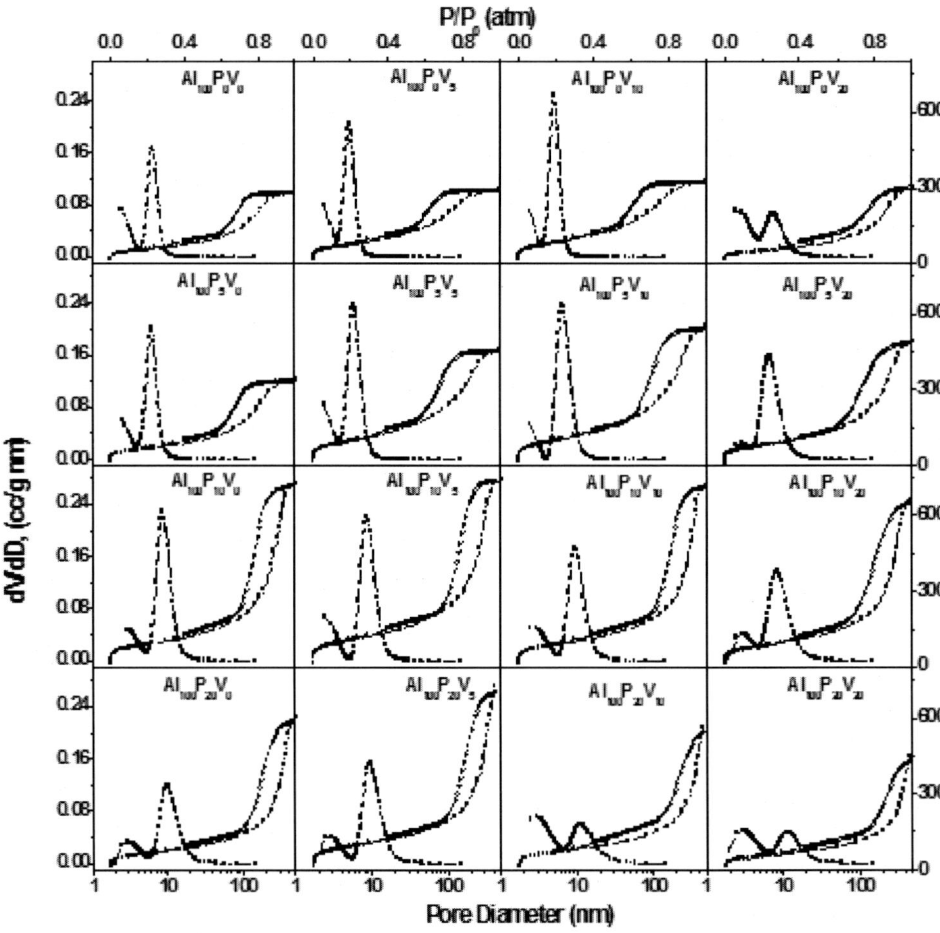

Figure 12. Nitrogen adsorption – desorption isotherms at 77 K for mesoporous $Al_{100}P_XV_Y$ (X, Y = 0, 5, 10, 20) solids and the corresponding pore size distribution curves.

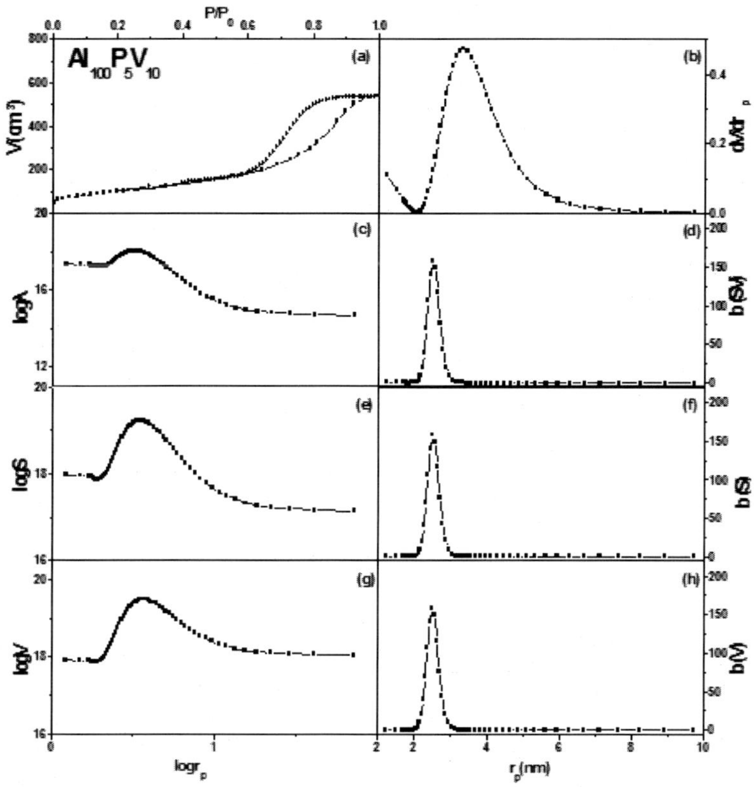

Figure 13. A typical example of the treatment of data for the $Al_{100}P_5V_{10}$ samples with relative high pore anisotropy *b*. (a) The N_2 adsorption – desorption isotherms, (b) the pore size distribution, (c) the plots $\log(S_i)$ *vs.* $\log(r_i)$, (d) the calculated values of $b_{i(S)}$, (e) the plots $\log(V_i)$ *vs.* $\log(r_i)$, (f) the calculated values of $b_{i(V)}$, (g) the plots $\log(S_i^3/V_i^2)$ *vs.* $\log(r_i)$, and (h) the calculated values of $b_{i(SV)}$.

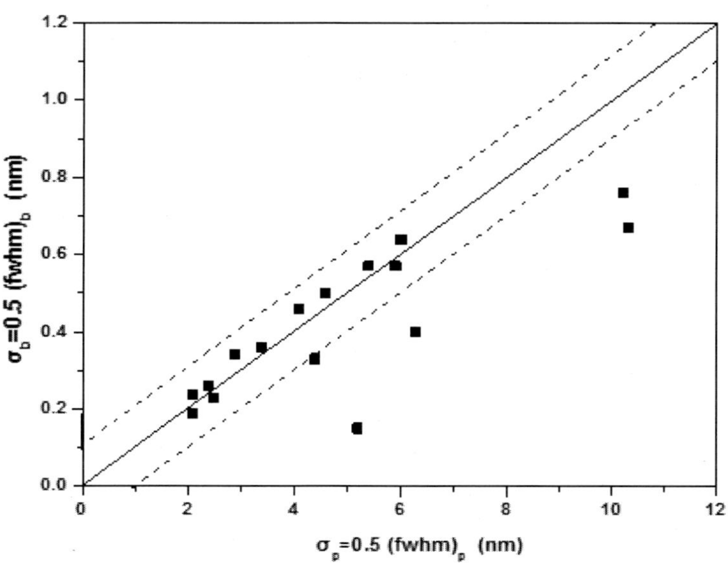

Figure 14. The relationship between the standard deviation of the pore size distribution $\sigma_p = 0.5(fwhm)_p$ and the standard deviation of the pore anisotropy distribution $\sigma_b = 0.5(fwhm)_b$. The line $\sigma_p = 10(\sigma_b \pm 0.1)$ is a guide for the eye.

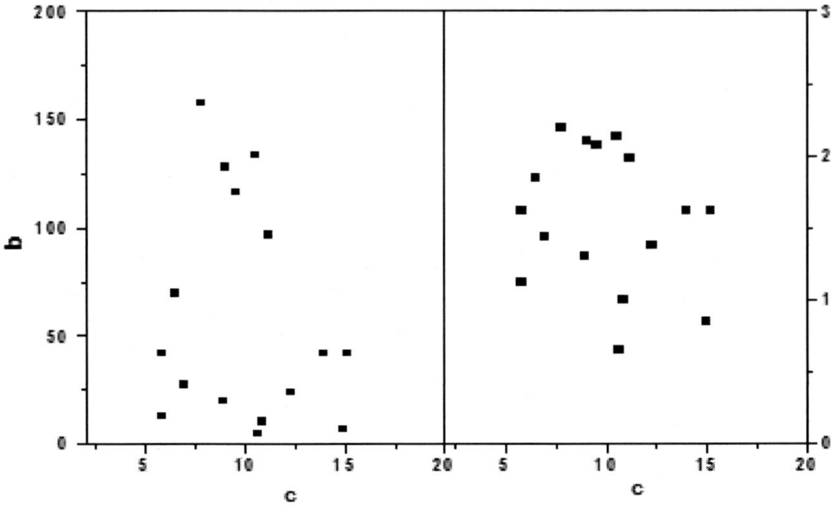

Figure 15. Plots of $b = f(c)$ (left) and $\log(b) = f(c)$ (right). No trends are apparent.

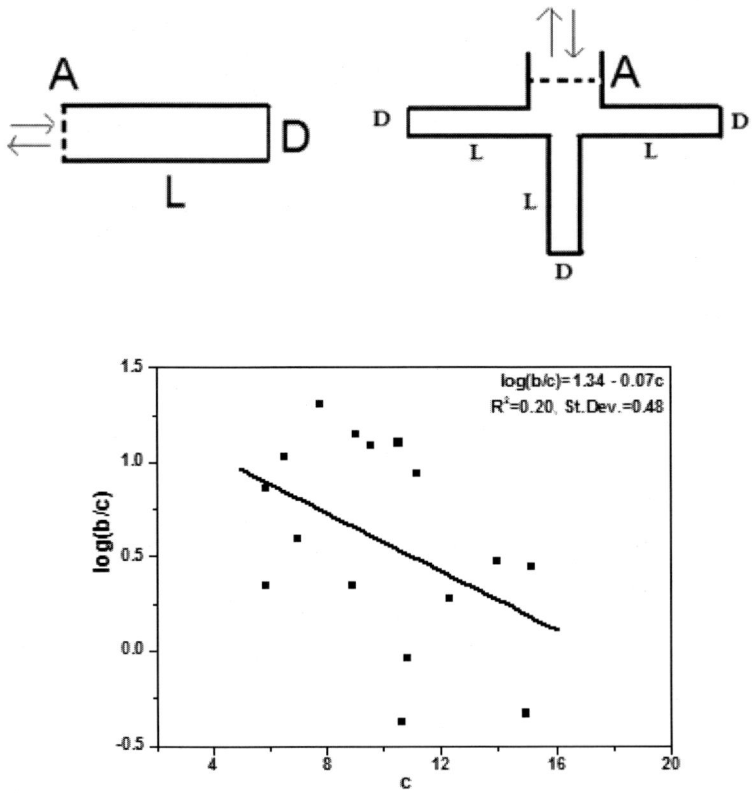

Figure 16. Illustration and plot of connected and unconnected pores. Upper part: An observer at the entrance A of an independent pore counts the desorbed amount of N_2 and then carries out the relevant calculation to find the pore anisotropy b = L/D (left) and an observer at a crossroad point A where three exactly similar pores are connected, based on the desorbed amount of N_2, may calculate a total pore anisotropy b_{total} = 3L/D, but the real anisotropy of each particular pore is actually $b_{real} = b_{total}/3$ = (3L/D)/3 = L/D (right). Lower part: Variation of the logarithm of the maximum pore anisotropy b reduced over the pore connectivity c, log (b/c), as a function of mean pore connectivity c.

But as shown in Figure 13d, f and h the *pore anisotropies* exhibit also some kind of distribution. This is expected since at each particular pore size, some pores will be longer while others will be shorter. Those anisotropy distributions possess also some particular values of variance $2\sigma_b$ which can be calculated from the $(fwhm)_b$ of the pore anisotropy distribution, $(fwhm)_b \approx 2\sigma_b$. These parameters for $Al_{100}P_{10}V_{20}$ samples are given in Table 3.

It is a crucial point that the values of standard deviation of the pore size distribution, σ_p, are related to the standard deviation of the *pore anisotropy distribution*, σ_b, more or less linearly as shown in Figure 14. Although, the relation between those two quantities seems almost linear, its exact form is not easy to be drawn, but as a rule $\sigma_p >> \sigma_b$ and in the majority of the cases $\sigma_p = 10(\sigma_b \pm 0.1)$. This means that the pores in the solids may have many different sizes but the apparent pore anisotropies cannot vary as much. In other words, of all the pore sizes in the studied solids, only very few (about one in ten) possess appreciable length.

Intuitively, it is expected that more connected pore networks exhibit smaller pore anisotropies and vice versa. To test this perception, we have drawn in Figure 15 (left) plots of the form $b = f(c)$. As it can be seen in this plot, there is hardly a relationship between those two quantities, the data are spread in a maverick way and no trend can be traced. In Figure 15(right), there is a plot of the form $\log(b) = f(c)$. Again the data are scattered and no trend is apparent. Perhaps the reason for this contradiction to our original intuition has to do with the way the pore anisotropy is viewed. When the pore anisotropy $b = L/D$ is defined and estimated as described above, what is actually meant is that the pores are cylindrical and independent to each other. In other words, they are not connected (see Figure 16, upper left). Then, an observer at the pore entrance A can easily count the desorbed amount of N_2 and, then, carry out the relevant calculations to find the pore anisotropy of that single pore only. The result is clear-cut and does not accept alternative interpretations. But if the observer finds himself at a crossroad point A where, let us say, three exactly similar pores are diverged and connected (see Figure 16, upper right) then the result (based again on the desorbed amount of N_2 and the relevant calculations) would refer to those three pores. So, although the observer may calculate a total pore anisotropy b_{total}, care must be taken that this value originates from three pores and the real anisotropy of each of them is actually $b_{real} = b_{total}/3$. In other words, the individual pore anisotropy is the estimated anisotropy divided by the connectivity.

Pore arrangements similar to that mentioned above and shown in Figure 16 (upper right) are often employed in theoretical models to simulate random pore networks. Such models are usually statistical pore networks consisting of a three-dimension lattice of nodes (sites) and bonds. The pore volume of a solid is represented by the nodes as well as the bonds connecting the nodes and forming a *Dual Site Bond Model*, DSBM. The lattice is constructed in such a way as to simulate satisfactorily the real pore structure of solids using nitrogen or mercury physisorption data. There are many such models of different sophistication. For instance, Portsmouth and Gladden [53] proposed such a model for a porous network with different connectivity. Nicholson and Petropoulos [45, 54, 55] employed similar network configurations to simulate two-phase flow in a three-dimensional network with *Gaussian radii distribution* as well the flow properties of random networks with various radius distributions. Armatas and Pomonis [56] performed similar simulations of pore network of silica materials using DSBM and *Monte Carlo* method, the N_2 desorption branch was considered whose fitting was achieved using the Monte Carlo method for selecting the proper location of each pore into the lattice.

So, the suggested arrangement shown in Figure 16 (upper right) is in line with previous studies and draws attention to the fact that desorption takes place via the nodes which are by necessity of extended size since various pores are converging at those points. Such a desorption mechanism is actually similar to “extrusion mechanism 2” in Hg porosimetry suggested by Portsmouth and Gladden [53]. This mechanism uses the nodes associated with the smallest pores in each individual pore cluster as starting sites for mercury withdrawal. In other words, the three small pores attached to the node with $c = 4$, in Figure 16 (upper right), will be emptied via the empty large cavity. Although the argument of these authors refers to the extraction of Hg, the physics behind the N_2 desorption should not be very different. To test this conjecture, we have plotted in Figure 16 (lower part) the logarithm of the pore anisotropy b reduced over the pore connectivity c, i.e. $\log(b/c)$, as a function of connectivity c for mesoporous $Al_{100}P_XV_Y$. As it can be seen, a loose trend becomes apparent obeying the relation $\log(b/c) = 1.34–0.07c$ but the least square correlation coefficient $R^2 = 0.20$ is poor.

Nevertheless, the pores connected to each node may possess *neither* the same anisotropy *nor* the same pore size. So, the (b/c) ratio, i.e. the pore anisotropies b reduced over the pore connectivity c, should be reduced *twice*: first over the standard deviation σ_b of the distribution of their anisotropies and second over the standard deviation σ_p of their pore sizes. To test this hypothesis, we have plotted the relationship $\log(b/c\sigma_p)$ *vs.* c in Figure 17(a) and the relationship $\log(b/c\sigma_b)$ *vs.* c in Figure 17(b). It is observed that the situation is impressively improved in both cases.

(i) The relationship $\log(b/c\sigma_p) = 1.61–0.13c$ exhibits square correlation coefficient $R^2 = 0.43$, which corresponds to a significance level P ~ 90%.
(ii) The relationship $\log(b/c\sigma_b) = 2.45–0.11c$ exhibits square correlation coefficient $R^2 = 0.38$, which also corresponds to a significance level slightly smaller than P ~ 90%.

So, it seems that both normalizations are important and significantly improve the linear regression between b and c parameters. However, the normalization over the standard deviation σ_p of the pore sizes is slightly more important compared to the normalization over the standard deviation σ_b of the distribution of anisotropies – at least for the present group of solids.

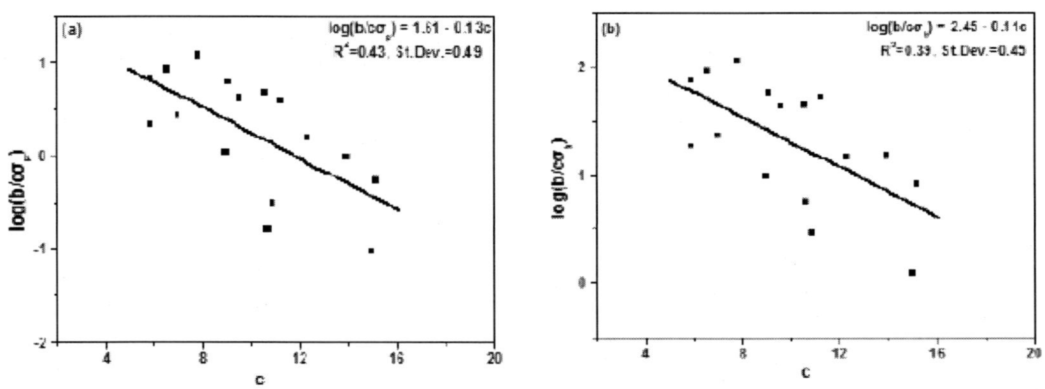

Figure 17. The relationship (a) $\log(b/c\sigma_p)$ *vs.* c, and (b) $\log(b/c\sigma_b)$ *vs.* c for mesoporous $Al_{100}P_XV_Y$ materials.

The next step is to check the combined effect of both σ_p and σ_b on the reduced (b/c) values and the results are shown in Figure 18.

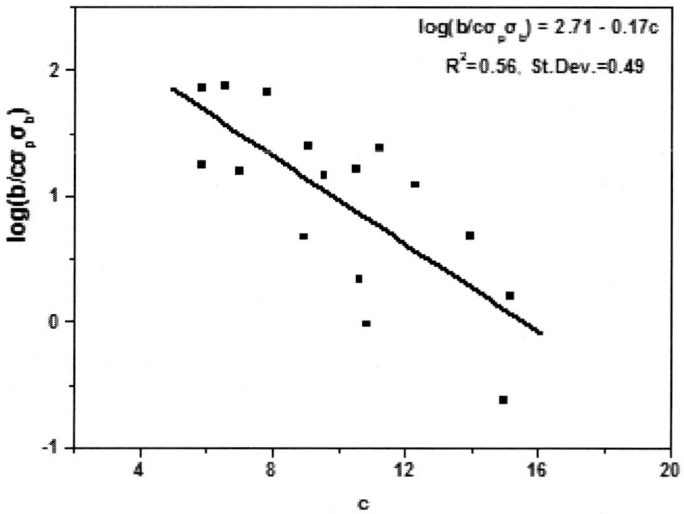

Figure 18. Variation of the logarithm of the maximum pore anisotropy b reduced over the mean pore connectivity c.

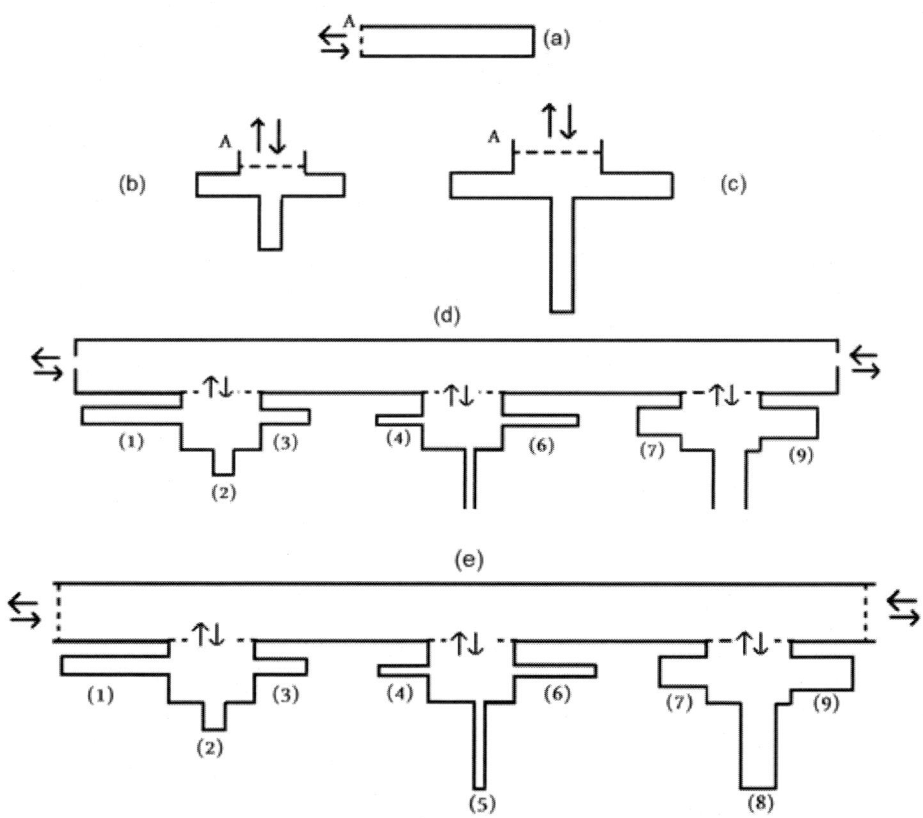

Figure 19. Dependence of pore anisotropy b on pore connectivity c, standard deviation of the pore size distribution σ_p and standard deviation of the pore anisotropy distribution σ_b.

A remarkable relationship becomes now apparent, from this figure, obeying the relation $\log(b/c\sigma_b\sigma_b) = 1.27 - 0.19c$ with a least square correlation coefficient $R^2 = 0.56$ corresponding to P ~ 99%. This relationship conveys the reasonable fact that in random porous networks, like the ones in the studied solids $Al_{100}P_XV_Y$, the pore anisotropy b reduced over the pore connectivity c, the standard deviation of the pore sizes distribution σ_p, and the standard deviation of the pore anisotropy distribution σ_b vary inversely to the pore connectivity c.

A model describing these effects and summarizing the discussion is shown in Figure 19.

(i) In Figure 19a): an observer at the entrance A of an independent pore counts the desorbed amount of N_2 and then carries out the relevant calculation to find the pore anisotropy b.

(ii) In Figure 19b): the observer at a crossroad point A, where c = 3 similar pores are connected, may calculate a total pore anisotropy b_{total} but the real anisotropy of each particular pore is actually $b_{real} = b_{total}/3 = b_{total}/c$.

(iii) In Figure 19c): the above c = 3 pores have the same size, but have different lengths, i.e. different anisotropies b. Then the term (b_{total}/c) should be reduced over the standard deviation of the pore anisotropy distribution σ_b, thus $b_{real} = (b_{total}/c\sigma_b)$.

(iv) In Figure 19d): the above c = 3 pores have both different size and different lengths. Then, the term $(b_{total}/c\sigma_b)$ should be reduced over the standard deviation of the pore size distribution σ_p, and therefore $b_{real} = (b_{total}/c\sigma_p\sigma_b)$. In this case, the collection of pores communicates with the outer environment via a large cavity with small openings, which is the reason for the observed hysteresis. It is also mentioned that any possible permutations between the pores (1), (2), (3), (4), (5), (6), (7), (8) and (9) in this figure or in Figure 19e will not affect the final outcome.

(i) Figure 19e): the large cavity shown in (d), where the various nodes are connected, may possess large openings. Then, no hysteresis loop is expected and no calculation of connectivity is possible according to the method of Seaton [47-51]. In this case the situation becomes similar to that described in the previous section where micropores are communicated with the outer environment via organized mesopores and the pore anisotropy is affected by the (%microporosity) according to the relationship $\log(b) = \log(b_o) - B(\%micro)/\Delta P_{micro})$ discussed in previous section.

8. Common Rules in the Development of Anisotropic Systems

As mentioned in the first section of this article, the most common example of anisotropy in nature is the development of plants which is favored along the vertical axis rather than along the horizontal direction [57-59]. This effect is generated by the anisotropic expansion of plant cells, which originates from their anisotropic mechanical construction of aligned cellulose microfibrils. This type of growth overcomes the prevailing *stress anisotropy*. The degree of cell's anisotropy may be controlled by factors in addition to the direction of microfibril alignment and the rates of expansion in maximal and minimal directions are regulated by *molecular mechanisms* [57]. It has been observed that disruption of the cell's ability to generate long cellulose microfibrils affects its ability to achieve growth anisotropy and also that shaping of plant organs may depend not only on the orientation but also on the relative length of cellulose microfibrils during axis establishment and growth [58]. The above well-established facts are depicted in Figure 20, which reveal a very crucial point: *The direction of structural elements, causing the cell anisotropy (e.g. cellulose microfibrils in this case), are aligned normal to the development of anisotropy.* This topic is huge and the reader can find more details in specialized texts [60].

It is remarkable that exactly the same arrangement is also observed in the structural elements, e.g. the C_nTAB surfactants, of micelles that cause the generation of extensive anisotropy of pores in MCM -type of solids (see TEM images in Figure 21, upper and middle part).

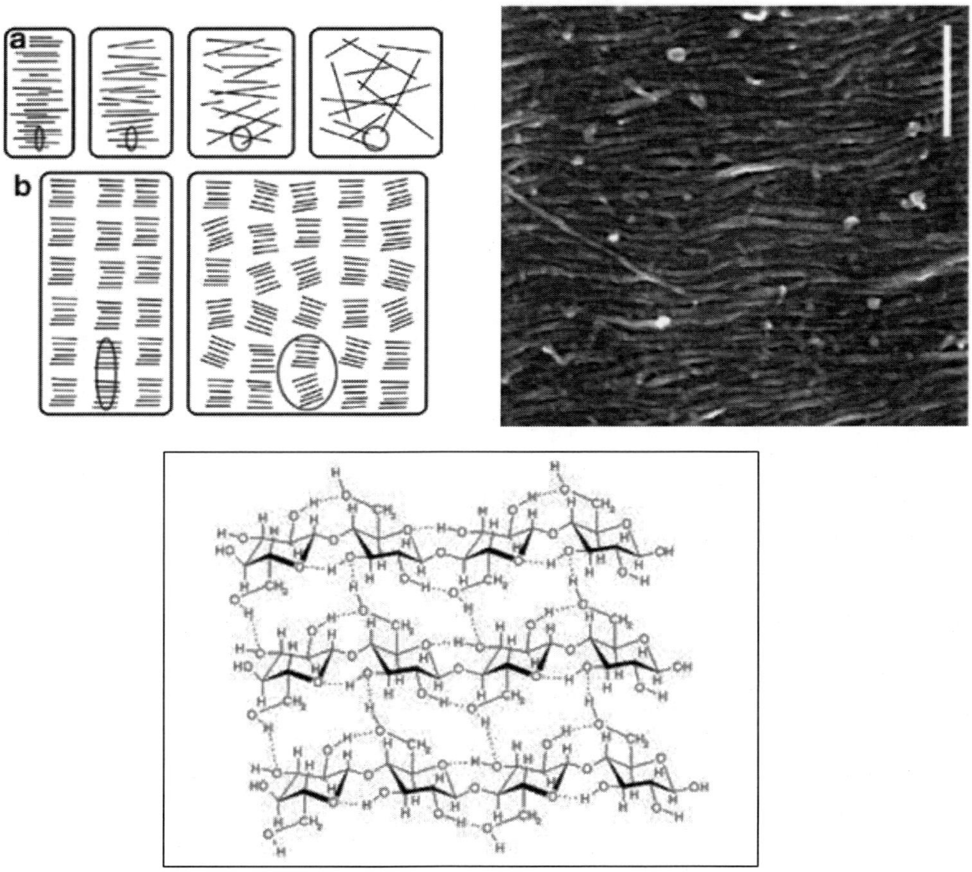

Figure 20. The role of cellulose microfibrils in the development of cell anisotropy in plants. Upper left: Models for controlling the degree of anisotropy by microfibril alignment. The degree of anisotropy is shown by the ellipses. (a) Green's hypothesis [61], where the degree of anisotropy is proportional to the degree of local alignment among microfibrils. (b) Modified hypothesis of Baskin [57] for a tissue (black rectangle with rounded corners) in which the degree of anisotropy is proportional to the uniformity of alignment in different cells or cell regions (cells are not shown). Upper right: Micrograph of the cell wall layer of a cucumber imaged with field-emission scanning electron microscopy [57]. The long axis of the stem runs upwards. The cell wall is structurally anisotropic. Scale bar = 250 nm. Lower: A strand of cellulose showing the hydrogen bonds (dashed) within and between cellulose molecules. The polymer is a rather stiff ribbon-like structure consisting of a 500-15000 units of a single type of sugar, glucose, and every other glucose residue is oriented 180 degrees relative to its neighbor.

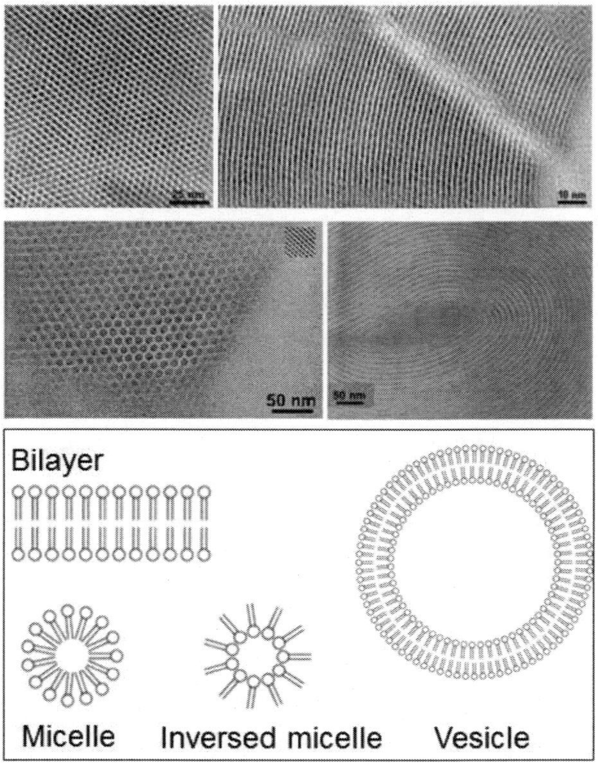

Figure 21. TEM images and mechanism of micelles formation. Upper part: TEM images of MCM-41 material parallel (left) and perpendicular (right) to the pore channel axis. Middle part: TEM images of SBA-15 parallel (left) and perpendicular (right) to the pore channel axis. Lower part: Mechanism for the formation of micelles, which serve as template and create internal porosity in the solid. The structural elements of those micelles are usually surfactant species consisted of long-chain alkyl quaternary ammonium ions, $C_nH_{2n+1}N^+(CH_3)_3$ with 8 < n < 16. Their orientation is always vertical to the long axis of micelle and therefore of the resulting long pores.

Under specific conditions, these molecules form micelles that serve as templates for the creation of internal porosity in solids, as shown in Figure 21 (lower part) and Figure 3. The concept of micelle template was originally developed for silica and silica-alumina based materials and later generalized for many other metal oxides, like Al_2O_3, TiO_2 and ZrO_2. In the original syntheses by Mobil researches, cationic surfactants – namely long-chain alkyl quaternary ammonium ions $C_nH_{2n+1}N^+(CH_3)_3$ with 8 < n < 16 – reacted with (alumino)silicate species in solution at high pH and moderate temperature. Depending on the surfactant to Si precursor ratio in the solution, various mesophases with 3D (cubic, MCM-48), 2D (hexagonal, MCM-41) or 1D (lamellar, MCM-50) pore structure can be obtained. A clear relationship has been observed between the arrangement of surfactant micelles and the structure of the corresponding mesophase.

On the basis of this relationship, a "*liquid crystal templating*" (LCT) mechanism was proposed in which positively charged micelles are formed in solution and interact with silicate species to generate silica rods and, then, long-range ordered nanostructures. As a consequence, the dimensionality and size of the pore system are dictated by the dimensionality and the size of micelles in the solution. In the particular case of the hexagonal member MCM-41, the pore diameter can be controlled between ca. 2 and 4 nm by changing the length of the alkyl chain of the surfactant from C_8 to C_{16}. *In all those cases the surfactant species are arranged normal to the long axis of micelle and eventually of pores, exactly like the cellulose fibrils in the anisotropic plant cells.*

What might be the reasons (physical, geometrical, physicochemical, biochemical or whatever) behind this common behavior? Following Baskin [57] closely, it is mentioned that for any material that expands anisotropically, fundamental laws of mechanics dictate that the direction in which the maximal rate occurs is perpendicular to the direction of the minimal rate (Figure 20). Therefore, anisotropy is characterized by two parameters: direction and degree. Direction specifies the direction in which the maximal strain rate occurs and degree specifies the relationship between the maximal and minimal strain rates. This corresponds to the parameter of anisotropy b = L/D in the case of pores.

In essence, plant cell growth in nature; as well micelle growth in a laboratory reflects the formation of particular water microvolume. If the forces acting to all directions are equal, then that microvolume is developed as a sphere. But, if the acting forces are stronger, let's say, along the x,y Euclidian level, then the microvolume will develop along the z axis. To use a metaphor, the situation is similar to the blowing-up of balloon, which will be spherical if no restraints are present. If the balloon is placed inside a solid "slinky" and then blown up, it will have a limited axis of expansion within the slinky and will enlarge towards a cylinder rather than a sphere.

Quantitatively, the divergence of the growth along the x,y level and the z axis represents different strains within the microvolume material. For the case of micelles formed during the organization of liquid crystals, one of the main reasons is the geometrical features of the surfactant molecule, which are the volume (V) occupied by the hydrophobic chain (i.e. C_nH_{2n+1}–), the length (l) of this chain and the cross sectional area (a_o) of hydrophilic head (i.e. $-N^+(CH_3)_3$). These three geometrical parameters are usually compared in the form of a dimensionless factor *g* suggested in 1976 by Israelachvili *et al.* [62].

$$g = (\text{cross section of the chain})/(\text{cross section of the head}) = (V/l)/a_o = V/a_o l$$

The value of the parameter *g*, which originates from simple geometry, reflects the shape of the surfactant as seen in Figure 22. The shape of the surfactant molecule in turn controls the shape of micelles formed as seen in Table 4.

It is observed that when $g < 1$, in other words when (cross section of the chain) < (cross section of the head), then the tight arrangement of surfactant molecules results by necessity to curved micelle surfaces, forming mesoporous structures like MCM-41 and MCM-48. But, when $g = 1$, i.e. when (cross section of the chain) = (cross section of the head), then the surfactant species are arranged parallel to each other forming a non-curved surface. Such an arrangement produces layered structure like MCM-50.

The situation is similar to the *cellulose fibrils* (i.e. the dominant architectural component of the cell walls of higher plants), which, via the cellulose synthase complex, are developed parallel to each other [60]. Therefore, it seems that the fundamental geometrical and physicochemical principles controlling the development of anisotropy in plant cells on one hand and the generation of micelle precursors, and eventually ordered mesoporous solids, on the other bear remarkable similarities which might be exploited for the generation of novel biomimetic materials.

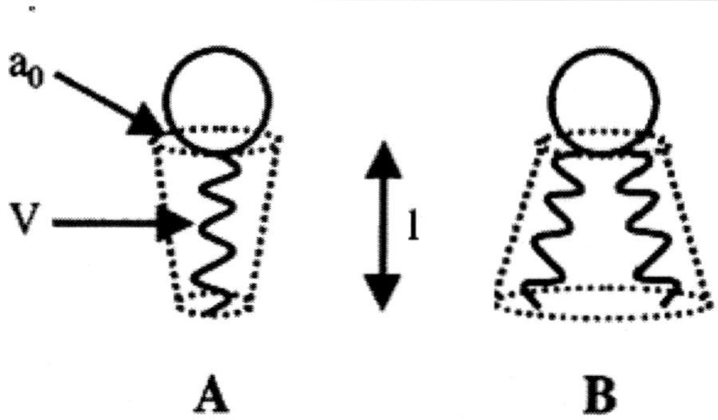

Figure 22. Surfactant molecules with different parameter g: (A) g < 1 and (B) g > 1.

Table 4. Effect of the dimensionless parameter g on the micelle arrangement and the final type of mesoporous material

Parameter *g*	Expected arrangement of micelles	Type of final mesoporous solids
1/3 to 1/2	Cubic e.g. MCM-48	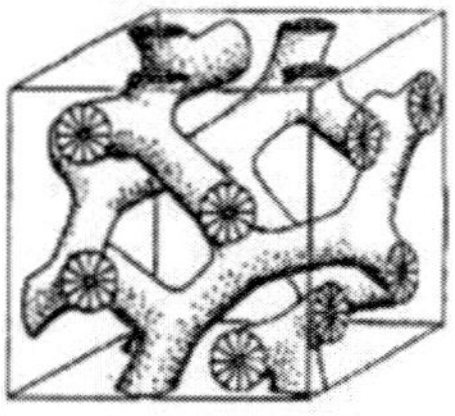
1/2	Hexagonal e.g. MCM-41	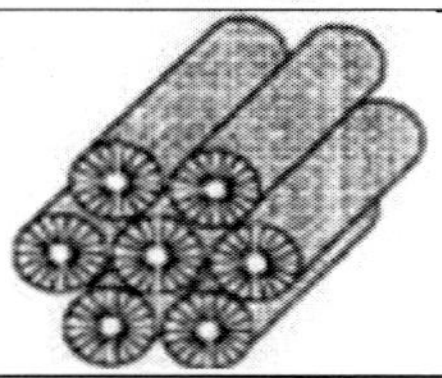
1	Layered e.g. MCM-50	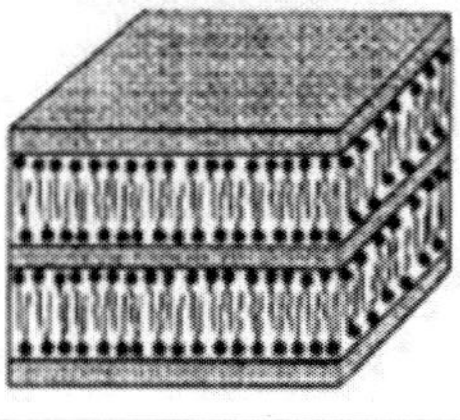

Conclusion

Anisotropy is a quite common phenomenon in nature where the geometrical or physical properties of a system (e.g. plants or clays) vary along different spatial axes. Anisotropic structural developments are also observed in porous inorganic materials like ordered mesoporous MCM-type materials.

In this chapter, the potential to calculate pore anisotropy in nanoporous solids from gas physisorption measurements has been demonstrated. This methodology is based on adsorption – desorption isotherm data and seems to be directly related to the fractal dimensionality of nanoporous surfaces. The materials described in this chapter have order or random pore structure. The anisotropy of pores is an intrinsic property of solid materials that is strongly related to the microporosity and mean pore connectivity. In general, high connected pore networks or porous domains with high microporosity exhibit smaller pore anisotropies and vice versa.

Comparison between the phenomenology describing the development of anisotropy in plants and the generation of organic-inorganic mesostructures in synthesis of ordered mesoporous solids revealed remarkable similarities.

These two systems, whereas entirely different in nature, are governed by similar fundamental geometrical and physisochemical principles. The exploitation of these characteristics will pave the way to understand the development and, thus, the synthesis of novel biomimetic materials.

ACKNOWLEDGMENTS

We wish to thank E. Tsaousi for her help. This work was supported by projects 09SYN-42-791 NANOMgO and THALES-NANOMESO, financed by ERDF and National Founds.

REFERENCES

[1] T. I. Baskin, *Ann. Rev. Cell Dev. Biol.,* 21, 203 (2005).

[2] C.T. Brett, *Int. Rev. Cytol.,* 199, 161 (2000).

[3] G. O. Wasteneys, *Curr. Opin. Plant Biol.* 7(6), 651 (2004).

[4] http://classes.mst.edu/ide120/lessons/wood/property_comparison/index.html.

[5] F. Bergaya, B.K.G. Theng and G. Lagaly, *Handbook of Clay Science*; Elsevier: Amsterdam, (2006).

[6] Gill, S.A. Korili, R. Trujilano, and M.A. Vicente, *Pillared Cays and related Catalysts*; Spinger (2010).

[7] F. Figueras, *Catal. Rev.* 30(3), 457 (1988).

[8] C.T. Kresge, M.E. Leonowicz, W.J. Roth, J.C. Vartuli and J.S. Beck, *Nature*, 359, 710 (1992).

[9] J.C. Beck, J.C. Vartuli, W.J. Roth, C.T. Leonowicz, C.T. Kresge, K.D. Schmitt, T.-W. C. Chu, D.H. Olson, E.W. Sheppard, S.B. McCullen,. J.B. Higgins, J.L. Schlenker, *J. Am. Chem. Soc.* 114, 10834 (1992).

[10] U. Ciesla and F. Schuth, *Micropor. Mesopor. Mater.* 27, 131(1999).

[11] G. Øye, J. Sjöblom and M. Stöcker, *Adv. Colloid Interface Sci.*, 89-90, 439 (2001).

[12] J.S. Gregg and K.S.W. Sing, *Adsorption, Surface Area and Porosity*, 2nd ed. Academic Press: New York (1982).

[13] F. Rouquerol, J. Rouquerol and K.S.W. Sing, *Adsorption by Powders and Porous Solids*; Academic Press: New York (1999).

[14] S. Lowell, *Introduction to Power Surface Area*; John Willey and Sons: Chichester, (1979).

[15] J.M. Thomas and W.J. Thomas, *Introduction to the Principles of Heterogeneous Catalysis*; Academic Press: London (1967).
[16] J.M. Thomas and W.J. Thomas, *Principles and Practice of Heterogeneous Catalysis*; VCH: New York (1997).
[17] E.P. Barrett, L. G. Joyner and P. P. Halenda, *J. Am. Chem. Soc.* (73), 373 (1951).
[18] F.A.L. Dullien, *Porous Media: Transport and Pore Structure*, 2nd ed., Academic Press: New York (1992).
[19] P. J. Pomonis and G. S. Armatas, *Langmuir*, 20, 6719 (2004).
[20] J.K. Knowles, G. Armatas, M.J. Hudson, P.J. Pomonis, *Langmuir*, 22, 410 (2006).
[21] R. Denoyel, J.M. Meneses, G.S. Armatas, J. Rouquerol, K.K. Unger, and P.J. Pomonis, *Langmuir*, 22, 5350 (2006).
[22] A.P. Katsoulidis, D.E. Petrakis, G.S. Armatas, P.N. Trikalitis and P.J. Pomonis, *Microp. Mesop. Mater.* 92, 71 (2006).
[23] A.P. Katsoulidis, D.E. Petrakis, G. S.Armatas and P. J. Pomonis, *J. Mater. Chem.*, 17, 1518 (2007).
[24] A.P. Katsoulidis, E.T. Tsaousi, G.S. Armatas and D.E. Petrakis, *Microp. Mesop. Mater.* 122, 175 (2009).
[25] P. J. Pomonis and E.T. Tsaousi, *Langmuir*, 25, 9986 (2009).
[26] J. Frenkel, *Kinetic Theory of Liquids*; Clarendon Press: Oxford (1946); (b) G. D. J. Halsey, *Chem. Phys.* 16, 31 (1948). (c) T. L. Hill, *Adv. Catal.*, 4, 211 (1952).
[27] P. Pfeifer, Y.J. Wu, M.W. Cole and Krim, *J. Phys. Rev. Lett.* 62, 1997 (1989). (b) P. Pfeifer, M.W. Cole, *New J. Chem.* 14, 221 (1990). (c) P. Pfeifer, J. Kenntner, and M.W. Cole, *In Fundamentals of Adsorption*; Mersmann, A. B.; Scholl, S. E.; Eds..; Am. Inst. Chem. Eng.: New York, 1991; p 689. (d) I. M. K. Ismail and P. Pfeifer, *Langmuir*, 10, 153 (1994).
[28] A.V. Neimark, *JETP Lett. 51*, 607 (1990). (b) A.V. Neimark, *Adsorpt. Sci. Technol. 7*, 210 (1991). (c) A.V. Neimark, and K.K. Unger, *J. Colloid Interface Sci. 150*, 4121993. (d) A.V. Neimark, M. Hanson and K.K. Unger, *J. Phys. Chem. 97*, 6011 (1993).
[29] Mandelbrot, *The Fractal Geometry of Nature*; Freeman: San Francisco (1982).
[30] D. Avnir, D. Farin, P. Pfeifer, *J. Chem. Phys. 79*, 3566 (1983). (b) D. Avnir, D. Farin, P. Pfeifer, *Nature*, 308, 261 (1984). (c) D. Avnir. *The Fractal Approach to Heterogeneous Chemistry: Surfaces, Colloids, Polymers*; John Wiley and Sons: Chichester (1989).
[31] M. Jaroniec, *Langmuir*, 11, 2316 (1995).
[32] B. Sahouli, S. Blacher and F. Brouers, *Langmuir*, 12, 2872 (1996).
[33] J. M. Drake, L.N. Yacullo, P. Levitz and J. Klafter, *J. Phys. Chem*., 98, 380 (1984). (b) F. Ehrburger-Dolle, M. Holz, and Lahaye, *J. Pure Appl. Chem.* 65, 2223 (1993). (c) Y. Lefebvre, S. Lacelle and C. Jolicoeur, *J. Mater. Res*. 7, 1888 (1992).
[34] K. Kaneko, M. Sato, T. Suzuki, Y. Fujiwara, K. Nishikawa, M. Jaroniec, *J. Chem. Soc.*, Faraday Trans., 87, 179 (1991). (b) H. Darmstadt, C. Roy, S. Kaliaguine, B. Sahouli, S. Blacher, R. Pirard and F. Brouers, *Rubber Chem. Technol.* 68, 330 (1995). (c) C. L. Wang, J. Krim, and M.F. Toney, *J. Vac. Sci. Technol. A*, 7, 2481 (1989).
[35] M.K. Wu, *Aerosol Sci. Technol.* 25, 392 (1996). (b) F. Meng, J.R. Schlup and L.T. Fan, *J. Colloid Interface Sci.* 197, 88 (1998). (c) E. Stathatos, P. Lianos and C. Tsakiroglou, *Microp. Mesop. Mater.* 75, 255 (2004).

[36] G.A. Niklasson, *Cem. Concr. Res.* 23, 1153 (1993). (b) M.J. Watt-Smith, K.J. Edler andS.P. Rigby, *Langmuir*, 21, 2281 (2005).
[37] R. Celis, J. Cornejo and M.C. Hermosin, *Clay Miner.* 31, 355 (1996). (b) F.J. Lee, C.-K. Lee and L.-C. Juang, *J. Colloid Interface Sci.* 217, 172 (1999). (c) A.D. Stefanis, A.A.G. Tomlinson, T.A. Steriotis, G.C. Charalambopoulou and U. Keiderling, *Appl. Surf. Sci.*, 253, 5633 (2007).
[38] A.K. Ladavos, P.N. Trikalitis and P.J. Pomonis, *J. Mol. Catal. A: Chem.*, 106, 241 (1996).
[39] F. Rubio, J. Rubio and J.L. Oteo, J. *Mater. Sci. Lett.* 16, 49 (1997). (b) G.M.S. El Shafei, *J. Colloid Interface Sci*, 250, 394 (2002).
[40] G.M.S. El Shafei, C.A. Philip and N.A. Moussa, *J. Colloid Interface Sci.*, 277, 410 (2004). (b) T.C. Vaimakis, C.S. Skordilis, and P.J. Pomonis, *J. Colloid Interface Sci.* 172, 311 (1995).
[41] D.E. Petrakis, I. Pasxalidis, C.R. Theoharis, M.J. Hudson and P.J. Pomonis, *J. Colloid Interface Sci.* 185, 104 (1997).
[42] C.-K. Lee and C.-S. Tsay *J. Phys. Chem. B*, 102, 4123 (1998).
[43] B. Sahouli, S. Blacher and F. Brouers, *Langmuir*, 13, 4391 (1997). (b) J.M. Drake, L.N. Yacullo, P. Levitz and J. Klafter, *J. Phys. Chem.* 98, 380 (1994). (c) P. Tang, N.Y.K. Chew, H.-K. Chan and J.A. Raper, *Langmuir*, 19, 2632 (2003).
[44] D.D. Do, *Adsorption Analysis: Equilibria and Kinetics*; Series on Chemical Engineering, Imperial College Press: Singapore (1998).
[45] P. J. Pomonis, K.M. Kolonia and G.S. Armatas, *Langmuir,* 17, 8397 (2001).
[46] G.S. Armatas, K.M. Kolonia and P.J. Pomonis, *Langmuir*, 18, 10421 (2002).
[47] A.N. Seaton, *Chem. Eng. Sci.* 46, 1895 (1991).
[48] H. Liu, L. Zhang, A.N. Seaton, *Chem. Eng. Sci.* 47, 4393 (1992).
[49] H. Liu, L. Zhang and A.N. Seaton, *J. Colloid Interface Sci.* 156, 285 (1993).
[50] H. Liu, L. Zhang and A.N. Seaton, *Langmuir*, 9, 2576 (1993).
[51] H. Liu and A.N. Seaton, *Chem. Eng. Sci.* 49, 1869 (1994).
[52] S. Kirkpatrick, In *Ill-Condensed Matter.*; Ballian, R.; Mayward, R.; Toulouse, G.; Eds.; North-Holland: Amsterdam (1979).
[53] R.L. Portsmouth and L.F. Gladden, *Chem. Eng. Sci.*, 46, 3023 (1991).
[54] D. Nicholson, J.H. Petropoulos, *J. Phys. D Appl. Phys.*, 8, 1430 (1975).
[55] D. Nicholson and J.H. Petropoulos, *J. Phys. D Appl., Phys.*, 4, 181 (1971).
[56] G. S. Armatas and P.J. Pomonis, *Chem. Eng. Sci.* 59, 5735 (2004).
[57] T.I. Baskin, *Ann. Rev. Cell Dev. Biol.* 21, 203 (2005).
[58] G.O. Wasteneys and M. Fujita, *J. Plant Res.* 119, 5 (2006).
[59] G.O. Wasteneys, *Curr. Opin. Plant Biol.* 7, 651 (2004).
[60] Alberts, A. Johnson, J. Lewis, M. Raff, K. Roberts and P. Walter, *Molecular Biology of the Cell*; 4th edition; Oxford University Press: New York (2003).
[61] P.B. Green, *Science* 138, 1404 (1962). (b) P.B. Green, *J. Cell Biol.* 27, 343 (1965).
[62] J.N. Israelachvili, D.J. Mitchell and B.W. Ninham, *J. Chem. Soc. Faraday Trans.*, 72, 1525 (1976).

In: Anisotropy Research
Editor: Hirpa Lemu

ISBN 978-1-62081-977-7

Chapter 4

EXPERIMENTAL STUDY OF DIFFUSION ANISOTROPY OF CLAY FORMATIONS

V. G. Rumynin[1,*], E. B. Pankina[2], A. M. Nikulenkov[1] and M. P. Glukhova[2]

[1]Saint Petersburg State University, Dept. of Geology;
Inst. of Env. Geology of the RAS, SPb Div, Russia
[2]A.P. Alexandrov Technology Inst., Sosnoviy Bor, Russia

ABSTRACT

The study presents experimental evidence of exhibition of the anisotropic behavior of Vendian and Cambrian clay formations (selected for long-term isolation of radioactive wastes at the Northwestern region of Russia) with regards to radionuclide (Cl-36, Sr-90, Cs-137 and Co-60) diffusion. Coupled diffusion and sorption processes in the clay sediments have been studied using different experimental set-up: 1D tests with packing clays (disturbed samples) and 1D and 3D tests with undisturbed rock matrix. Experiments have shown that the studied clays exhibit fairly strong diffusion anisotropy, at that $D_{aL} > D_{aT}$: the longitudinal apparent diffusion coefficient is appreciably larger than the transverse coefficient (by a factor of up to 6). This effect is caused by the stratification of clay formations, the preferred orientation of the microlayers at the mm-cm scale and clay particles. The obtained results agreed with the published results of laboratory tests with clay samples taken from clay formations considered as potential host rocks for the underground repository for radioactive disposal in the Western Europe.

1. INTRODUCTION

Recent studies in many countries have been focused on the diffusion properties of regionally persisting clay formations with a large (>100 m) thickness, which are of interest in the context of their possible use for long-term isolation of radioactive wastes (RW).

* Corresponding author: rumynin@hge.pu.ru.

Molecular diffusion is regarded as the main radionuclide transport mechanism in the estimation of the risk associated with RW burial. Thus, a number of laboratory and in situ tests have been performed to study diffusion in Aalenian Opalinus clay (Switzerland), Rupelian Boom clay at Mol (Belgium), Callovo-Oxfordian clayey siltstone/silty clay at Bure (France), Vendian/Kothlin, Cambrian/Blue clays (the north-western region of Russian Federation).

The studies show all these clay strata to feature anisotropic physical properties due to the microstratification, i.e., alternation of beds (with thickness varying from several millimeters to a few centimeters) with different content of clay minerals. Such effect can be caused, for example, by intercalation of *argillite* and *aleurolite* (siltstone). In addition to intercalation, there are other factors that may influence the anisotropy of clays amongst which is consolidation.

Studies of the structure of clays have shown that clay particles tend to become oriented perpendicular to the major principal stress during one dimensional consolidation. This commonly results in horizontal bedding planes having physical properties different in horizontal and vertical directions [1, 2].

Experiments with clay samples from all mentioned above sites of the planned RW burial demonstrate anisotropy in flow and diffusion properties: the *hydraulic conductivity*, k_L, and *effective-diffusion coefficient*, D_{eL}, along the bed are many times greater than the same coefficients, k_T and D_{eT} governing the solute transfer across the beds [3–11]. The anisotropy of the effective permeability and diffusion coefficients is mostly due to stratification, the preferred orientation of the microlayers at the mm-cm scale and clay particles. The main focus of this chapter will be on *anisotropy of diffusion*.

Thus, the tortuosity of the diffusion pathways is larger when the direction is normal rather than parallel to the preferred orientation of the layers [12]. So, diffusion anisotropy is a consequence of rock heterogeneity and bedding.

Tracer transport in an anisotropic clay domain is described by the following diffusion equation [3, 7 – 9, 13]:

$$\frac{\partial C}{\partial t} - \nabla \cdot (\mathbf{D}_m \cdot \nabla C) = 0 \tag{1}$$

where $\mathbf{D}_m$ is the orthotropic pore diffusion tensor ($\mathbf{D}_m = \mathbf{D}_e/\phi$). The expression for the latter is as follows:

$$\mathbf{D}_m = \begin{pmatrix} D_{mL} & 0 & 0 \\ 0 & D_{mL} & 0 \\ 0 & 0 & D_{mT} \end{pmatrix} \tag{2}$$

where D_{mL} and D_{mT} are the principal components of the tensor. In this study, D_{mL} is the pore diffusion coefficient parallel to the horizontal bedding while D_{mT} is the pore diffusion coefficient perpendicular to the bedding.

2. Geological Settings and Physical Properties of the Sediment

2.1. Location

Clay formations are widespread in the Northwestern part of the Russian Federation. In St. Petersburg region southeast of the Gulf of Finland and Ladoga Lake, they occur close to the surface. The sediments are represented by two formations, which formed in Vendian and Cambrian geological periods (650–500 Ma). The degree of the clayey sediment consolidation is rather high and therefore they can also be referred to the mudstone lithological type of rock [14].

In the northwestern Russian Federation, Vendian and Cambrian clays have local names, Kothlin (associated with a geological suite) and Blue (associated with the characteristic color), respectively.

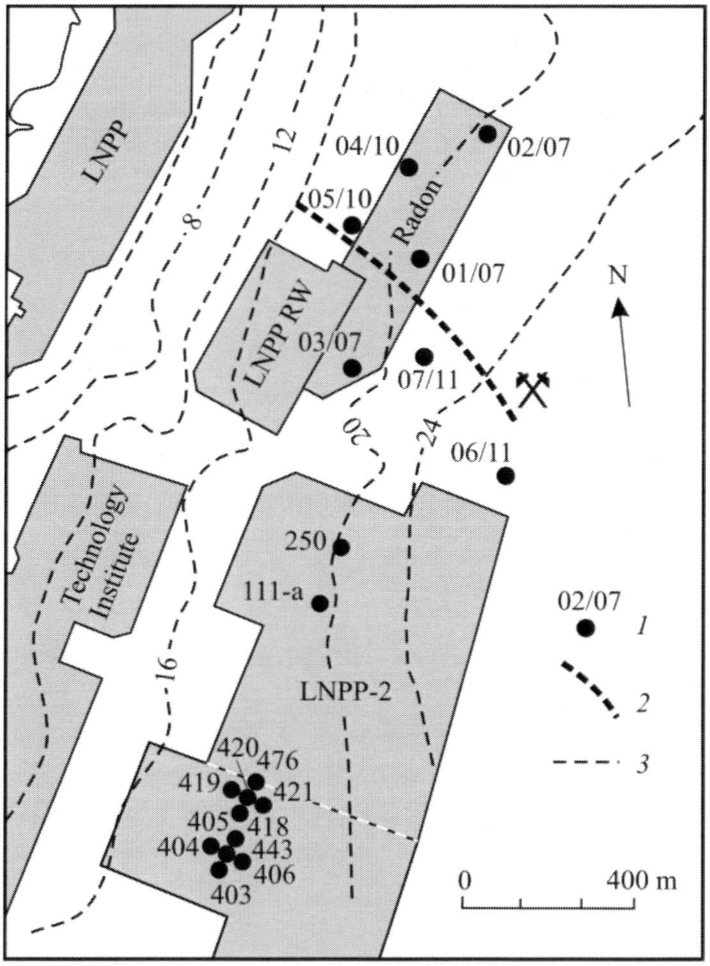

Figure 1. Layout of engineering-geological boreholes at Radon and LNPP-2 sites. (1) Engineering-geological borehole and its number; (2) the route of underground mine (Engineered URW); (3) isoline of hydraulic head (Cambrian aquifer).

The clays having low permeability and high sorption capacity are considered to be a suitable host environment for toxic or radioactive waste (RW) repositories. To quantify the barrier and geotechnical properties of these geological units and to assess the potential impact of engineered RW repository on the environment, field observations and laboratory experiments as well as modeling studies have been carried out in 2006–2011.

Data on the physicochemical, radionuclide-retention, diffusion, and geomechanical (geotechnical) properties of Kothlin (Vkt) and Blue (CBlue) mudstones obtained elsewhere as well as data on the relevant properties of clay formations in the Western Europe were collected and analyzed. In particular, a preliminary comparative analysis of the Kothlin and Blue clay properties against the results of the study of Rupelian Boom clay at Mol (BC) (Belgium), Callovo-Oxford argillaceous rocks (siltstone-silty clay) at Bure (COx) (France) and Aalenian Opalinus clay (OPA) at Mont Terry (Switzerland) has been performed [11, 14-16]. These formations were selected as potential host formations for the disposal of HLW (high level and long lived radioactive waste).

Unlike the BC, Cox, and OPA formations in Western Europe, where clays are overlain by thick (hundreds of meters) sedimentary rock strata, the clays in the northwestern region of the Russian Federation lay near the surface at depth not more than a few tens of meters. As compared with BC, Cox, and OPA clays, the Vkt and CBlue clays are more loosely and subject to exogenous processes. Therefore, these clays show high spatial variations in physical properties and the sensitivity to the distribution of external load.

Currently, the most likely option for the construction of underground RW repository in northwestern Russian Federation is the area coinciding with the existing Radon surface RW repository (Figure 1). This area is close to the construction site of the Leningrad NPP (LNPP-2), which has been a subject of detailed geological and engineering-geological studies.

2.2. The Structure, Chemical and Mineral Composition and the Physical Properties of the Clays

Microscale structure: The samples taken from the key sites demonstrate a micro-layered structure (Figure 2): silt layers from a few mm to 2 cm in thickness alternate with argillite layers of increased thickness that is illustrated by core-samples measurements.

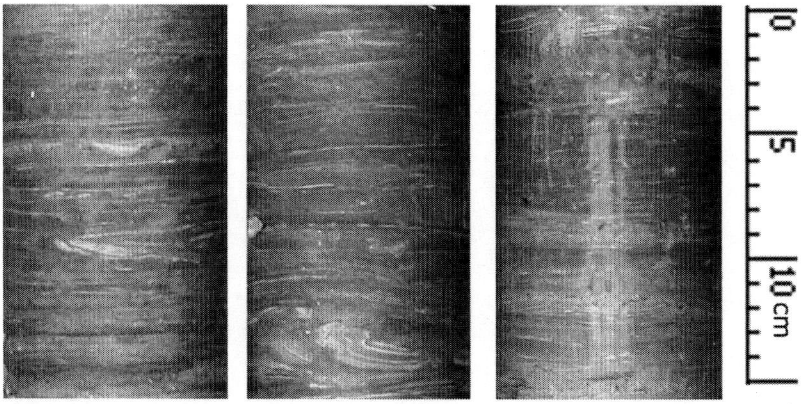

Figure 2. Micro-layered structure of the Vkt clay.

So, one may expect some exhibition of anisotropy in the physical properties of the studied formations due to their sedimentary origin. The clay formation anisotropy is also a first important characteristic of the Boom, Callovo-Oxfordian, and Aalenian Opalinus clays.

Discontinuities in the clay formation: Field observations of geological outcrops, shallow (near-surface) excavations, underground mines and boreholes have shown that

(1) the upper part of clay formations, which outcrops near the surface, was found to be disintegrated by a fracture network of different genesis; the joints generated by a release of natural stress (decompression) and the effect of weathering are predominant;
(2) the density of joints decreases when the clay is overlain by other geological layers;
(3) only few fractures were identified by examining rock samples from boreholes; some fractures are sealed with secondary minerals; no significant shifts of lithological microlayers along fractures were found;
(4) three sets of natural joints were revealed in mines related to the St. Petersburg subway tubes under construction: two of them are nearly vertical and can be associated with movements of the Archean and Proterozoic crystalline base of the territory; the third set, including horizontal joints, is connected with the clayey deposit formation. The walls of the natural joints are smooth and dry. The joints are closed, and no secondary minerals covering the walls were observed. They differ from artificial, *decompression joints*, which have rough walls; a typical distance between joints is around 20 to 30 cm.

Overall, the performed studies confirm the conclusions of M. Arnould [14], who analyzed data on the fracturing of clay formations Boom Clay, Opalinus Clay and Callovo-Oxfordian Silty Clay.

Thus, the comparison of data obtained by studying fracturing in near-surface open-pit mines, on the one hand, and the results of drilling and examining deep mine workings (immediately after their driving), on the other hand, shows fractures to exist only in the near-surface zones of sections and their formation to be caused only by weathering processes and the release of stress on the examined rock massifs.

Only single fractures, if any, were recorded at large depths; this fact is in agreement with the results of drilling key wells at the Radon site. Not denying the existence of sparse fracture network at large depths, M. Arnould [14] stresses that the external compressive stresses in clays are quite sufficient to cause complete healing of fractures, thus ensuring the high water resistance of clay strata at the macroscopic level.

Grain size distribution: Most Vkt and CBlue clays contain > 40–45% of clay particles (<0.005 mm). This is close to the content of fraction from 0.005–0.05 mm size-interval (Table 1; the number of experiments with different clays varies between 20 and 40).

Chemical composition and mineralogical constituents: The Vkt and CBlue clays consist of oxides of silicium (>50%), aluminum (>15%), iron (>5), sodium (up to 2%), and magnesium (up to 6%). The pelitic fraction (<0.01 mm) of the clay shows the presence of principal clay minerals, quartz, and orthoclase. Illite dominates among the clay minerals.

Cation exchange capacity (CEC): The Vkt clay shows rather low (for a clayey type of sediments) values of CEC: the average value is 6.6 mg-equiv/100 g (Table 2). CEC of the CBlue clay varies between 10.2 and 13.4 mg-equiv/100 g.

Table 1. The statistics of grain size distribution

	D [mm]	1–0.5	0.5–0.25	0.25–0.10	0.10–0.05	0.05–0.01	0.01–0.005	<0.005
				Radon (V_2kt)				
Content%	Min	–	–	–	2.7	12.7	10.6	15.9
	Max	6.8	11.0	8.5	17.9	56.0	21.5	58.7
	Avr	1.2	1.9	2.5	8.3	29.9	15.7	40.5
				LNPP-2 (V_2kt)				
Content %	Min	0	0	0.1	1.8	14.3	8.4	17.6
	Max	1.2	1.9	8.5	49.3	47.9	26.9	63.9
	Avr	0.1	1.2	1.6	12	23.1	16.5	45.5

Table 2. CEC variations (Radon site)

Interval, m	CEC [a]	W [b]	Interval, m	CEC	W
Borehole 01/07					
20.50–20.65	6.0	12.4	77.00–77.15	6.8	10.5
36.80–36.95	5.3	13.9	88.30–88.45	8.3	10.4
48.45–48.60	9.0	13.9	97.90–98.05	5.3	9.6
58.20–58.35	7.5	12.2	111.45–111.60	3.0	9.8
65.95–66.10	5.3	12.1	117.95–118.05	0.8	13.3
			133.25–133.40	4.5	10.0
Borehole 02/07			Borehole 03/07		
16.20–16.35	7.8	13.5	15.95–16.10	6.0	13.8
23.35–23.50	6.0	13.0	23.25–23.40	5.3	11.8
36.85–37.00	10.7	12.6	44.95–45.10	10.5	12.2
45.80–45.95	10.3	13.3	48.50–48.65	9.0	13.7
59.35–59.50	8.3	12.0	58.45–58.60	5.3	11.0
68.60–68.80	5.3	11.6	65.75–65.90	6.0	10.7

[a] Cation exchange capacity, mg-equiv/100 g; [b] moisture content, %.

Summary of physical properties: The following clay physical properties were tested: field moisture content, W; bulk density, ρ_b; bulk density of dry sediment, ρ_{bd}; density of particles, ρ_s; porosity, ϕ; moisture at the limits of yielding flow (liquid limit), W_L; moisture at the limit of plasticity (plastic limit), W_P; plasticity index, PI; saturation, S_r. Thorough laboratory examinations of 77 samples from five boreholes (1/07, 2/07, 3/07, 4/10 and 5/10) at the Radon site have established statistically representative variation ranges of physical parameters of the Vendian clay as shown in Table 3.

The mechanical properties of the clay formation have been characterized by triaxial (under drained and undrained loading conditions) and uniaxial compression tests.

The results obtained made it possible to determine a complete set of mechanical parameters, e.g., modulus of general deformation, E_0; modulus of deformation, E_d; shearing resistance, C; the angle of shearing resistance, φ; uniaxial compression strength, σ_c, and Poisson's ratio, ν (Table 4).

Table 3. Physical properties of the Vkt clay (Radon site)

Characteristics	Min	Max	Mean	CV[a]
W, %	7.1	22.5	13.62	0.15
ρ_b, g/cm^3	2.09	2.42	2.24	0.02
ρ_{bd}, g/cm^3	1.71	2.26	1.97	–
ρ_s, g/cm^3	2.65	2.81	2.73	0.012
ϕ	0.19	0.31	0.25	–
W_L, %	20.7	40	30.65	0.14
W_P, %	14.4	27.0	20.55	0.14
PI, %	4.1	17.6	11.87	–
S_r,	–	–	0.95	–

[a] Coefficient of variation.

Table 4. Geomechanical characteristics of the clayey formations

Parameters	Unit	CBlue, OPs[a]	Vkt, Radon	Vkt, Metro-1
Cohesion, C	MPa	0.15–0.3	0.2–0.35	2.5–3
Friction angle, φ	Degree	10–20	20–35	20
Uniaxial strength, σ_c	MPa	0.3–0.4	0.9	2.5–7.5
Modulus of deformation, E_d	MPa	50–60	40–120	970–1300
Poisson's ratio, ν		0.35–0.4	0.2–0.3	0.2–0.3

[a] Open pit sites (Nikolsky and Krasny Bor open pits).

When analyzing the laboratory data, one has to take into account that when the core assigned for testing is lifted up to the surface, the decompression tends to exaggerate the micro-fracture openings, while they are more close under natural stress conditions. This may result in erroneous geotechnical parameters.

2.3. Anisotropy in the Mechanical Properties

This anisotropy is primarily due to the layered structure of the sediment and the process of sedimentation followed by one-dimensional consolidation. Laboratory geotechnical tests show that *physical and mechanical properties* are dependent on orientation of microlayers with respect to the loading direction.

The clay is more compressible in the vertical direction in comparison with the horizontal direction: modulus of deformation across layering can exceed up to two times the same coefficients along layering. The experimental results demonstrate also the direction-variability of shear strength: factor of *cohesion anisotropy* can reach 0.7–0.6.

Anisotropy of the hydraulic conductivity: The clay formation has been characterized by core flow tests. They were performed under elevated-pressure conditions using modified triaxial equipment. Two samples 50 mm in diameter were cut out from each block sample intended for tests in such a manner that flow was directed along the bedding in one sample and across it in the other sample.

Table 5. Hydraulic conductivity of the tested clay (Vkt, Radon site)

Borehole number (Figure 1)	Depth, [m]	Along layering, [m/s]	Across layering, [m/s]	Anisotropy factor
5/10	48.6–49.0	$6.94 \cdot 10^{-11}$	$1.50 \cdot 10^{-11}$	4.6
4/10	51.8–52.1	$4.63 \cdot 10^{-11}$	$6.94 \cdot 10^{-12}$	6.7
1/07	55.2–55.35	N.d. [a]	$5.79 \cdot 10^{-12}$	N.d.
1/07	65.8–65.95	N.d.	$5.79 \cdot 10^{-12}$	N.d.
4/10	69.2–69.5	$2.20 \cdot 10^{-10}$	$1.62 \cdot 10^{-11}$	13.6
7/11	70.3–70.6	$3.38 \cdot 10^{-10}$	$1.54 \cdot 10^{-11}$	22
7/11	84.7–85.0	$4.12 \cdot 10^{-8}$	$2.99 \cdot 10^{-10}$	138
7/11	94.0–94.25	$2.36 \cdot 10^{-9}$	$2.66 \cdot 10^{-10}$	8.9

[a] No data.

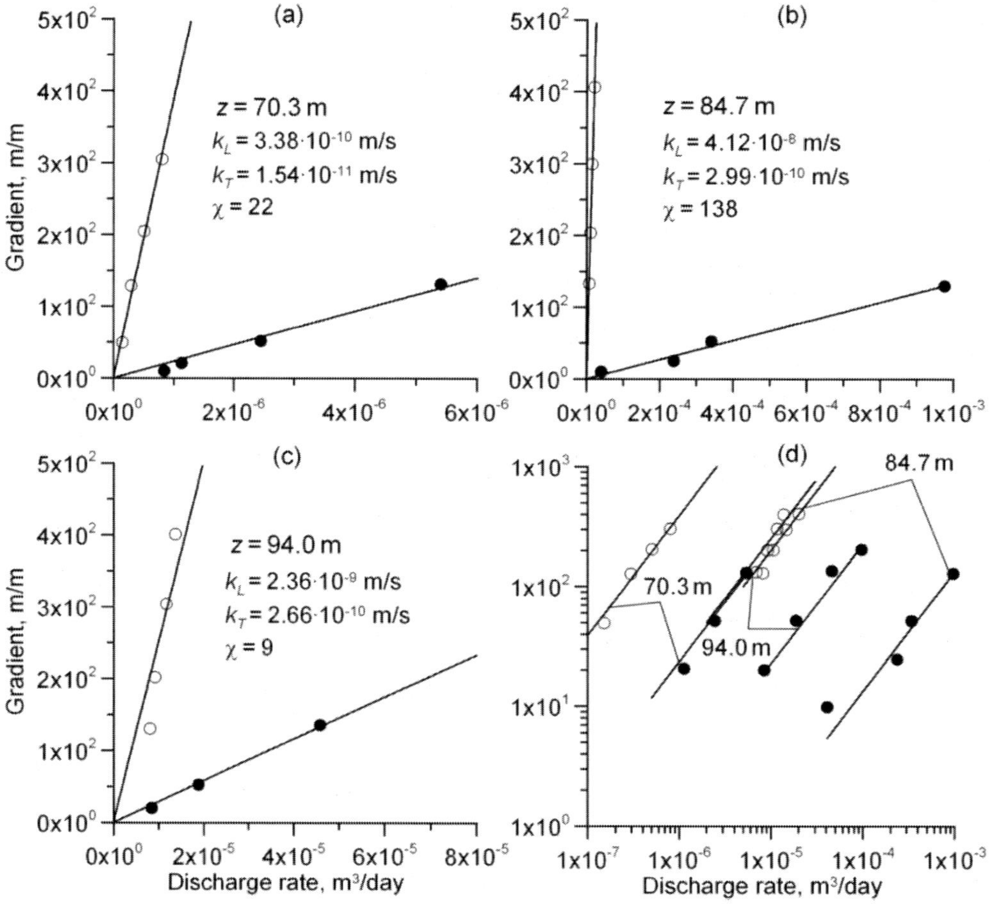

Figure 3. Relationship between discharge rate and hydraulic gradient in the flow tests with clay samples taken from borehole 7/11 at depth of sampling (z) of (a) 70.3 m, (b) 84.7 m and (c) 94 m (built with a normal coordinate system). (d) – log-log coordinates. Empty circles – tests across layering, filled circles – tests along layering.

Overall, eight experimental series were carried out. The obtained values of hydraulic conductivity (Table 5) varied from $4.6 \cdot 10^{-11}$ to $4.1 \cdot 10^{-8}$ m/s along the bedding and from

$5.8 \cdot 10^{-12}$ to $3.0 \cdot 10^{-10}$ m/s across it. Thus, the clay massif shows an appreciable anisotropy in permeability. It was identified that flow in the studied samples is governed by linear Darcy's law, and the initial hydraulic gradient was not observed in the specified range of pressure drop (Figure 3).

3. Study of Molecular Diffusion

Molecular diffusion is the predominant radionuclide-transport mechanism in low-permeability clay formations. Therefore this process was studied in detail using different experimental approaches. The studies included two types of experimental activities: 1D tests with packing clays (disturbed samples) and 1D and 3D tests with undisturbed rock matrix.

3.1. Single-Chamber Diffusion Cell (in-Diffusion) Tests

Experimental Setup and Interpretation

Experimental setup: Before the experiments, clay samples were disintegrated to the size of individual aggregates less than 1 mm. The pasted samples were conditioned using $SrCl_2$ solution of 0.1 N or 3.0 N at room temperature of 20–25° C.

The excessive solution was separated from the solids by centrifuging. These conditioned samples were used in tests carried out under isotonic conditions. To create nonisotonic conditions, pore water was desalted after saturation of the exchange complex of clay with stable strontium (pasted samples were conditioned with distilled water, until chloride ion concentration dropped below 0.001 N).

After conditioning, the pasted samples were thoroughly mixed and placed into chemical columns in small portions with subsequent packing (Figure 4). The height of the clay layer in all experiments was 20–25 cm. Next the column with clay was filled with a stable solution of $SrCl_2$ with appropriate concentration and exposed for 3–5 days. Just before the diffusion experiment, the solution of stable $SrCl_2$ was removed from the prepared cylindrical columns, the columns were weighted, and the volume filled by clay was measured.

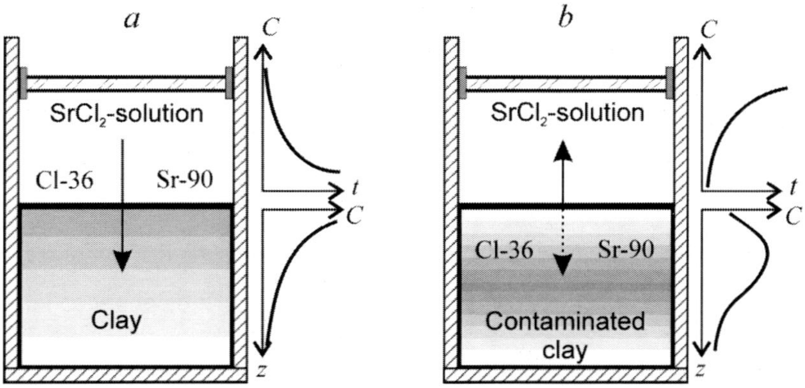

Figure 4. Experimental scheme: (*a*) diffusion from solution into sample, (b) diffusion from sample into solution.

Radioactive solutions were prepared by adding Cl-36 and Sr-90 isotopes to previously prepared and conditioned solutions of stable $SrCl_2$ (0.1 and 3.0 N). The experiments involved two stages: in-diffusion and out-diffusion. At the last stage of diffusion experiment, the vertical distribution of residual activity of *radionuclides* in clay was examined. At the end of the out-diffusion experiment, the clay core was sliced and the diffusion profile in the clay was determined.

Mathematical model: The first stage of parameter identifications was based on the measurement of the concentration decrease in the injection chamber. The process can be described by a partial solution of the one-dimensional differential equation

$$R\frac{\partial C}{\partial t} - D_m \frac{\partial^2 C}{\partial x^2} = 0 \tag{3}$$

combined with the condition of *isotope balance* in a limited volume of solution in the chamber

$$h\frac{\partial C}{\partial t} = -\phi D_m \left.\frac{\partial C}{\partial x}\right|_{x=0} \tag{4}$$

This solution, written for the concentration function in the chamber $C \equiv C(x=0,\ t)$ under the conditions $C(x,t=0)=C_0$, $C(x\to\infty,\ t)=0$, can be represented as given in [17, 18, 19] as follows.

$$\bar{C} = \exp(a^2 t)\,\mathrm{erfc}(a\sqrt{t}) \tag{5}$$

where $a=\phi\sqrt{D_m R}/h$, $\bar{C}=C/C_0$, C_0 is the initial concentration in the chamber, D_m is the *pore diffusion coefficient*, ϕ is the *porosity*, h is the thickness of the solution layer in the chamber, $R=1+K_d\rho_b/\phi$ is the *retardation factor*.

Unfortunately, as mentioned above, the processing of the experimental breakthrough curve, characterizing the drop in concentration in the registration chamber, gives little information, since only a value for the combination $\phi^2 R D_m^{\mathrm{in}}$ can be obtained.

The interpretation potential radically increases in the presence of data on the spatial distribution of indicators (both adsorbed and inert) in the pore space, $C=C(x,t)$, and the record of function $C^{\mathrm{out}} \equiv C(0,t>t_0)$, obtained in the period of their *reverse diffusion* from the sample into solution (in the case of change in its composition at some moment $C^{\mathrm{out}}=(0,t=t_0)=0$); here, t_0 is the duration of diffusion into the sample. In this case, the solution of the problem for heterogeneous initial conditions can be obtained numerically.

Summary of Tests with Packing Cambrian Clay

The obtained values of molecular diffusion coefficients for chlorine and strontium ions were found to be close to one another (Table 6): in the solution with a low ionic strength, D_m

= (1.3–1.7)·10^{-9} m^2/s, while in the concentrated solution D_m = (9.7–9.9)·10^{-10} m^2/s, i.e., the diffusion coefficient decreases by a factor of about one and a half. The values of effective porosity, established for chlorine ions, are close to the total clay porosity ($\phi^0 \approx 0.6$). Hereafter, we assume $\phi = 0.5$.

The experiments revealed appreciable adsorption of Sr-90, which was maximal in pore waters with low *ionic strength* (0.1 N): the appropriate coefficient of adsorption distribution K_d = 4.2 cm^3/g is 6 times as large as K_d (0.7 cm^3/g), evaluated for a concentrated (3.0 N) solution.

It is important to note that the processing of the results of experiments with radiostrontium yielded a considerable difference between the calculated (fitted) values of R for the stages of diffusion from solution into the sample (R^{in}), when *adsorption* of this radionuclide takes place, and for the stage of diffusion from this sample (R^{out}), when desorption takes place: $R^{out} >> R^{in}$. The calculated values of K_d were about 11.4 and 2.4 cm^3/g (in the rocks saturated with low-salinity and concentrated solutions, respectively).

Single-Chamber Diffusion Cell Tests with Undisturbed Vendian Clay Samples

In this experimental set-up samples for single-chamber diffusion cells (Table 3 *a*) were cut from the borehole core material without its crushing. Samples of cylindrical-shape (ranging in length from 3 to 5 cm and 5 cm in diameter) were drilled out using special core-drilling bit. For each experimental series with Sr-90, Cs-137 and Co-60 three pairs of the cylindrical-shape samples were prepared.

Each pair of the samples characterized a certain depth of the Vkt formation (Table 7). One out of each pair of samples was drilled out in such a way that its axis was directed parallel to bedding planes. The other pair-sample was reoriented towards a direction that is perpendicular to bedding planes.

Table 6. Calculated (model) parameters of diffusion and adsorption (desorption) distribution (isotonic conditions)

Test parameters		Isotopes and solutions			
		Cl-36		Sr-90	
		0.1 N $SrCl_2$	3.0N $SrCl_2$	0.1 N $SrCl_2$	3.0 N $SrCl_2$
a, $d^{-1/2}$		0.153	0.145	0.616	0.278
$\phi^2 D_m R^{in}$, m^2/s		5.60·10^{-10}	2.43·10^{-10}	6.44·10^{-9}	8.39·10^{-10}
D_m, m^2/s		1.3·10^{-9}	9.7·10^{-10}	1.7·10^{-9}	9.9·10^{-10}
D_a, m^2/s		1.2·10^{-9}	9.7·10^{-10}	1.1·10^{-10}	2.9·10^{-10}
R	In	1.1	1.0	15.0	3.4
	Out	1.1	1.0	39.0	9.0
K_d, cm^3/g	In	≈ 0.03	0.0	4.19	0.71
	Out	≈ 0.03	0.0	11.38	2.38

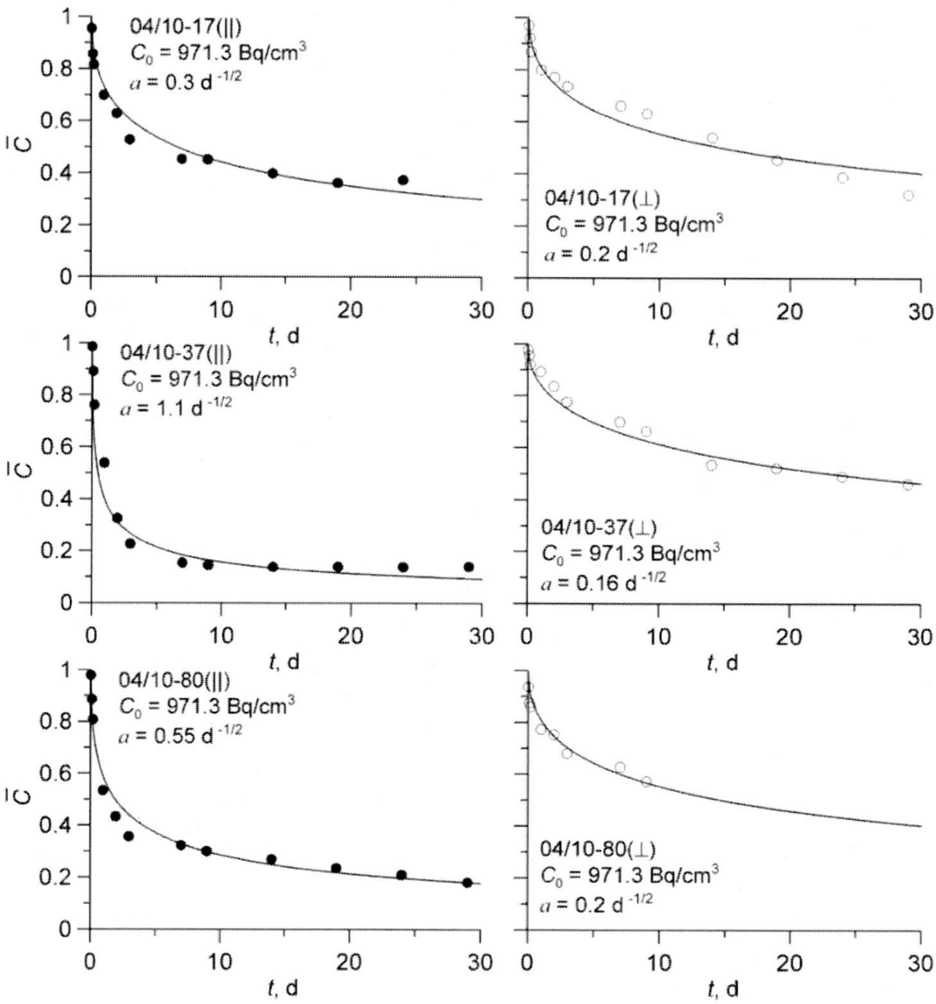

Figure 5. Breakthrough curves for Sr-90 diffusion-in experiments with undisturbed samples.

Table 7. Results of interpretation of single-chamber diffusion cell tests with undisturbed clay samples

Code of sample [a]	a, d^{-1} [b]	$D_m\phi^2R \times 10^9$, m^2/s [c]	Code of sample	a, d^{-1}	$D_m\phi^2R \times 10^9$, m^2/s
4/10-17(| |)	0.3	3.88	4/10-65(⊥)	0.25	2.70
4/10-17(⊥)	0.2	1.73	4/10-68(| |)	0.55	13.1
4/10-37(| |)	1.1	52.0	4/10-68(⊥)	0.3	3.88
4/10-37(⊥)	0.16	1.11	4/10-80(| |)	0.55	13.1
4/10-53(| |)	0.3	3.88	4/10-80(⊥)	0.2	1.73
4/10-53(⊥)	0.3	3.88	4/10-35(⊥)	0.7	21.1
4/10-65(| |)	0.13	0.73	4/10-83(⊥)	0.4	6.91

[a] First number corresponds to borehole number (Sect. 4.2), depth (in m) of sampled interval follows after hyphen, (| |) is a sign for a diffusion test parallel to bedding planes; (⊥) the same for a diffusion test perpendicular to bedding planes.

[b] As in solution (5). [c] For $h = 6.11$ cm.

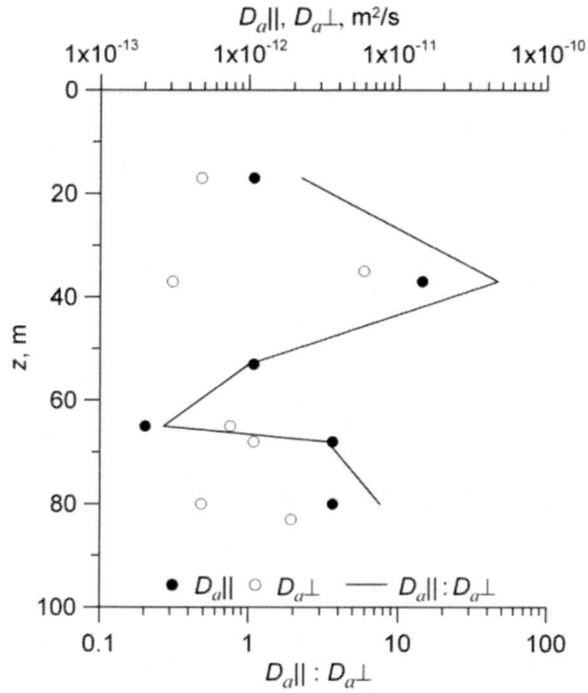

Figure 6. Longitudinal (parallel to bedding planes) and transverse (perpendicular to bedding planes) apparent diffusion coefficients for Sr-90 versus depth (from the data in Table 7, assuming $R = 299.5$, $\phi = 0.2$). Factor R was calculated assuming $K_d = 30$ cm^3/g, $\rho_b = 1.99$ g/cm^3.

The laboratory diffusion experiments were analyzed following the mathematical model presented above, Eq. (3)–(5). The selected experimental set-up and sampling technique allow us to determine the values of the combined parameter, $D_m\phi^2R$, reflecting *radionuclide diffusion* and retention parallel and perpendicular to clay bedding planes, in different intervals of the studied cross-section. Typical breakthrough curves are presented in Figure 5. Table 7 summarizes interpretation of the experimental data with Sr-90.

Results of experiments with Sr-90 demonstrate that combined parameter, $D_m\phi^2R$, varies widely over depth that can be explained by the changes in lithologic composition and mineralogy of the strata, and variability in clay to silt-sand particle ratio. Obviously micro-layered character of the rock is responsible for the measured apparent diffusion anisotropy (Figure 6): calculated values of $D_a(\|)/D_a(\perp)$ [D_{aL} / D_{aT}] ratio exceed (with one exception) 1.

Unfortunately, the experiments with two other radionuclides, Cs-137 and Co-60, could not be successively interpreted in the frameworks of model (3)–(5). Selected experimental time for these strongly adsorbable species was too short to allow them to diffuse inside the core samples more than in a few tens to a few hundreds of micrometers from the liquid/solid surface interface, and observed decrease in concentration of Cs-137 and Co-60 in the injection chamber was connected mostly with *sorption kinetics* on this interface.

In general, it seems that single-chamber experimental set-up is not appropriate option for studying diffusion and retention of strongly adsorbable radionuclides: the sensitivity of the breakthrough curves to the diffusion process is very low in comparison to the surface sorption kinetics.

3.2. A 3D Diffusion Test with a Clay Sample of Natural Structure

In this section, we give the results of experimental studies of diffusion transport of *radionuclides* in a block sample of a rock that has a microlayered structure. This type of *rock heterogeneity* can lead to different values of molecular diffusion coefficient in different directions – along and across bedding planes. This is a manifestation of *anisotropy in clay properties* for this mass transport mechanism. The block sample was taken at a depth of 30 m from the roof of clays at the lower bench of Nikol'skii Quarry (Leningrad province), which strips Cambrian blue clays.

Experimental Setting

The dimensions of the sample were 46×20×28 cm. The duration of experiment was 17 months (510 days). To preserve the natural moisture content, the sample was paraffined according to a standard procedure (Figure 7*a*). A hole 4.2 cm in depth was drilled at the flat end of the sample (Figure 7*b*) and filled by a radioactive "cocktail" of radionuclides Cl-36, Sr-90, Cs-137 and Co-60 (Table 8).

The hole was sealed by clay powder and the exposed surface was sealed by paraffin. Considering the small volume of the introduced liquid, we can assume that the experiment was carried out at constant moisture content. To minimize the external impact, the container with clay sample was stored in a basement room at constant temperature (10° C). After the exposition, the block sample was taken from the container, divided into layers 4 cm in height perpendicular to the direction of the drilled hole. After marking each layer, it was cut into cubes with a side of 2 cm (Figure 7*c*). Immediately after cutting, the cubes were dried until constant mass at $T = 105°$ C and homogenized into powder.

Table 8. Characteristic of the concentration source

Radionuclide	Activity (Bq) [a]
Cl-36	$1.68{\cdot}10^{+6}$
Sr-90	$3.70{\cdot}10^{+5}$
Cs-137	$5.14{\cdot}10^{+5}$
Co-60	$3.43{\cdot}10^{+5}$

[a] The activities of radionuclides 2.5 ml in volume in the sample.

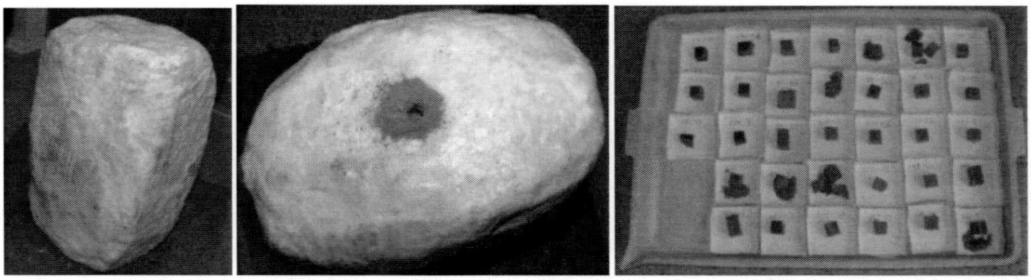

(a) (b) (c)

Figure 7. (*a*) A sample of clay in protected paraffin coating, (*b*) a hole for radionuclide cocktail injection, and (*c*) rectangular samples of clayey material prepared for isotope detection.

The clay sample treated in this manner was placed on measuring backings with appropriate shape, on which the composition and specific activity of radionuclides in individual disintegrated cubes were determined by β- and γ-spectrometry. The measured weight concentrations (activities) obviously reflect the total concentrations of radionuclides in pore solution and on the mineral matrix at the moment when sample drying started.

Mathematical Model of the Test

Diffusion of a stable nonreactive component in an anisotropic porous medium is described by Eq. (1). The migration of a sorbable unstable component can be described by the following equation:

$$R\frac{\partial C}{\partial t} - D_{mT}\frac{\partial^2 C}{\partial x^2} - D_{mL}\frac{\partial^2 C}{\partial y^2} - D_{mL}\frac{\partial^2 C}{\partial z^2} + \lambda CR = 0 \quad (6)$$

where D_{mL} [$D_m(\|)$] and D_{mT} [$D_m(\perp)$] are the principal components of the orthotropic pore diffusion tensor (2), parallel to the bedding and perpendicular to the bedding, respectively; C is the concentration (activity) of the radionuclide in solution [Bq/cm^3].

As applied to a stratified system, equation (6) is effective since the phenomena induced by "rithmic" heterogeneity of the medium are described by it in an indirect manner by incorporating generalized parameters D_{mT} and D_{mL} in the model. However, when the averaging of the process is not sufficient, scale effects may appear. These effects are due to the fact that the part of the medium involved in the process within the experiment time includes a limited number of heterogeneity elements (layers).

Of interest for us is the solution of equation (6) with the following boundary conditions:

$$C(x \to \infty, y, z, t) = 0,\ C(x, y \to \infty, z, t) = 0,\ C(x, y, z \to \infty, t) = 0, \quad (7a)$$

$$C_i(x, y, z, 0) = \frac{M}{nR}\delta(x, y, z) \quad (7b)$$

here $\delta(x, y, z)$ is Dirac delta function [cm^{-3}], M is the activity of the radionuclide that entered the porous medium in point $x = 0,\ y = 0,\ z = 0$ at moment $t = 0$ [Bq].

The exact solution of equation (6) with boundary conditions (7) has the form (Ogata and Banks 1961):

$$C(x, y, z, t) = \frac{M}{8\phi\sqrt{\pi^3 t^3 D_{aT} D_{aL} D_{aL}}} \cdot \exp\left[-\frac{x^2}{4D_{aT}t} - \frac{y^2}{4D_{aL}t} - \frac{z^2}{4D_{aL}t} - \lambda t\right] \quad (8)$$

where $D_a = D_m / R$ is the *apparent diffusion coefficient* [cm^{2}d^{-1}]; x, y, z are the distances from the source along coordinate axes [cm]. The dimensions used here assume that the volumetric activity of radionuclides in solution is measured in Bq/cm^3.

As mentioned above, the activity of radionuclides measured in a rock charge is associated with the total (both in pore water and rock exchange complex) activity of radionuclides converted to unit rock mass ΣC_g (Bq/g). Given the density ρ_b, the volumetric activity becomes $\Sigma C = \rho_b \Sigma C_g$ (Bq/cm^3). Therefore, the adaptation of the calculation scheme to the above measurement procedure can be described by the equation

$$\Sigma C = \phi C + N = \phi R C \tag{9}$$

where ΣC is the total activity of the radionuclide in a unit rock volume [Bq/cm^3]; C is *radionuclide activity* in pore water [Bq/cm$^3_{\text{liquid}}$]; $N = K_d \rho_b C$ is the activity on the rock [Bq/cm$^3_{\text{solid}}$].

We rewrite equation (8) for the total activity function (9), with the assumption that the matter source is located near an impermeable (reflecting) boundary

$$\Sigma C(x,y,z,t) = \frac{MR}{4\sqrt{\pi^3 t^3 D_{aL}^{\;3}/\chi}} \cdot \exp\left[-\frac{\chi x^2 + y^2 + z^2}{4D_{aL}t} - \lambda t\right] \tag{10}$$

where $\chi = D_{aL}/D_{aT} > 1$ is the *anisotropy coefficient* for parameter D_a.

In principle, Eq. (10) can be used directly to calculate the 3D concentration profiles for different values of the parallel and perpendicular diffusion coefficients in the tests. These theoretical profiles can be compared with the experimental results in order to identify which values of the diffusion coefficients provide the best fitting of the experimental results. An alternative approach is based on linearization of Eq. (10). Taking the logarithms of the left and right parts of solution (10), we obtain its linearized representation:

$$\ln\left(\Sigma C(x,y,z,t)\right) = G - B(\chi \cdot x^2 + y^2 + z^2) \tag{11}$$

where $B = \dfrac{1}{4D_{aL}t}$; $\quad G = \ln\left(\dfrac{MR}{4\sqrt{\pi^3 t^3 D_{aL}^{\;3}/\chi}}\right) - \lambda t$.

Equation (11) is linear in coordinates $\ln\left(\Sigma C\right) \div (\chi \cdot x^2 + y^2 + z^2)$. Therefore, it makes sense in the quantitative analysis of experimental material to use the following procedure:

(1) to fit the value of the anisotropy coefficient χ, such that the experimental data are best (with the least scatter, i.e., minimal mean-square deviation) approximated by the straight line (11);
(2) to evaluate the slope of this line, i.e., the coefficient *B*, and the shift *G*;
(3) to evaluate the apparent diffusion coefficient D_{aL} from the slope *B* using the formula $D_{aL} = 1/4Bt$;
(4) to evaluate the second component of the diffusion coefficient: $D_{aT} = D_{aL}/\chi$;

(5) to use D_{aL} and G to calculate the retention factor: $R = (4\exp(G+\lambda t)/M)\sqrt{\pi^3 t^3 D_{aL}^{\ 3}} / \chi$;

(6) to evaluate diffusion coefficients for each direction: $D_{mL} = D_{aL} R$, $D_{mT} = D_{aT} R$.

Since measurements of Cs-137 and Co-60 concentrations show considerable scatter, we will focus our further analysis on interpretation of measurements of 3D distribution of Cl-36 and Sr-90 within the clayey block.

4. RESULTS OF TEST DATA INTERPRETATION

Figure 8 gives the results of interpretation of experimental data for Cl-36 and Sr-90, which was carried out in accordance with the above scheme. Each calculation variant is illustrated by (a) the complete set of samples with the approximating solution line (11) and the fitted anisotropy coefficient, χ; (b) a set of points that characterizes the radionuclide concentration distribution in the main plane (0, *Y*, *Z*). The construction of the plot in Figure 8*c* was required for minimizing the effect on the accuracy of evaluation of averaged parameters caused by the distribution of experimental points in different lithological layers (see below).

The best agreement of experimental and calculated concentration distributions with Eq. (10) was obtained as shown in Table 9. It is seen that Vkt clay exhibits fairly strong diffusion anisotropy due to stratification, and, as could be expected, $D_{aL} > D_{aT}$. Moreover, it can be seen that the appropriate values form the series: Cl-36 > Sr-90. By the degree of their sorption, the radionuclides form another series: Sr-90 > Cl-36.

Table 9. Summary table of the values of apparent diffusion coefficients $D_a = D_m / R$ (m²/s)

Radionuclide	B	G	D_{aL}	D_{aT}	R [a]	D_{eL}	D_{eT}	χ
Cl-36	18	19.3	$1.7\cdot10^{-10}$	$2.9\cdot10^{-11}$	1.05	$3.7\cdot10^{-11}$	$6.4\cdot10^{-12}$	5.8
Sr-90	150	23.2	$3.8\cdot10^{-11}$	$6.3\cdot10^{-12}$	20.6*	$1.6\cdot10^{-10}$	$2.7\cdot10^{-11}$	6

[a] $K_d \approx 2$ cm³/g (for $\phi = 0.21$ and $\rho_b = 2.1$ g/cm³).

At the same time, the scatter of values around the theoretical line of Eq. (11) is quite significant. This is due to the *lithological heterogeneity* (stratification) of the sample. i.e., by the alternation of layers (mm-scale heterogeneity) with different clay material content: a sampling point lying in a clay seam yields extremely high radionuclide activity, corresponding to its adsorbed state, while the sampling of an adjacent silt/sand-enriched layer yields lower activity. For Cs-137 and Co-60, which are the most sorbed radionuclides, this effect proved to be critical, since the localization of these radionuclides near the concentration source prevented the reaching of the required averaging level of the properties of clay rock with microlayer structure. Overall, the experiment is unique in some sense, since it simulates the three-dimensional migration of radionuclides in a rock sample with natural structure under nearly real conditions.

The effective parameters of diffusion and sorption coefficients obtained for different radionuclides make it possible to simulate their migration in clay strata in assessing the safety of construction of underground repository for RW disposal.

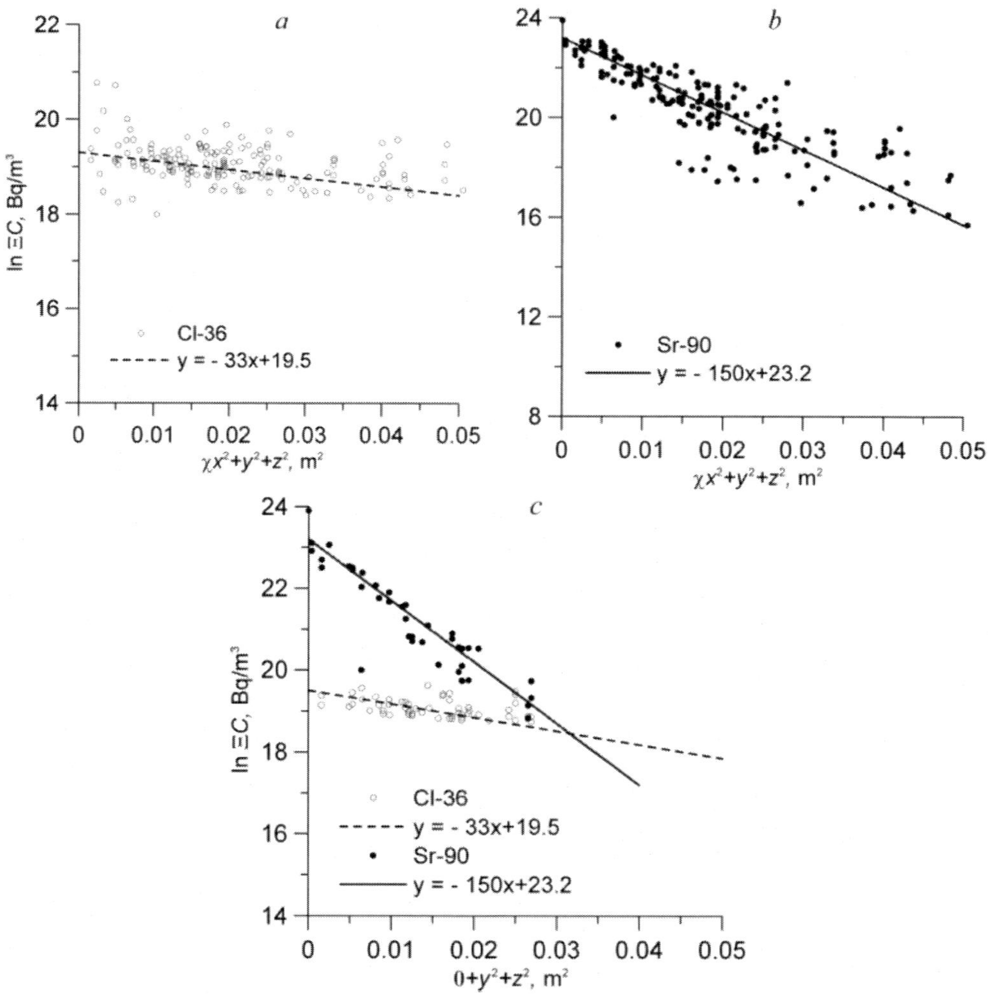

Figure 8. Linearization of experimental data: (*a*) Cl-36: complete sample of points *C*(*x*, *y*, *z*) with fitting the anisotropy coefficient χ ; (*b*) Sr-90: complete set of points *C*(*x*, *y*, *z*) with fitting the anisotropy coefficient χ ; (*c*) set of points along the bedding plane (*Y–Z* plane).

5. A Comparative Analysis

Tables 10 and 11 summarize the published results of laboratory tests with undisturbed clay samples taken from clayey formations considered as potential host rocks for the underground repository for RW disposal in the Western Europe.

Analyzing these results one may come to the conclusions as follows.

(1) Because of anion exclusion effects [21], anions (such as $^{36}Cl^{-}$ and $^{125}I^{-}$) diffuse slower and exhibit smaller *diffusion-accessible porosities*, ϕ than the uncharged HTO: the values of the effective diffusion coefficient of HTO are 3–10 times higher than those of $^{36}Cl^{-}$ and $^{125}I^{-}$ (Table 10).

(2) Similar values of D_e for the studied neutral and anionic radionuclides were measured in both OPA and COx clays. These values are 3–4 times smaller than the values of effective diffusion coefficients obtained for less consolidated Boom clay (BC), Table 10.

(3) Diffusion coefficients are larger in the direction parallel to the bedding plane than perpendicular to it.

(4) The values of effective diffusion coefficients D_{eL} and D_{eT} for $^{36}Cl^{-}$ obtained in the 3D laboratory experiment with Vkt clay (Table 9) are similar to those obtained in OPA and COx clays (Table 10).

(5) In situ diffusion experiments (carried out at Mont Terri, OPA clay) with nonsorbable tracers (HTO, HDO, $^{36}Cl^{-}$ and $^{125}I^{-}$) confirmed, in general, all tendencies established under laboratory conditions [3, 4, 7, 9]. In particular, the field effective diffusion coefficients, D_{eL} and D_{eT}, predominantly fall within the average range of values found in laboratory tests. The anisotropy factor $\chi = D_{eL} / D_{eT}$ varies within the average range from 2 to 6.

(6) The comparison of the D_{aT} value for Sr-90 obtained in the 3D laboratory experiment (Vkt clay, Table 9) with previously measured D_{aT} values (COx, OPA and BC clays, Table 10) indicates that migration process in all studied clays is characterized by almost the same values of the apparent diffusion coefficient.

(7) The obtained values of *sorption parameters*, R and K_d, for Sr-90 in the 3D experiment with Vkt clay are rather close to the relevant parameters characterizing the sorption of Sr-90 in diffusion experiments with compacted COx and OPA clay samples: the coefficient R (Table 9) is within the range of 14.9–22 measured in COx and OPA clays (Table 9), and $K_d \approx 2\ cm^3/g$, which is about twice as large as the average value from Table 11.

(8) The comparison of data from Table 9 and Table 6 shows that the diffusion of sorbed radionuclides in a compacted (undisturbed) material, characterized by the apparent diffusion coefficients, is much slower relatively to the diffusion of the same radionuclides in packing (disturbed/disintegrated) clay samples: the apparent diffusion coefficients for both Cl-36 and Sr-90 in the first set of experiments with clay pastes were significantly higher (up to an order of magnitude) than in the 3D experiment. From the latter it follows that modeling of *radionuclide transport* based on data from experiments with packing clays can overestimate migration process intensity, and the resulting forecasts can be too conservative.

(9) The batch sorption experiments with Vkt clay resulted in the distribution coefficients, K_d, which are about 20 times higher (the average value is about 45 cm^3/g) than K_d ($\approx 2\ cm^3/g$) from diffusion experiments with compacted Vkt clay.

Table 10. Effective diffusion coefficients, D_{eT} (m^2/s), for conservative (non-sorbed) tracers obtained in laboratory experiments with BC, Cox, and OPA, undisturbed clay samples

Index	BC (Mol)	Cox (Bure)			OPA (Mont Terri)		
Ref	[1]	[2] [a]	[3]	[4] [b]	[5]	[4] [b]	[6] [b]
HTO, 10^{-11}	8.5	2.6	1.8–4.2	6–9	1.2–1.5	4–8	1.5
ϕ	0.37	0.18	0.15	0.16	0.1	0.16	0.15
$^{36}Cl^-$, 10^{-12}	24	5.8	4.4–8.9		4.1–5.7		1.0
ϕ	0.16	0.09			0.07		0.08
$^{125}I^-$, 10^{-12}	22	2.8			3.4–4.8		
ϕ	0.16	0.13			0.08–0.11		

[a] Coefficient D_{eT} calculated using D_{eL} data from through-diffusion laboratory experiments, $\chi = 1.56$; [b] large scale laboratory experiment, $\chi = 3-4$. Ref.: [8, 6, 22, 23, 24, 25].

Table 11. Apparent diffusion coefficients, D_{aT} (m^2/s), for Sr-90 (89) obtained in laboratory experiments with BC, Cox, and OPA undisturbed clay samples

Index	BC (Mol)	Cox (Bure)		OPA (Mont Terri)		
Ref	[8][a]	[3] [a]	[3,4] [b]	[7] [a]	[4] [b]	[9][c]
D_{aT}, 10^{-12}	7.8±1.5	7.06±0.26	0.6–2	7.4–8.5	1–3	22
ϕ	0.37	0.15		0.09–0.11		0.17
K_d, cm^3/g	257	0.87±0.45		0.64–0.75		1.4
ρ_b, g/cm^3	1.7 (dry)	2.4		~ 2.0		~ 2.0
R	1182	14.9		17-22		17.5

[a] Small-scale laboratory experiment; [b] large-scale laboratory experiment, D_{aL} value; [c] small-scale laboratory experiment, D_{aL} value.

CONCLUSION

The obtained experimental results indicate that diffusion anisotropy is a common property of the clay formations. The coefficient of diffusion is often greater in the direction of the layering when compared to values in the orthogonal direction. Diffusion anisotropy is accompanied by anisotropy of other physical and mechanical properties of clay materials (flow, deformation etc.). The establishment of relationship between different physical phenomena and parameters governing the anisotropy is one of the major questions for future investigations.

REFERENCES

[1] L. Barden, G. *Journal of the Soil Mechanics and Foundations Division,* ASCE I970. 96(SM4):1171–200 (I970).
[2] A.A. AI-Karni, M.A. Al-Shamrani. *Computers and Geotechnics* 26, 83–103 (2000).
[3] J.M. Palut, P. Montarnal, A. Gautschi. *J. Contam. Hydrol.* 61, 203–218 (2003).
[4] Yllera, A. Hernández, M. Mingarro. *Appl. Clay Sci.* 26, 181–196 (2004).
[5] J.-Y. Boisson, Clay Club Catalogue of Characteristics of Argillaceous Rocks Compiled. Nuclear Energy Agency. *Organisation for Economic Co-operation and Development.* NEA No 4436. OECD (2005).
[6] M. García-Gutiérrez, J.L. Cormenzana, T. Missana et.al. *Phys. Chem. Earth* 31, 523–530 (2006).
[7] J. Samper, S. Dewonck, L. Zheng et al. *Phys. Chem. Earth* 33, 1000–1008 (2008).
[8] J. Samper, C. Yang, A. Naves et al. *Phys. Chem. Earth* 31, 531–540 (2006).
[9] J.M. Soler, J. Samper, A. Yllera et al. *Phys. Chem. Earth* 33, 196–207 (2008).
[10] J.L. Cormenzana, M. García-Gutiérrez, T. Missana. *Phys. Chem. Earth* 33, 949–956 (2008).
[11] V.G. Rumynin, E.B. Pankina, G. Volckaert et al. "Geotechnical, flow and transport properties of Kotlin (Vendian age) and Blue (Cambrian age) clays with respect to design of underground storage facilities for radioactive waste disposal in the north-west region of Russia". *The IV International Nuclear Forum (2009).* St. Petersburg. Pp 95–210.
[12] S. Suzuki, H. Sato, T. Ishidera et al. *J Contam Hydrol* 68, 23–37 (2004).
[13] J. Bear, *Dynamics of fluids in porous media*, Dover Publ. Inc, New York (1972).
[14] M. Arnould. *Bull. Eng. Geol. Envir.* 65, 413–42 (2006).
[15] H. Bock, P. Blümling, H. Konietzky. *Bull. Eng. Geol. Envir.* 65,195–207 (2006).
[16] M. Huysmans, A. Dassargues. *Hydrogeol. J.* 14, 1094–1106 (2006).
[17] K.C.J. Van Rees, E.A. Sudicky, P.S.C. Rae et al. *Environ. Sci. Technol.* 25, 1605–1611 (1991).
[18] G.J. Moridis. *Water Resour. Res.* 35, 1729–1740 (1999).
[19] G. Giu, L Barbour, B.C. Si. *Environ. Sci. Technol.* 43, 2412–2416 (2009).
[20] Ogata, R.B. Banks. U.S. *Geol. Survey Prof. Paper* 411-A (1961).
[21] L.R. Van Loon, B. Baeyens, M.H. Bradbury. *Appl. Geochem.* 20, 2351–2363 (2005).
[22] M. Aertsens, M. Van Gompel, P. De Cannibre. *Phys. Chem. Earth* 33, 61–66 (2008).
[23] N. Maes, M. Aertsens, S. Salah et al. *External Report of the Belgian Nuclear Research Centre*, SCK•CEN-ER-98 09/NMa/P-108 (2009).
[24] L.R. Van Loon, J.M. Soler, A. Jakob et al. *Appl. Geochem.* 18, 1653–1662 (2003).
[25] P. Wersin, J.M. Soler, L. Van Loon. *Appl. Geochem.* 23, 678–691(2008).

In: Anisotropy Research
Editor: Hirpa Lemu

ISBN 978-1-62081-977-7

Chapter 5

WHITE MATTER ANISOTROPY AND THEORY OF MIND DEFICIT IN PATIENTS AT ULTRA HIGH RISK OF PSYCHOSIS: A TRACT-BASED SPATIAL STATISTICS STUDY

Daniel Serrani*
Faculty of Psychology, Department of Psychobiology
Universidad Nacional de Rosario, Argentina

ABSTRACT

Theory of mind (TM) refers to the ability to infer one's own and other persons' mental states. Few studies have attempted to investigate these deficits in ultra-high-risk of psychosis (UHRP) individuals and its association with white matter microstructure and connectivity abnormalities. UHRP patients and age- and IQ-matched healthy control (HC) group were administered first and second order TM, False Belief (FB) and Strange Story (SS) tasks and emotional recognition was assessed using the Emotional Quotient. DTI data were acquired on a 3.0-T MRI unit using a recently developed technique called tract-based spatial statistics (TBSS). TBSS maps each subject's DTI onto a common white matter tract center ("skeleton"). UHR and HC groups were compared using independent t-test for continuously distributed variables and chi-square test for categorical ones. A 2 × 2 two-way analyses of variance for TM and EQ, and effect size with Cohen's D were also performed. UHRP performed significantly worse compared to the HC group for TM, SS and EQ tasks, with effect size at an intermediate level (0.81, 0.78 and 0.79 respectively). MANOVAs main effects for TM and EQ scores were also found. A cluster of FA reductions in inferior longitudinal fasciculus (ILF), left uncinate fasciculus (UF) and left anterior cingulate (LAC) was associated with deficits in TM, SS and EQ tasks, respectively. On the other hand, no significant correlation of fractional anisotropy with any of the tasks was found in HC. In this chapter, TBSS DTI is used to demonstrate an association between reduced white matter integrity and TM and EQ deficits in UHRP subjects with positive association between selective white matter FA reduction in ILF, UF, LAC and disturbed performance on TM, SS and EQ tasks.

* Daniel Serrani, MD, E-mail: danielserrani@argentina.com.

These results indicate that impaired structural connectivity among frontal networks may contribute to TM and emotional recognition deficits in UHRP.

1. Introduction

Theory of Mind (TM) is a term used to explain a related set of intellectual abilities that enable to understand that others have beliefs, desires, plans, hopes, information, and intentions that may differ from our own, enabling to explain and predict behavior in order to adaptively function in a social environment. TM deficits have been disclosed in several psychiatric conditions ranging from *autism* [1], *obsessive compulsive disorder* [2], attention deficit *hyperactivity disorder* [3], *developmental language disorders* [4], *and schizophrenia* [5] as well as acquired disorders of the right brain and traumatic brain injury [6] or multiple sclerosis [7]. TM appears to be a composite function, involving memory, joint attention, and complex perceptual recognition (such as face and gaze processing), language, executive functions (such as tracking of intentions and goals and moral reasoning), emotion processing-recognition, empathy and imitation. Hence, TM is dependent on the maturation of several brain systems and is shaped by parenting, social relations, training, and education, standing as an example of the complex interaction that occurs between brain development and social environment. Such deficits have critical implications for the patients, as they contribute to impairment in *social functioning* [8] and play a role in positive and negative psychotic symptoms [9]. Some specific brain regions associated with TM have been identified in *Orbital prefrontal cortex* (PFC) [10], temporal lobes [11], the temporal poles and posterior superior temporal sulcus [12]. In addition, several imaging studies have confirmed that brain regions involved in TM exhibit abnormal hemodynamic responses in such areas as medial and lateral portions of the PFC, relative to healthy controls [13]. There is a growing body of evidence suggesting that TM abnormalities may represent a trait character of schizophrenia [14], although it is not clear yet if TM deficits are primary, that is, independent of general intellectual disabilities and executive function [15]. Impaired TM performance can reflect not only in a cognitive but also in a behavioral domain, as observed in first episode patients [16], patients on remission [17], early stages [18], at high genetic [19] or psychometric risk of psychosis [20]. In a recent meta-analysis review on TM deficits in remitted and non-remitted patients with schizophrenia [21] the authors concluded that even if a trait-related TM impairment can be found in schizophrenia, more sound research is needed to put into evidence this same finding in symptom-free patients and people at high risk for developing psychosis.

A cohort of at risk children was studied employing neuroimaging techniques such as fMRI (functional magnetic resonance imaging) assessments of response inhibition and error-related processes, whole brain voxel-based morphometry of gray matter and diffusion tensor imaging to probe *white matter integrity* [22]. The results showed reduced activity within right frontal and bilateral temporal cortex for response inhibition and reduced activity within the anterior cingulate, insula and middle frontal gyrus for error-related processing. Also morphometry revealed increased gray matter within middle and superior temporal gyri, angular gyrus, orbitofrontal gyrus and decreases within the inferior temporal gyrus. At last, DTI analysis identified decreased white matter along the inferior fronto–occipital fasciculus, cingulum and inferior longitudinal fasciculus. An aberrant prefrontal-temporal dysfunction in addition to cingulated and insular dysfunctions was also found, but relationship with functional measures or social cognition was not assessed. It has been argued that functional disability in schizophrenia, particularly impairment in social and role functioning are linked to cognitive deficits.

Retrospective analyses demonstrate that functional decline precedes the onset of *psychosis* [23]. Recent investigations reveal that also individuals at *clinical-high-risk* (CHR) for psychosis show impairments in social relationships, work/school functioning and daily living skills [24-25]. Youth at high risk of psychosis (HRP) also exhibit a pattern of impairment across a wide range of cognitive domains, including social cognition, which is qualitatively similar to that of individuals with schizophrenia [26-27]. While many studies have sought to elucidate predictors of clinical deterioration, specifically the development of schizophrenia in HRP samples, few have reviewed factors related to psychosocial measures, including deficit in *theory of mind* [28-30] Impaired theory of mind has been found in several investigations performed on schizophrenia [31-33], subjects at high risk of developing psychosis [34-36]. In general, these studies have observed that neural basis of mentalizing is characterized as a network including the dorsomedial prefrontal cortex (dmPFC), the superior temporal sulcus extending into the *temporo-parietal junction* (TPJ) and was sometimes found to include the temporal poles [37].

Although the accumulated findings of TM deficits in patients at all stages of schizophrenia seem to suggest that a trait-related TM impairment exists, more research is still lacking in this area regarding people at high risk for developing psychosis. This prodromal phase of the illness has gained considerable attention among researchers and the criteria defining this population has been described in detail [38-39], suggesting that prior to the onset of psychosis, UHR youths already have brain abnormalities similar to those present in patients with schizophrenia [40-43]. They also exhibit neuropsychological deficits that match, in many features, those found in patients with schizophrenia, although to a lesser degree [44-45], including impaired social functioning and related problems with social skills [46-47], which in turn predict the onset of positive and negative symptoms that could account for a rate transition to full blown psychosis of more than 20% [48]. As a whole, HRP subjects perform significantly worse during TM tasks [49-51], and this fact has been consistently associated with neuro-anatomical and functional correlates [52-57].

Brain structural abnormalities appear to progress along the frontotemporal connections, disturbing the cortico-striato-thalamic loop. An investigation which analyzed structural MRI in HRP patients concluded that prefrontal, anterior cingulate, medial temporal, insular and cerebellar areas showed decreased gray matter volume, with medium effect sizes in HRP who transitioned to psychosis compared to those without transition (HR-NT). In functional imaging studies they showed reduced brain activation in prefrontal cortex, reduced neuronal density, and increased membrane turnover in frontal and cingulate cortex [58-59]. Insofar, functional and structural abnormalities within brain regions dedicated to self and other-referential processing have been implicated in the pathophysiology of the disorder [60]. Despite these findings, to date no study has examined white matter structure abnormalities associated with TM deficits in HRP patients.

There are several strengths to this research, as it allows for the study of potential structural markers of risk transition in a vulnerable group. Here we used *tract-based spatial statistics* (TBSS) of white matter *fractional anisotropy* (FA) to examine abnormalities associated with ToM in a group of individuals at high risk of psychosis, comparing them with a healthy control group. With the advent of diffusion tensor imaging (DTI), a complex picture of white matter abnormalities has emerged. In fact, diffusion tensor imaging is an MRI modality that examines the self-diffusion of water which depends on the Brownian motion of molecules [61]. In the absence of structural boundaries, white matter diffuses in all directions equally (isotropic diffusion). Structural barriers in the brain including axonal and nerve myelin impose definite margins and directionality to the diffusion ensues (*anisotropic diffusion*). Water diffusion within brain's white matter follows axons direction which allows DTI to identify the position and direction of white matter tracts [62]. DTI can provide among a measure of the directionality of diffusion known as fractional anisotropy (FA) [63-64], which ranges from 0 to 1, with 0 indicating completely random diffusion and 1 indicating completely directional diffusion, as would be exhibited by a water molecule traveling along an infinitesimally narrow path. FA is higher in white matter than gray matter, although the main determinant of anisotropic diffusion appears to be axonal integrity, with myelin playing a modulatory role [65]. For this reason, FA is considered as an index of white matter integrity. Greater FA is associated with white matter fibers longitudinal alignment, myelinization and increased volume averaging from differently oriented fibers [66], while reduced FA results from edema, disrupted myelin sheath or changes fiber bundles distribution [67]. Pathologic changes in WM tracts associated with FA abnormalities can be measured using eigenvalues that represent averaged values of main diffusivities in the longitudinal, transverse and oblique fibers directions. Increased longitudinal diffusivity suggests greater numbers of longitudinally aligned fibers, and decreased radial diffusivity suggests reduced numbers of obliquely aligned fibers.

On the other hand, ROI (regions of interest) analysis are subjective, manually intensive, requires pre-specification of spatial location of regions of interest, and sometimes is compromised by the use of standard registration algorithms, so alignment of FA images from multiple subjects in voxelwise analyses remains a problem that can be overcome by tract-based spatial statistics (TBSS). TBSS is an automated observer-independent method of aligning FA images from multiple subjects to allow groupwise comparisons of DTI data [68]. Tract-based spatial statistics investigates the whole brain estimating a groupwise FA skeleton. Individual data are then projected into this skeleton so each voxel takes the value from the nearest local tract center, solving alignment issues using nonlinear registration [69].

2. Subjects and Method

Theory of Mind is associated with a wide variety of tasks that tap into a range of mental state attribution abilities which comprise attribution of knowledge or beliefs, intentions and/or affects such as desire, fear or anger; using stories or pictures. To explore TM skills there are a number of tests, including *Happé's* "Strange Stories" task [70], Faux Pas Recognition [71], "false belief tasks" based on verbal, pictorial or puppets acted stories [73], eye gaze direction as a measure of the ability to infer mental states from others [74], the ability to infer deceptive intention [75] or whether something is intentional or accidental [76]. Seventy-two age- and IQ-matched subjects were recruited from an outpatient community mental health service and consisted of 36 HC and 36 UHR individuals. All subjects were aged 17–29 years. The UHRP group was recruited according to The *Structured Interview* for *Prodromal Syndromes* SIPS criteria [77] which emphasize onset or worsening in the past 12 months of attenuated positive symptoms in 1 or more of 5 possible categories for a prodromal syndrome: unusual thought content, suspicion/paranoia, perceptual anomalies, grandiosity, and disorganized communication.

A 7-point severity scale is used for each symptom, reflecting its frequency, duration, impact on functioning, and degree of loss of insight. Levels of 0 to 2 (none, questionable, or mild) indicate normal to subprodromal functioning; levels of 3 to 5 (moderate, moderately severe, or severe), a prodromal state; and a level of 6, a fully psychotic state. For example, a prodromal level of unusual thought content corresponds to an idea of reference or an odd belief that is worrisome or becomes meaningful because it will not go away and may be accompanied by an emerging sense that the event is caused by an external source, but doubt in this notion can be induced by contrary evidence. By contrast, a psychotic level of unusual thought content is an idea of reference or odd belief that is accompanied by full conviction for a specified period or that is acutely disruptive or disabling. The instrument is used to rate the severity of symptoms and to derive a categorical determination of prodromal status. A subject may also qualify for a prodromal syndrome on the SIPS because of onset in the past 3 months of brief intermittent psychotic symptoms, which are positive symptoms of psychotic intensity but below the threshold required for a DSM-IV Axis I psychotic disorder diagnosis, or by having a genetic risk (defined as having a first-degree relative with a psychotic disorder or as having a diagnosis of schizotypal personality disorder) for psychosis and deterioration of 30% or greater on the General Assessment of Functioning Scale in the past 12 months. A genetic relationship between *schizophrenia* and *schizotypal personality disorder* has been detected in samples of families, twins, and adoptees [78].

The HC group consisted of 36 subjects who were recruited via the social networks of community mental health center staff members, and were excluded if meet criteria of any first- or second-degree biological relatives with a lifetime history of a psychotic disorder, lifetime diagnosis of psychotic illness, substance dependence, or neurological disease, history of head injury or medical illness with documented cognitive sequelae, sensory impairment, current use of psychotropic medication or a full-scale IQ estimate of below 70. Subjects aged $\geq$ 21 years gave informed consent, whereas subjects below 18 years of age were subject to informed consent provided by parents. All procedures were carried out in accordance with the current version of the *Declaration of Helsinki*. The study was approved by the local Institutional Ethics Review Board.

2.1. Clinical Assessments

At intake the SIPS, an expanded version of the *Brief Psychiatric Rating Scale* [79], the *Structured Clinical Interview* for DSM-IV Axis I Disorder (SCID) and the Global *Assessment of Functioning* (GAF) were administered.

2.2. Neuropsychological Assessment

The intelligence profile was assessed using the WAIS-III [80]. Four factors were derived: Verbal Comprehension, Perceptual Organization, Freedom from Distractibility and

Processing Speed, as factor analytic studies indicate that these offer the best estimates of intelligence [81]. Other measures assessed were Executive function using Winsconsin Card Sorting Test [82], Trail Making Test B [83], Stroop Test Color-Word [84], Controlled Oral Word Association Test [85]; Working Memory using Wechsler Adult Intelligence Scale Digit Span and Spatial Location [86]; Memory and Learning using Rey-Osterrieth Complex Figure copy, immediate and delayed recall [87], Califormia Verbal Learning Test immediate and delayed recall [89]) and processing speed by means of Stroop Test Color and Trail Making Test A.

To assess theory of mind we used a variety of neuropsychological tasks. All subjects were tested alone in avoiding distracting stimuli. Theory of mind tests were presented on paper and, in case of the Strange Stories and the Faux pas test, they were read out by the experimenter. Reading the Mind in the Eyes Test was developed [90] to measure individual differences in sensitivity to social perceptions. The revised version consists of 36 photographs of the region around the eyes of both genders, and participants have to match the thought or feeling depicted in each image to one out of four words best describing them. The selected eye region extends from the forehead to the nose, and each picture has two opposite words describing mental states printed on it, such as "concerned/unconcerned" or "observing /daydreaming" to test inferences about affective o mental states respectively. Participants were asked to voice the one word best describing the person feeling or thought. This task was designed to have low executive function load because it does not require sequencing or set-shifting [91]. Even though this test is less than a verbal one, may be results on it could be affected due to poor comprehension of mental state terms used in it, which could be overcome explaining the words such as reflective/unreflective, happy reflection/sad reflection, dominant /submissive, flirtatious/uninterested, indecisive/decisive. The False Belief (FB) task was composed of first-order and second-order tasks [92]. First-order FB tasks tests the recognition of a story character's false belief about the world, whereas second order FB tasks assess the subject's mental ability to infer a third character's mental state. A first question about the reality of the situation was asked first, while a second one was aimed to uncover the subject's knowledge of the mental state of one of the characters, reflecting the character's FB about the situation. The score was rated as 1 if subjects answered the reality question correctly and 2 if they passed the second question. The maximum sum score of both FB tasks taken together was 12 points. The *Strange Story task* (SS) [93] was composed of 24 short passages, each followed by a test question. These materials were adapted for adults. The story vignettes were of two types: a TM and a Physical Story. The TM stories concerned double bluff, mistakes, persuasion, and white lies. These stories were followed by questions requiring an inference about the characters' thoughts and feelings, and in most cases, an inference about the speakers or actors intentions.

As a non-TM task, the *Physical Story* was composed of eight short stories, and required inferences to be made in order to avoid involving mental states in the correct answer. Faux-Pas Test (FP) is a task in which participants were required to detect if anyone in the story said something that could be considered odd and subsequently to find out what the cover motive could. The experimenter reads out each story, which was also positioned facing the participants allowing reading them as well to reduce memory load. After each story subjects were asked questions searching the detection of the faux-pas (did anyone said something odd?), the person identification (who said something odd?), the content (what was odd?), the explanation (why was it odd?), the false belief itself (Did they recall that?), and an empathy question (what feeling did it elicited?). The adult version of the Faux pas test [93]. The test includes twenty stories, ten with and ten without a faux-pas. After each story the participant was prompted to detect the faux, asking if anyone had said something they shouldn't had said. If the participant succeeded identifying the right person it was assumed to be a correct answer. Follow-up questions such as why shouldn't the subject in the story had said what they did, or why the participant believed that they had said that to test inferences about emotionality and intentionality. At the end, to test for comprehension of the story, a question regarding some detail of the story was issued. Participants unable to answer correctly to the first question were asked the control question as the next step. As Faux Pas task doesn't seem to be affected by changes in working memory, it doesn't seem to place a load on it. To score and interpret the answers the following instructions were used. All subjects get questions scores, even when answering negatively to the first one, but in that case he will get 0 points for the whole story. Besides on the control stories, the score reach 2 points if they made it clear that what was said should in fact be said, but 0 points if they affirm that someone said something that shouldn't have been said, ranging a total of 20 points on the control stories. Control questions on these stories received 1 point. The *Empathy Quotient* (EQ) [94] was developed to examine empathizing tendencies which involves two elements: ability to attribute mental states to oneself and others and to show an emotional reaction that is appropriate to the other person's mental state. The instrument comprises 60 questions, 20 of which are filler items and 40 items examine empathizing. Scores range between 0 and 80, being the lower scores associated with worse empathizing abilities.

2.3. Psychological Data Analysis

Statistical analyses for FB, SS, FP and EQ data were performed in SPSS version 15.0 (SPSS Inc., Chicago). Scores of both groups were compared using two-tailed t-tests. Group comparisons of performance in each of the tasks were performed using 2 × 2 two-way analyses of variance (ANOVAs), with group as a between-subjects factor and mind (theory of mind) and emotion (EQ) inferences as a within-subjects factor.

For analyses assessing repeated measures effects, violations of sphericity assumptions were accounted for by correcting the degrees of freedom according to Greenhouse–Geisser, including *Mauchly's Test of Sphericity* [95]. The statistical significance level was set at corrected $p < 0.05$ (cluster level) or uncorrected $p < 0.001$ (voxel level).

2.4. MRI Acquisition

Diffusion tensor images (DTI) were acquired on a 3.0-T X series Philips Achieva MRI system (Philips Medical Systems, Best, The Netherlands) using single-shot echo-planar imaging (EPI), SENSE parallel imaging factor = 2.5, b values = 0, 1200 s/mm^2, with SPIR fat suppression and dynamic stabilization in an image acquisition time of 17 min, TR = 12561 ms, TE = 59 ms, flip angle = 90°, voxel dimensions = 2.00 mm × 257 mm × 360 mm, Field of View (FOV) = 224, number of slices = 60, slice gap = 0 mm, spatial resolution = 1.8 x 1.8 x 2.1 mm DTI images were acquired in 8 directions together with a baseline image having no diffusion weighting. Total scan time was approximately 2 minutes for the acquisition of one diffusion weighted data set.

2.5. Data Preprocessing

Data were analyzed using tractography in Explore DTI [96] and Tract-Based Spatial Statistics (TBSS) implemented in FSL. All data were processed using FSL (FMRIB Software Library, FMRIB Center from Oxford) [97]. Images were skull stripped using BET (Brain Extraction Tool) with fractional threshold, f = 0.1, and vertical gradient, g = 0 [98]. Finally, a diffusion tensor model was fitted on the data to determine the level of anisotropy for each voxel independently by calculating the tensor eigen values using FDT (FMRIB Diffusion Toolbox) describing the diffusion strength in the primary, secondary and tertiary diffusion directions.

To allow voxelwise analysis of FA data across subjects, individual maps were aligned, target image extracted and affine-aligned to MNI-152 standard space [99]. All other images were transformed into MNI-152 space by combining the nonlinear transform to the target FA image with the affine transform from that target to MNI-152 space, resulting in a standard space version of every subject's FA image. Mean alignment-invariant tract representation was created (mean FA skeleton), onto which participant's FA data were projected, using cluster-size thresholds defined by $t > 3$ and a level of $P < 0.05$.

2.6. DTI Data Analysis

Processing of DTI data was performed using FSL ver. 4.1. (hwww.fmrib.ox.ac.uk/fsl) and raw images were corrected for head motion and eddy currents using DTIFit part of FMRIB's Diffusion Toolbox (www.fmrib.ox.ac.uk/fsl/fdt) with affine alignment to its corresponding b0 image before the extraction procedure of non-brain voxels using a factor of 0.35. *Fractional anisotropy* (FA), *mean diffusivity* (MD), *axial diffusivity* (AD-$\lambda 1$ eigenvalue) and *radial diffusivity* (RD-$\lambda 2$ and $\lambda 3$ eigenvalues) maps were generated using the FDT module. For voxelwise statistical analysis was used TBSS ver. 1.2, part of FSL [100]. The FA data of participants were aligned into a common space using the nonlinear registration tool FNIRT, averaged onto a mean FA image, and thinned to create a mean FA skeleton, taking the centers of white matter tracts common to all subjects.

On a next step, Voxel values of each participant FA map were projected onto the skeleton by searching the local maxima along the perpendicular direction from the skeleton and Voxelwise permutation-based nonparametric inference was performed on skeletonized FA data, using *FSL Randomize* ver 2.1.

Group comparisons between UHRP and HC were performed using an analysis of covariance ANCOVA, with gender, age and instruction level as covariates, which were fed into the matrix and after that were removed from the data before accomplishing permutation inferences. Multiple repeated comparisons and 10,000 permutations between UHRP and HC were tested. In order to keep benefits of cluster-based tresholding without having to arbitrarily a) define initial cluster forming threshold or b) work on a scale-space basis for a voxel signal smoothing we took advantage of the *threshold-free cluster enhancement* (TFCE) in which voxelwise values represent the weighted sum of all spatial supporting cluster-like signals, without the need to recur to hard *cluster-forming thresholding* [101] and optimizing detection of a wide range of signals shape, so control of family-wise error and voxelwise accuracy become both possible, using a significance level of $p < 0.01$. Voxelwise simple regression analyses of UHRP patients were performed using TBSS. Design matrices were created for each task, modeling total task scores as explanatory variables.

Seed-masks were used to include only those voxels with significant group differences. Permutation-based non-parametric inference on unsmoothed statistical maps with 10,000 permutations of the data were undertaken to test against using randomize v2.1 [102]. Statistical significance level was thresholded for correction of multiple comparisons using a false discovery rate (FDR) [103]. The fiber tracts corresponding to the clusters were identified with reference to the Johns Hopkins University DTI-based White Matter Atlas (http://cmrm.med.jhmi.edu) [104].

3. Results

The demographic and clinical data are shown in Table 1. Patients and healthy controls differed only on GAF and WAIS scores. None of the patients were taking psychotropic medications.

Table 1. Demographic characteristics and clinical symptoms in UHRP and HC groups

	HC	UHRP [e]	Analysis [f] F or x^2	Df	P
Male/female			0.02	1	0.76
Right/left handed			4.12	1	0.35
age (years)	28.46 (9.03)	26.81 (6.91)	0.46	2.79	0.69
Education (years)	14.3 (1.5)	13.14 (1.98)	2.94	2.79	0.35
GAF [a]	89.72 (3.56)	58.83 (3.58)	12.51	2.79	0.001 **
WAIS scores [b]					
Full scale	98.26 (5.68)	116.34 (11.30)	2.41	2.79	0.03 *
Verbal comprehension	101.2 (10.42)	116.7 (10.41)	3.91	2.79	0.04 *
Perceptual organization	103.4 (11.45)	111.4 (10.56)	5.27	2.79	0.035 *
Freedom from distraction	102.4 (9.42)	112.5 (4.61)	4.36	2.79	0.046 *
Processing speed	103.8 (10.34)	112.6 (11.49)	5.27	2.79	0.002 **
BPRS		41.37 (5.91)			
CAARMS		34.15 (11.52)			

* $p < 0.05$; ** $p < 0.01$;

[a] GAF = global assessment of functioning; [b] WAIS = Wechsler Adult Intelligence test; [c] BPRS = Brief Psychiatric Rating Scale; [d] CAARMS = Comprehensive Assessment of at risk mental states; [e] Assessment of UHRP based on state+trait risk factors, attenuated symptoms and/or brief limited intermittent psychotic symptoms (BLIPS); [f] Independent t-test or Welch's t-test if both groups have heterogeneous variances, and chi-square analysis for categorical data.

3.1. Psychological Data

Results of neuropsychological tests are exhibited in Table 2, while differences between groups in WAIS and subcategories scores are shown in Table 3. Results of the faux pas test, TM and EQ are also displayed in Table 4, 5 and 6 respectively.

There were significant differences between the two groups in terms of TO and emotional tasks performance. Repeated measures ANOVAs revealed significant main effects of group in all four tasks: F(1,70) = 7.16, p = 0.006 in FP task; F(1,70) = 11.96, p = 0.002 in Strange Story task; F(1,70) = 13.38, p < 0.001 in FB task; and F(1,70) = 17.48, p < 0.001 in EQ task, with UHRP patients performing worse than healthy controls.

Table 2. Performance in Neuropsychological tests of UHRP and HC group

Measures [a]	HC [b]	UHRP [b]	Analysis		
	Mean (SD)	Mean (SD)	F	df	P
Executive function					
WCST correct	65.36 (12.32)	54.67 (12.38)	0.15	2.32	0.03
Preservative response	9.34 (4.25)	14.69 (4.58)	1.34	2.32	0.05
TMT-B (sec)	55.23 (12.60)	71.25 (15.47)	4.63	2.32	0.01 *
Stroop C-W (sec)	87.15 (14.52)	119.34 (25.83)	8.04	2.32	0.001 **
COWAT	41.51 (9.02)	36.61 (10.11)	1.43	2.32	0.04 *
Working Memory					
WAIS Digit Span	13.48 (3.26)	10.76 (4.12)	3.62	2.32	0.04 *
Spatial location	46.27 (3.27)	40.62 (5.28)	8.92	2.32	0.003 **
Memory/learning					
ROCF copy (score)	35.27 (3.51)	30.71 (4.62)	1.48	2.32	0.07*
Immediate recall (score)	24.51 (4.91)	18.34 (5.12)	4.90	2.32	0.087
Delayed recall (score)	22.11 (5.52)	15.72 (3.83)	6.23	2.32	0.054
CVLT immediate recall	13.97 (2.93)	11.59 (3.14)	2.73	2.32	0.72
Delayed recall	13.52 (3.62)	12.60 (4.11)	0.34	2.32	0.97
Processing speed					
Stroop C (sec)	57.34 (10.17)	61.39 (11.58)	1.83	2.32	0.002 **
TMT-A (sec)	30.58 (10.42)	33.68 (11.91)	4.11	2.32	0.03 *

* $p < 0.05$. ** $p < 0.01$.

Abbreviations: WCST = Wisconsin Card Sorting Test; TMT-B = Trail making Test B; Stroop C-W = stroop color-word; COWA = Controlled Oral Word Association test; WAIS digit span = Weschler Adult Intelligence Test digit span; ROCF = Rey-Osterrieth Compex Figure Test; CVLT = California Verbal Learning test; Stroop C = Stroop Test color; TMT-A = Trail Making Test A; HC = healthy controls; UHRP = ultra high risk of psychosis.

Table 3. Differences between HC and UHRP in WAIS and subcategories scores

	HC	UHRP	Statistic	p-Value
Gender (male:female)				
Mean age			F(2.68) = 2.42	0.32
WAIS scores				
Full scale	98.26 (5.68)	116.34 (11.30)	F(2.68) = 1.27	0.02
Verbal comprehension	101.2 (10.42)	116.7 (10.41)	F(2.68) = 4.17	0.01
Perceptual organization	103.4 (11.45)	111.4 (10.56)	F(2.68) = 3.17	0.02
Freedom from distraction	102.4 (9.42)	112.5 (4.61)	F(2.68) = 4.12	0.03
Processing speed	103.8 (10.34)	112.6 (11.49)	F(2.68) = 3.57	0.01

Table 4. Performance of UHRP and HC on faux pas test

	HC	UHRP	Fp
Score on faux pas questions media (DE)	36.7 (6.43)	11.4 (2.52)	0.003 *
Score on control question media (DE)	8.59 (2.24)	7.39 (2.41)	0.76

* $p < 0.05$.

Table 5. Performance of UHRP and HC on Reading the Mind in the Eyes task

	HC	UHRP	Fp
Overall (media/N)	24.4/36	16.7/36	0.005 *
Score on mental states items (media/N)	7.9/11	5.9/11	0.006 *
Scores on affective mental items (media/N)	16.3/25	11.7/25	0.0035 **

* $p < 0.05$; ** $p < 0.01$.

Table 6. EQ scores for UHRP and HC groups

	Total scores on EQ		
	HC Media (SD)	UHRP Media (SD)	F p
Male	39.57 (10.28)	35.31 (10.12)	3.69 = 0.03 *
Female	42.23 (11.52)	38.48 (9.72)	5.42 = 0.005 **

* $p < 0.01$; ** $p < 0.001$.
HC = healthy controls, UHRP = ultra high risk of psychosis.

Table 7. Effect size based on differences in raw means divided by pooled post test standard deviation

	HC	UHRP	Fp
Score on FB task (DE) †	8.7 (1.25)	11.4 (1.39)	0.001 *
Score on SS task (DE) ‡	12.28 (2.11)	18.22 (2.35)	0.006 *
Score on EQ task (DE) ‡‡	9.14 (3.01)	5.21 (1.99)	0.004 *

* $p < 0.001$.
† FB = False Belief tests consisted of first-order and second order tasks.
‡ SS = Strange Stories tests required the participants to either interpret another's behavior or understand another's emotion.
‡‡ EQ= emotional quotient test.

Table 8. Mixed-model 2x2 ANOVA source table for mind (TM) and emotion (EQ) inference scores

	Within-subjects test (T)			T x G			Between-subjects Group (G)		
	df	MS	F (p)	df	MS	F (p)	df	MS	F (p)
TM [1]	70,2	214.5	54.5(p<0.01)	70,2	118.1	12.7(p<0.01)	69,3	227.3	5.36(p<0.01)
EQ [2]	70,2	147.2	24.5(p<0.01)	70,2	63.2	5.6(p<0.01)	69,3	89.8	12.6(p<0.01)

[1] TM = theory of mind tasks, [2] EQ = emotion quotient tasks, MS = Mean square.

Effect size for differences in TM tasks between UHRP and HC groups are shown in Table 7, and Cohen's D effect size for differences between groups were 0.78 for FB, 0.81 for SS and 0.79 for EQ.

There was also significant group by emotion and mind inference interaction in the EQ and TM tasks (Table 8).

3.2. Imaging Data

Patients showed a cluster of significant FA reductions compared with controls including bilateral deep white matter in the frontal lobes and CC (Figure 1). Because there was no significant cluster of FA increase in patients, the following correlation analysis was performed only within the FA reduction mask.

Patients' performance in TM task showed a strong positive correlation with three white matter regions, which were the *inferior longitudinal fasciculus* (Figure 1) that represents a main association tract connecting frontal with parietal and temporal areas, the *left uncinate fasciculus* (Figure 2) that connects the *orbitofrontal lobe* (OFL) to the *temporal pole* (TP), and the *left anterior cingulum* (Figure 3). Correlation coefficients at the peak voxels in these regions were 0.79, 0.77 and 0.72, respectively. The EQ tasks showed significant correlations with FA reduction in the *left anterior cingulum* that runs between the frontal and temporal lobes, interconnecting parts of the limbic system. On the other hand, the correlation analyses in healthy controls revealed no significant correlation of FA with any of the TM or EQ tasks.

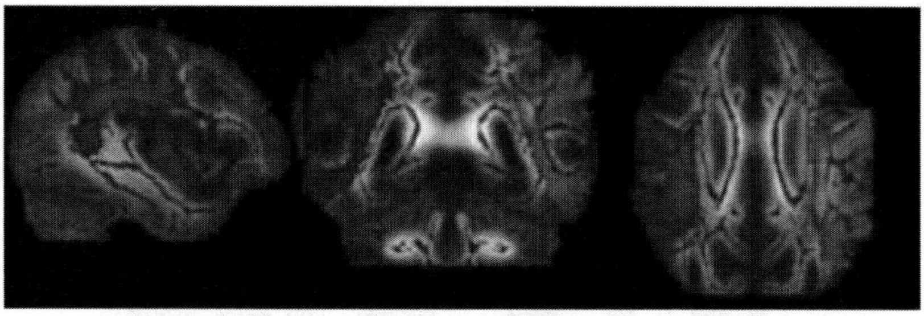

(a) Sagittal slice (b) Coronal slice (c) Axial slice

Figure 1. TBSS Inferior Longitudinal Fasciculus.

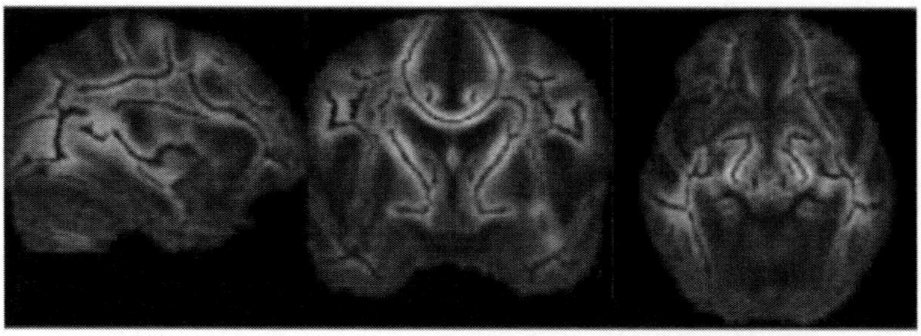

(a) Sagittal slice (b) Coronal slice (c) Axial slice

Figure 2. TBSS Left Uncinate Fasciculus.

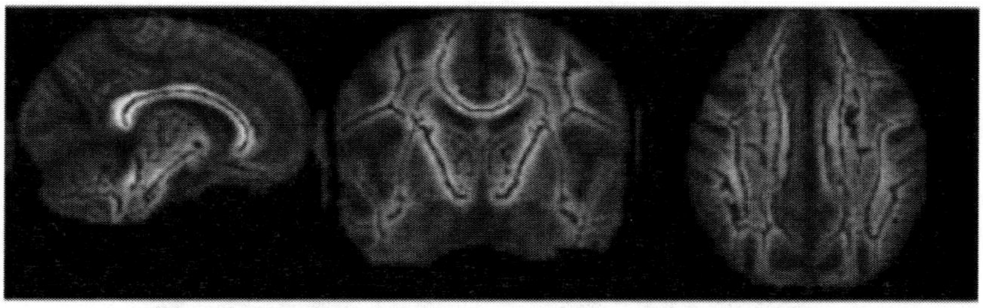

(a) Sagittal slice (b) Coronal slice (c) Axial slice

Figure 3. TBSS Left Anterior Cingulate.

Figure 1 shows regions of significant FA reduction in UHRP patients relative to healthy controls. To aid visualization, regions showing significant FA reductions ($p < 0.01$ corrected by TFCE) are thickened using the tbss_fill script implemented in FSL (red-green). Results are shown overlaid on the mean FA map and the FA skeleton (blue). Left-right orientation is according to the radiological convention.

Significant correlation between TM task performance and FA in UHRP patients is observed from the images given in Figure 2. To aid visualization, regions showing significant FA reductions ($p < 0.01$, corrected by FDR) are thickened using the tbss_fill script implemented in FSL (red-green). The left uncinate region is shown overlaid on the mean FA map and the FA skeleton (blue).

As can be seen from the images in Figure 3, significant correlation is also observed between EQ task performance and FA in UHRP patients. To aid visualization, regions showing significant FA reductions ($p < 0.01$, corrected by FDR) are thickened using the tbss_fill script implemented in FSL (red-green). The left anterior cingulate region is shown overlaid on the mean FA map and the FA skeleton (blue).

CONCLUSION

In this chapter evidence in favor of an association between reduced white matter integrity and both TM and emotional cognition deficits in UHRP is offered, providing evidence of disturbed frontal and cingulated callosal interhemispheric connectivity supporting impaired emotional and mind inferences in UHRP patients. Voxelwise correlation analysis in UHRP patients revealed a positive correlation between three white matter areas of FA reduction and performance in mental inference task, as well as a positive correlation between one white matter areas of FA reduction in emotional inference attribution task. These results indicate that, in addition to the previously demonstrated regional gray matter deficits [105-106], structural disconnection among gray matter networks may contribute to theory of mind and socio-emotional cognition impairment in UHRP patients. An Interaction between diagnosis, emotion and mentalizing categories was found, indicating that the deficit was of both areas, rather than being specific to any one of the categories.

The result of TBSS group comparison revealed a relatively large cluster of reduced FA in UHRP patients. Similar patterns of FA reduction have been robustly reported in earlier voxel-based DTI studies [107-108]. In the correlation analyses of patients' EQ, an emotional recognition task, a strong association was observed with the left anterior CC region, comprised mainly of callosal interhemispheric fibers and the right anterior cingulum. Nevertheless, as multiple fiber tracts are contained in this region, caution must be taken in interpretation. Recent studies indicate a critical role of uncinate fasciculus and right anterior cingulum as a predictor of social functioning at follow-up in ultra high of psychosis subjects [109-111]. In a study using a multimodal approach, reduced white matter integrity was found in the uncinate fasciculus in 11-13 year old schoolchildren presenting with subclinical psychotic symptoms, together with the cingulum and inferior longitudinal fasciculus, suggesting aberrant prefrontal-temporal dysfunction in psychosis proneness subjects [112-113]. On the whole, reduced FA values were found in those brain areas implicated in theory of mind tasks, including ventromedial prefrontal area, anterior cingulate, temporoparietal junction and superior temporal lobe [114].

It has been described that posterior and anterior cingulate are involved in emotional awareness as expressed in facial gestures [115-116], including bilateral temporo-parietal junction, but only orbitofrontal, medial prefrontal and anterior cingulate areas have been described as capable to tag deceptive intentions to another subject [117]. Reported deficits in oligodendrocyte and myelin gene expression may be associated with reduction in the number of oligodendrocytes [118-119], which in turn could lead to disruption or insufficient myelin development contributing to axonal disorganization and altered frontotemporal connectivity [120]. Reduced FA and asymmetry disruption in *the left uncinate fasciculus* was found in accordance with others [121]. It has been suggested a putative role in executive measures and high order processes as assessed in TM evaluation for this bundle, including self-awareness, goal-directed behavior and social cognition. Brain frontotemporal networks connected by these fibers seem to support this hypothesis [122]. The observed decreased FA of the left anterior cingulum bundle in UHRP subjects compared with controls is also consistent with the disconnection process theory and may explain the challenge imposed on them by TM tasks [123]. Reduced FA in the right white matter temporal portion of the Inferior longitudinal fasciculus, the major white matter tract connecting temporal and occipital lobes up to the temporal–parietal border, is consistent with impairment role functioning, predicting according to some authors have further deterioration in community functioning in UHRP participants at 15-month follow-up. This may extend well beyond prodromal period as similar findings have been described in schizophrenia patients [124]. Inferior longitudinal fasciculus is a ventral associative bundle with long and short fibers connecting the occipital and temporal lobes. The long fibers are medial to the short fibers and connect visual areas to the amygdala and hippocampus [125].

Brain areas connected by these fibers may explain clinical consequences on deficits in social cognition ranging from deficits in theory of mind assessing ability to comprehend other's intentions to deficits in social knowledge, referred to as skillfulness in using social cues like gestures, or employing social rules to solve relationships situations. Contrary to our findings other studies failed to demonstrate reduced FA in the arcuate, uncinate and cingulum bundle [126] but it must be taken into account that UHRP subjects exhibit great heterogeneity regarding gender, age and developmental stage at time of examination which could account for differences between investigated groups in the location of FA impairments [127]. White matter FA abnormalities were found in subjects with genetic high risk of psychosis, having twins or first-degree relative with schizophrenia, including anterior and posterior cingulate and angular white matter [128]. This suggests that white matter medial frontal FA reduction could be associated with deficit in social reasoning and ulterior full psychosis development as has been suggested by some authors [129]. Performance in FP, on the other hand, failed to show any significant association with FA reduction [130]. This could be due to the fact that abnormalities associated with TM deficits were located in white matter regions adjacent to cortical gray matter, not in deep white matter tracts, making it possible that these regions were excluded from the skeleton in TBSS.

One limitation of the study could be accounted by the fact that several studies have revealed correlations between cognitive functions and FA in healthy subjects [131]. Thus, it would seem reasonable to hypothesize that individual differences in a TM and EQ task performance were correlated with individual variations of FA values in healthy people. Correlation analyses in healthy controls revealed no association between FA values and any of the FP, SS, EQ or FB tasks. However, this in turn could be due to small variances in healthy controls, leading to weak associations between the measured variables. In conclusion, direct association between regional FA reduction in three white matter areas and impaired Theory of Mind and Emotional Quotient as well in UHRP patients is remarked.

REFERENCES

[1] J.M. Moran, L.L. Young, R. Saxe, S.M. Lee, D. O'Young, P.L. Mavros and J.D. Gabrieli. *PNAS 2011*, 108 (7), 2688 (2011).

[2] G.S. Chasson, K.R. Timpano, J.L. Greenberg, A. Shaw, T. Singer and S. Wilhelm, *Clinical Psychology Review*, 31(4), 653 (2011).

[3] J. Uekermann, M. Kraemer, M. Abdel-Hamid, B.G. Schimmelmann, J. Hebebrand, I. Daum, J. Wiltfang and B. Kis, *Neuroscience and Biobehavioral Reviews*, 34(5), 734 (2010).

[4] N. Rakhlin, S.A. Kornilov, J. Reich, M. Babyonyshev, R.A. Koposov and E.L. Grigorenko, *Journal of Neurolinguistics*, 24(4), 476 (2011).

[5] P.E. Bailey and J.D. Henry, *British Journal of Clinical Psychology*, 49, 43 (2010).

[6] S. McDonald, A. Saad, C. James. *Journal of Clinical and Experimental Neuropsychology*, 33(6), 619 (2011).

[7] M. Banati, J. Sandor, A. Mike, E. Illes, L. Bors, A. Feldmann, R. Herold, Z. Illes. *European Journal of Neurology*, 17, 426 (2010).

[8] S.M. Couture, E.L. Granholm, S.C. Fish. *Schizophrenia Research*, 125(2-3),152 (2011).

[9] M. Abdel-Hamid, C. Lehmkamper, B.C. Sonntag, G. Juckel, I. Daum, M. Brune. *Psychiatry Research*, 165(1-2),19 (2009).

[10] J.L. Powell, P.A. Lewis, R.I.M. Dunbar, M. Garcia-Finana, N. Roberts. *Neuropsychologia*, 48(12), 3554 (2010).

[11] A.C.H. Lee, V.L. Scahill, K.S. Graham. *Cerebral Cortex*, 18, 683 (2008).

[12] B.C. Vander Wyk, C.M. Hudac, E.J. Carter, D.M. Sobel, K.A. Pelphrey. *Psychological Science*, 20, 771 (2009).

[13] D. Vistoli, E. Brunet-Gouet, A. Lemoalle, M.C. Hardy-Baylé, C. Passerieux. *Social Neuroscience*, 6(3), 316 (2011).

[14] M. Yücel, C. Pantelis. *Acta Psychiatrica Scandinavica*, 120(4), 253 (2009).

[15] E. Bora, M. Yucel, C. Pantelis. *Schizophrenia Research*, 109(1-3), 1-9 (2009).

[16] K. Koelkebeck, A. Pedersen, T. Suslow, K.A. Kueppers, V. Arolt, P. Ohrmann. *Schizophr. Res.*, 119(1-3), 115 (2010).

[17] E. Bora, S. Gökçen, B. Kayahan, B. Veznedaroglu. *J. Nerv. Ment. Dis.*, 196(2), 95 (2008).

[18] R. Herold, A. Feldmann, M. Simon, T. Tényi, F. Kövér, F. Nagy, E. Varga, S. Fekete. *Acta Psychiatrica Scandinavica*, 119(3),199 (2009).

[19] C.M. Gibson, D.L. Penn, M.J. Prinstein, D.O. Perkins, A. Belger. *Schizophr. Res.*; 122(1-3),179 (2010).

[20] M. Barragan, K.R. Laurens, J. Blas Navarro, J.E. Obiols. *Psychiatry Research*, 186(2-3), 225 (2011).

[21] A.J. Fett, W. Viechtbauer, M. Dominguez, D.L. Penn, J. van Os, L. Krabbendam. *Biobehavioral Reviews*, 35(3), 573 (2011).

[22] S. Jacobson, I. Kelleher, M. Harley, A. Murtagh, M. Clarke, M. Blanchard, C. Connolly, E. O'Hanlon, H. Garavan, M. Cannon. *Neuroimage*, 15; 49(2),1875 (2010).

[23] T.A. Niendam, M. Jalbrzikowski, C.E. Bearden. *Neuropsychol. Rev.*, 19, 280 (2009).

[24] G. Shim, D.H. Kang, Y.S. Chung, S.Y. Yoo, N.Y. Shin, J.S. Kwon. *Aust. N. Z. J. Psychiatry*; 42(8), 678 (2008).

[25] P. Fusar-Poli, A. Meneghelli, L. Valmaggia, P. Allen, F. Galvan, P. McGuire, A. Cocchi. *Br. J. Psychiatry*, 194(2), 181 (2009).

[26] S.M. Couture, D.L. Penn, J. Addington, S.W. Woods, D.O. Perkins. *Schizophrenia Research*, 100(1-3), 237 (2008).

[27] B.A. Cornblatt, A.M. Auther, T. Niendam, C.W. Smith, J. Zinberg, C.E. Bearden, T.D. Cannon. *Schizophr. Bull.*, 33(3), 688 (2007).

[28] C.U. Correll, M. Hauser, A.M. Auther, B.A. Cornblatt. *Journal of Child Psychology and Psychiatry*, 51(4), 390 (2010).

[29] M.F. Green, D.L. Penn, R. Bentall, W.T. Carpenter, W. Gaebel, R.C. Gur, A.M. Kring, S. Park, S.M. Silverstein, R. Heinssen. *Schizophrenia Bulletin*, 34(6), 1211 (2008).

[30] D.L. Penn, L.J. Sanna, D.L. Roberts. *Schizophr. Bull.,* 34(3), 408 (2008).

[31] M. Sprong, P. Schothorst, E. Vos, J. Hox, H. van Engeland. *Br. J. Psychiatry*, 191, 5 (2007).

[32] C. Gibson, D.L. Penn, M.J. Prinstein, D.O. Perkins, *Schizophrenia Research*, 122(1-3), 179 (2010).

[33] W.H. Schnell, S. Erk, C. Arnold, P. Kirsch, C. Esslinger, D. Mier, M.M. Schmitgen, M. Rietschel, S.H. Witt, M.M. Nöthen, S. Cichon, A. Meyer-Lindenberg. *Mol. Psychiatry*, 16(4), 462 (2011).

[34] W.H. Jung, J.S. Kim, J.H. Jang, J.S. Choi, M.H. Jung, J.G. Park, J.Y. Han, C.H. Choi, D.H. Kang, C.K. Chung, J.S. Kwon. *Schizophr. Bull.,* 37(4), 839 (2011).

[35] C.D. Frith, U. Frith. *Neuron*, 50(4), 531 (2006).

[36] T.J. Miller, T.H. McGlashan, J.L. Rosen, L. Somjee, P.J. Markovich, K. Stein, S.W. Woods. *Am. J. Psychiatry*, 159, 863 (2002).

[37] A.R. Yung, H.P Yuen, P.D. McGorry, L.J. Phillips, D. Kelly, M. Dell'Olio, S.M. Francey, E.M. Cosgrave, E. Killackey, C. Stanford *et al. Aust. N. Z. J. Psychiatry*, 39, 964 (2005).

[38] K.S. Shin, J.S. Kim, D.H. Kang, Y. Koh, J.S. Choi, B.F. O'Donnell, C.K. Chung, J.S.S Kwon. *Biol. Psychiatry*, 65, 1071 (2009).

[39] A.E. Pinkham, D.L. Penn, D.O. Perkins, K.A. Graham, M. Siegel. *Cogn. Neuropsychiatry*, 12, 198 (2007).
[40] J. Addington, D. Penn, S.W. Woods, D. Addington, D.O. Perkins. *Schizophr. Res.*, 99, 119 (2008).
[41] G. Shim, D.H. Kang, Y.S. Chung, Y.S. Yoo, N.Y. Shin, J.S. Kwon. *Aust. N. Z. J. Psychiatry*, 42, 678 (2008).
[42] A.R. Yung, L.J. Phillips, H.P. Yuen, S.M. Francey, C.A. McFarlane, M. Hallgren, P.D. McGorry. *Schizophr. Res.*, 60, 21 (2003).
[43] K.A. Woodberry, L.J. Seidman, A.J. Giuliano, M.B. Verdi, W.L. Cook, W.R. McFarlane. *Schizophrenia Research*, 123(2-3), 188 (2010).
[44] M. O'Connor, J.M. Harris, A.M. McIntosh, D.G. Owens, S.M. Lawrie, E.C. Johnstone. *Psychol. Med.*; 39(10), 1649 (2009).
[45] C.M. Corcoran, D. Kimhy, M.A. Parrilla-Escobar, V.L. Cressman, A.D. Stanford, J. Thompson, S.B. David, A. Crumbley, S. Schobel, H. Moore, D. Malaspina. *Psychol. Med.*, 41(2), 251 (2011).
[46] A.E. Pinkham, D.L. Penn, D.O. Perkins, K.A. Graham, M. Siegel. *Cogn. Neuropsychiatry*, 12, 198 (2007).
[47] S. Ruhrmann, F. Schultze-Lutter, R.K. Salokangas, M. Heinimaa, D. Linszen, P. Dingemans, M. Birchwood, P. Patterson, G. Juckel, A. Heinz, A. Morrison, S. Lewis, H.G. von Reventlow, J. Klosterkötter. *Arch. Gen. Psychiatry*; 67(3), 241 (2010).
[48] Y.S. Chung, D.H. Kang, N.Y. Shin, S.Y. Yoo, J.S. Kwon. *Schizophr. Res.*, 99, 111 (2008).
[49] P. Moller, R. Husby. *Schizophr. Bull.*, 26, 217 (2000).
[50] G. Modinos, R. Renken, S.G. Shamay-Tsoory, J. Ormel, A. Aleman. *Neuropsychologia*, 48(13), 3715 (2010).
[51] R. Pukrop, J. Klosterkötter. *Neurotox. Res.*; 18(3-4), 272 (2010).
[52] Rosso, N. Makris, H.W. Thermenos, S.M. Hodge, A. Brown, D. Kennedy, V.S. Caviness, S.V. Faraone, M.. Tsuang, L.J. Seidman. *Schizophrenia Research*, 123(1), 15 (2010).
[53] W.H. Jung, J.W. Jang, M.S. Byun, S.K. An, J.S. Kwon. *J. Korean Med. Sci.*; 25, 1700 (2010).
[54] M. Brune, S. Ozgurdal, N. Ansorge, H.G. von Reventlow, S. Peters, V. Nicolas, M. Tegenthoff, G. Juckel, S. Lissek. *NeuroImage*, 55(1), 329 (2011).
[55] H.L. Gallagher, C.D. Frith. *Trends in Cognitive Sciences*, 7, 77 (2003).
[56] E. Brunet, Y. Sarfati, M.C. Hardy-Bayle, J. Decety. *Neuropsychologia*, 41, 1574 (2003).
[57] R.C.K. Chan, X. Di, G.M. McAlonan, Q. Gong. *Schizophr. Bull.*, 37(1), 177 (2011).
[58] D. Marjoram, D.E. Job, H.C. Whalley, V.E. Gountona, A.M. McIntosh, E. Simonotto *et al. Neuroimage*, 31(4), 1850 (2006).
[59] R. Smieskova, P. Fusar-Poli, P. Allen, K. Bendfeldt, R.D. Stieglitz, J. Drewe, E.W. Radue, P.K. McGuire, A. Riecher-Rossler, S.J. Borgwardt. *Neuroscience and Biobehavioral Reviews*; 34(8), 1207 (2010).
[60] B. Nelson, A. Fornito, B.J. Harrison, M. Yücel, L.A. Sass, A.R. Yung *et al. Neuroscience and Biobehavioral Reviews*, 33(6), 807 (2009).
[61] S. Chanraud, N. Zahr, E.V. Sullivan, A. Pfefferbaum. *Neuropsychology Review*, 20(2), 209 (2010).

[62] M. Wahl, Y.O. Li, J. Ng, S.C. Lahue, S.R. Cooper, E.H. Sherr, P. Mukherjee. *Neuroimage*, 51(2), 531 (2010).

[63] M. Hori, K. Ishigame, N. Shiraga, H. Kumagai, S. Aoki, T. Araki. *Eur. Radiol.*, 18(4), 830 (2008).

[64] P. Pierpaoli, P. Basser. *Magn. Reson. Med.*, 36, 893 (1996).

[65] C. Beaulieu. *NMR Biomed.*, 15, 435 (2002).

[66] P. Mukherjee, J.I. Berman, S.W. Chung, C.P. Hess, R.G. Henry. *Am. J. Neuroradiol.*, 29, 632 (2008).

[67] K. Arfanakis, V.M. Haughton, J.D. Carew, B.P. Rogers, R.J. Dempsey, M.E. Meyerand. AJNR *Am. J. Neuroradiol.*; 23(5), 794 (2002).

[68] S.M. Smith, H. Johansen-Berg, M. Jenkinson, D. Rueckert, T.E. Nichols, K.L. Miller, M.D. Robson, D.K. Jones, J.C. Klein, A.J. Bartsch, T.E. Behrens. *Nat. Protoc.*, 2(3), 499 (2007).

[69] J. Miyata, M. Yamada, C. Namiki, K. Hirao, T. Saze, H. Fujiwara, M. Shimizu et. al. *Schizophrenia Research*, 119(3), 232 (2010).

[70] F. Happé. *Journal of Autism and Developmental Disorders*; 24:129–54 (1994).

[71] Ziv, D. Leiser, J. Levine. *Cogn. Neuropsychiatry*, 16(1), 71 (2011).

[72] S. Baron-Cohen, M. O'Riordan, V.E. Stone, R. Jones, K. Plaisted. *Journal of Autism and Developmental Disorders*, 29, 407 (1999).

[73] S. Baron-Cohen, S. Wheelwright, J. Hill, Y. Raste, I. Plumb. *Journal of Child Psychology and Psychiatry*; 42, 241 (2001).

[74] H. Walter, K. Schnell, S. Erk, C. Arnold, P. Kirsch, C. Esslinger, D. Mier, M.M. Schmitgen, M. Rietschel, S.H. Witt, M.M. Nöthen, S. Cichon, *Molecular Psychiatry,* 16, 462 (2011).

[75] J. Call, M. Tomasello. *Journal of Comparative Psychology*; 112, 192 (1998).

[76] T.J. Miller, T.H. McGlashan, J.L. Rosen, K. Cadenhead, T. Cannon, J. Ventura, W. McFarlane, D.O. Perkins, G.D. Pearlson, S.W. Woods. *Schizophr. Bull.*; 29(4), 703 (2003).

[77] R.K.R. Salokangas, T.H. McGlashan. *Nordic Journal of Psychiatry*; 62(2), 92 (2008).

[78] D. Lukoff, K.H. Nuechterlein, J. Ventura. *Schizophrenia Bulletin*, 12, 594 (1986).

[79] Wechsler D. *Wechsler abbreviated scale of intelligence.* San Antonio, TX: The Psychological Corporation, (1999).

[80] J.J. Ryan, A.M. Paolo. *Archives of Clinical Neuropsychology*, 16(2), 151 (2001).

[81] R.K. Heaton. *Wisconsin Card Sorting Test Manual.* Psychological Assessment Cooperation Press, Odessa, FL. (1981).

[82] R.M. Reitan, D. Wolfson. *The Halstead–Reitan Neuropsychological Test Battery.* Neuropsychology Press, Tucson. (1985).

[83] J.R. Stroop. *J. Exp. Psychol.*, 18, 643 (1935).

[84] D. Stuss, D. Benson. *The Frontal Lobes.* Ravens Press, NY. (1986).

[85] Wechsler D. *Wechsler Adult Intelligence Scale—Revised.* New York: The Psychological Corporation (1981).

[86] K.B. Boone, I.M. Lesser, E. Hill-Gutierrez, N.G. Berman, L.F. D'elia. *Clinical Neuropsychologist*, 7(1), 22 (1993).

[87] D.C. Delis, J.H. Kramer, E. Kaplan, B.A. Ober. *California Verbal Learning Test-Second Edition California Verbal Learning Test: Second Edition*. San Antonio, TX: Psychological Corporation (2000).

[88] A.D. Stanford, J. Messinger, D. Malaspina, C.M. Corcoran. *Schizophr. Res.*, 131(1-3), 11 (2011).
[89] A.D. Thompson, C. Bartholomeusz, A.R. Yung. *Early Intervention in Psychiatry*, 5(3), 192 (2011).
[90] J. Call, A. Tomasello. *Child Development*, 70, 381 (1999).
[91] N. Kaland, A. Møller-Nielsen, L. Smith, E.L. Mortensen, K. Callesen, D. Gottlieb. *European Child and Adolescent Psychiatry*; 14(2), 73 (2005).
[92] C. Gregory, S. Lough, V.E. Stone, S. Erzinclioglu, L. Martin, S. Baron-Cohen, J. Hodges. *Brain*, 125, 752 (2002).
[93] S. Baron-Cohen, S. Wheelwright. *J Autism Dev Disord*, 34, 163 (2004).
[94] J.W. Mauchly. *The Annals of Mathematical Statistics*, 11, 204 (1940).
[95] Leemans, B. Jeurissen, J. Sijbers, D.K. Jones. In: *17th Annual Meeting of Intl. Soc. Mag. Reson. Med.* 3537 (2009).
[96] Ellison-Wright, E. Bullmore, *Schizophrenia Research*, 108(1-3), 3 (2009).
[97] S.M. Smith. *Hum. Brain Mapp.*, 17, 143 (2002).
[98] S.B. Eickhoff, T. Paus, S. Caspers, M. Grosbras, A.C. Evans, K. Zilles, K. Amunts. *NeuroImage*, 36(3), 511 (2007).
[99] S.M. Smith, M. Jenkinson, H. Johansen-Berg, T.E. Rueckert, T.E. Nichols, K.L. Miller, M.D. Robson *et al. Neuroimage*; 31, 1487 (2006).
[100] S.M. Smith, T.E. Nichols. *Neuroimage*; 44, 83 (2009).
[101] T.E. Nichols, A.P. Holmes. *Hum. Brain Mapp.*; 15, 1 (2002).
[102] Y. Benjamini, T. Hochberg. *Journal of Educational and Behavioral Statistics*; 26, 60 (2000).
[103] S. Wakana, H. Jiang, L.M. Nagae-Poetscher, P.C.M. van Zijl, S. Mori. *Radiology*; 230, 77 (2004).
[104] S.J. Borgwardt, A. Riecher-Rössler, P. Dazzan, X. Chitnis, J. Aston, M. Drewe *et al. Biological Psychiatry*, 61(10), 1148 (2008).
[105] H. Witthaus, C. Kaufmann, G. Bohner, S. Özgürdal, Y. Gudlowski, J. Gallinat *et al. Psychiatry Research: Neuroimaging*, 173(3), 163 (2009).
[106] B.D. Peters, N. Schmitz, P.M. Dingemans, T.A. van Amelsvoort, C.B. Majoie, G.J. den Heeten, D.H. Linszen, L. de Haan. *Schizophrenia Research*, 111(1-3), 192 (2009).
[107] B.D. Peters, P.M. Dingemans, N. Dekker, J. Blaas, E. Akkerman, T.A. van Amelsvoort *et al. Neuroimaging*, 181(1), 44 (2010).
[108] P.D. Peters, J. Blaas, L. de Haan. *Journal of Psychiatric Research*, 44(15), 993 (2010).
[109] J.I. Friedman, C. Tang, D. Carpenter, M. Buchsbaum, J. Schmeidler, L. Flanagan, S. Golembo *et al. Am. J. Psychiatry*; 165(8), 1024 (2008).
[110] J. Tang, Y. LiaO, B. Zhou, C. Tan, T. Liu, W. Hao, D. Hu, X. Chen. *Brain Res.*; 1343, 199 (2010).
[111] S. Jacobson, I. Kelleher, M. Harley, A. Murtagh, M. Clarke, M. Blanchard *et al. NeuroImage*, 49(2),1875 (2010).
[112] T. Niendam, M. Jalbrzikowski, C. Bearden. *Neuropsychology Review*, 19(3), 280 (2009).
[113] N. Barnea-Goraly, H. Kwon, V. Menon, S. Eliez, L. Lotspeich, A.L. Reiss. *Biol Psychiatry,* 55, 323 (2004).
[114] M. Jehna, C. Neuper, A. Ischebeck, M. Loitfelder, S. Ropele,. Langkammer, F. Ebner *et al. Brain Research*, 1393, 73 (2011).

[115] Fornito, A.R. Yung, S.J. Wood, L.J. Phillips, B. Nelson, S. Cotton, D. Velakoulis *et al. Biological Psychiatry*, 64(9), 758 (2008).

[116] S. Lissek, S. Peters, N. Fuchs, H. Witthaus, V. Nicolas *et al. PLoS ONE*, 3(4), e2023 (2008).

[117] S.N. Mitkus, T.M. Hyde, R. Vakkalanka, B. Kolachana, D.R. Weinberger, J.E. Kleinman, B.K. Lipska. *Schizophrenia Research*, 98 (1-3), 129 (2008).

[118] K. Barley, S. Dracheva, W. Byne. Subcortical oligodendrocyte- and astrocyte-associated gene expression in subjects with schizophrenia, major depression and bipolar disorder. *Schizophrenia Research*, 112 (1-3), 54 (2009).

[119] S. Muñoz Maniega, G. Lymer, M. Bastin, D. Marjoram, D. Job, T. Moorhead, D. Owens, E. Johnstone, A. McIntosh, S. Lawrie. *Schizophrenia Research*, 106(2-3), 132 (2008).

[120] Voineskos, N. Lobaugh, S. Bouix, T. Rajji, D. Miranda, J. Kennedy, B. Mulsant, B. Pollock, M. Shenton. *Brain*, 133(5), 1494 (2010).

[121] F. Benedetti, A. Bernasconi, M. Bosia, R. Cavallaro, S. Dallaspezia, A. Falini, A. Poletti *et al. Schizophrenia Research*, 114(1-3), 154 (2009).

[122] H. Walter, M. Adenzato, A. Ciaramidaro, I. Enrici, L. Pia, B.G. Bara. *J. Cogn. Neurosci.*, 16, 1854 (2004).

[123] *M. Kubicki, R. McCarley, C.F. Westin, H.J. Park, S. Maier, R. Kikinis et al. J. Psychiatr. Res.;* 41, 15 (2007).

[124] M. Ashtari, J. Cottone, B. Ardekani, K. Cervellione, P. Szeszko, J. Wu, S. Chen, S. Kumra. *Arch. Gen. Psychiatry*, 64, 1270 (2007).

[125] K.H. Karlsgodt, T.A. Niendam, C.E. Bearden, T.D. Cannon. *Biol. Psychiatry*; 66(6), 562 (2009).

[126] M.J. Hoptman,, J. Nierenberg, H. Bertisch, D. Catalano, B. Ardekani, C. Branch, L. DeLisi. *Schizophrenia Research*, 106(2-3), 115 (2008).

[127] H. Fujiwara, C. Namiki, K. Hirao, J. Miyata, M. Shimizu, H. Fukuyama *et al. Schizophr. Res.*; 95, 215 (2007).

[128] B.A. Völlm, A.N. Taylor, P. Richardson, R. Corcoran, J. Stirling *et al. Neuroimage*, 29, 90 (2006).

[129] D.H. Salat, D.S. Tuch, N.D. Hevelone, B. Fischl, S. Corkin, H.D. Rosas, A.M. Dale. *Ann. N. Y. Acad. Sci.*, 1064, 37 (2007).

[130] Giorgio, K.E. Watkins, G. Douaud, A.C. James, S. James, N. De Stefano, P.M. Matthews, S.M. Smith, H. Johansen-Berg. *NeuroImage*, 39(1), 52 (2008).

[131] D. Qiu, L. Tan, K. Zhou, P. Khong. *NeuroImage*, 41(2), 223 (2008).

In: Anisotropy Research
Editor: Hirpa Lemu

ISBN 978-1-62081-977-7

Chapter 6

ANGLE AND POLARIZATION ANISOTROPY OF PHONONS AND PHONON-PLASMON MODES IN GaAs/AlAs SUPERLATTICES: RAMAN SPECTROSCOPY STUDIES

V. A. Volodin*

A.V. Rzhanov Institute of Semiconductor Physics, Siberian Branch of Russian Academy of Sciences, Novosibirsk, Russia

Novosibirsk State University, Novosibirsk, Russia

ABSTRACT

In this chapter Raman spectroscopy technique was applied to study phonons and phonon-plasmon coupled modes in undoped and Si doped set of GaAs/AlAs superlattices (SLs), grown on (100), (311)A and (311)B substrates. The observed phonon anisotropy (splitting of TO-like phonons) for (311)A corrugated superlattices (CSLs) can be indirect evidence of anisotropic structure of formed on (311)A GaAs surface quantum objects. Anisotropy in dispersion of phonon-plasmon modes with wave-vector normal to SL layers and in SL layers was observed for (100) SLs. Additional lateral anisotropy of the phonons and phonon-plasmon modes was observed for SLs grown on nano-corrugated (311)A surface. The observed anisotropy is supposed to be result of structural anisotropy of corrugated (311)A CSLs.

1. INTRODUCTION

GaAs/AlAs superlattices (SLs) and corrugated superlattices (CSLs) grown on (311)A,B and (001) surfaces were studied using *Raman spectroscopy*. Thickness of GaAs layers was varied from 1 monolayer (ML) to 14 MLs, thickness of AlAs barriers was varied from 8 to 14

* E-mail: volodin@isp.ncs.ru.

MLs in (311) direction. The observed *phonon anisotropy* of (311) A CSLs can be indirect evidence of anisotropic structure formed on (311) A GaAs surface quantum objects.

Effects of interaction of phonons with electrons in bulk semiconductors and semiconductor superlattices (SLs) have long been under study by wide community of researchers [1] but remain actual up today. The analysis of dispersion of *phonon-plasmon modes* in SLs can be expressed by non-destructive tools for measurements of charge carrier concentration and effective mass anisotropy [2, 3].

According to calculations [2, 3], in the case of the effective mass anisotropy of charge carriers, the phonon–plasmon modes propagating in the directions of "heavy" and "light" masses should have different frequencies. In this case, the mass difference is not as important as the wave-number of the *phonon–plasmon modes*. For the frequency difference to be experimentally observable (e.g., at least a few inverse centimeters), the wave-number of the modes must be on the order of one million of reversed centimeters. Since the semiconductor SLs are optically opaque, they may be Raman studied only using *backscattering geometry*. To reach a necessary wave-number difference, it is insufficient to only vary the angle of incidence of light relative to normal of a SL. It is necessary to implement the normal incidence geometry and the geometry in which the wave vectors of incident and scattered light lie in a SL plane. This is possible if the backscattering occurs at the side face of the SL as shown in Figure 1. Since the thickness of the SLs is usually less than a micron, one has to use a Raman scattering setup with micro-Raman setup.

2. EXPERIMENTAL

The studied SLs were grown using molecular beam epitaxy technique on (311)A-(1x8) reconstructed GaAs substrate and (001)-(2x4) reconstructed GaAs substrate at temperature 550° C. The control of reconstruction was carried out *in situ* with the use of reflection of high energy electron diffraction. In the case of (311) substrates the angle disorientation of the substrates from (311) direction was less than 15'. The flux of As_4 molecules was $5*10^{14}$ molecules/cm^2 per second the growth rate was about 0.3 μm/h for GaAs and 0.32 μm/h for AlAs.

All SL-satellites on (001) and (311)B GaAs substrates were grown in the same growth procedures – "side by side". The (311)A and B surfaces were detected using anisotropy of chemically etching. The conditions of growth are described in more detail elsewhere [4]. The data of the grown set of $GaAs_n/AlAs_m$ SLs are shown in Table 1. Really two sets of SLs were grown, first is not doped, second is doped by Si. To achieve electron concentration at room temperature $2*10^{18}$ cm^{-3}, the impurity (Si) concentration in doped SLs was $2.5*10^{19}$ cm^{-3}. Impurity of Si can be both acceptor and donor, but in chosen growth condition it gives the n-type material.

The Raman spectra were registered at room temperature in *backscattering geometry* in various polarization geometries. The line 514.5 nm of Ar^+ laser was used as a light source. The triple spectrometer T64000 (Horiba Jobin Yvon) equipped with cooled CCD camera and micro-Raman set-up was used. The incident laser light has always linear polarization; in some cases the polarization of scattered light was not analyzed. Polarization resolved Raman spectra were registered using polarization optic elements.

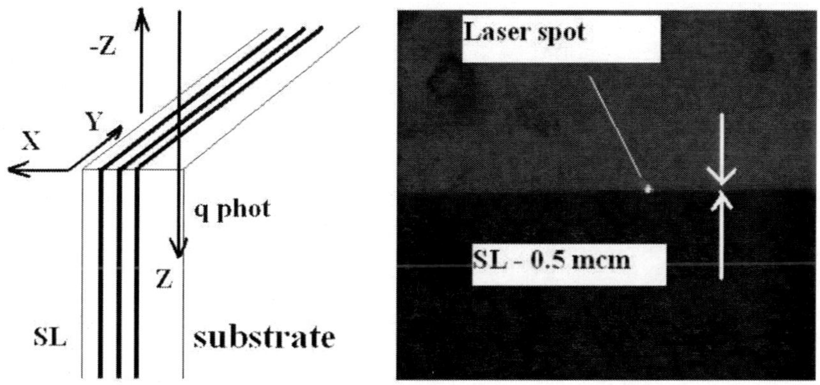

Figure 1. Scheme of "in-plane" Raman geometry (Left) and micro-photography of edge of a sample with a laser spot (Right).

Table 1. Specifications of the studied SLs $GaAs_nAlAs_8$. 1 ML in 311 direction = 1.7 angstrom

Substrate			GaAs thickness, Å.	AlAs thickness, Å.	Periods
311A	311B	100			
1A	1B	1	1.7	13.6	400
2A	2B	2	3.4	13.6	300
3A	3B	3	6.8	13.6	200
4A	4B	4	10.2	13.6	100
5A	5B	5	17	13.6	100
6A	6B	6	20.4	13.6	100

3. Results and Discussion

3.1. Not Doped SLs – Polarization Anisotropy of Phonons

Raman tensors for back-scattering geometry from (311) surface for *longitudinal optical* (LO) and *transverse optical* (TO) phonons are expressed as follows [4]:

$$\text{LO-}\ \frac{1}{\sqrt{11}}\begin{pmatrix} 0 & d & d \\ d & 0 & 3d \\ d & 3d & 0 \end{pmatrix};\ \text{TO}_x\text{-}\ \frac{1}{\sqrt{2}}\begin{pmatrix} 0 & -d & d \\ -d & 0 & 0 \\ d & 0 & 0 \end{pmatrix};\ \text{TO}_y\text{-}\ \frac{1}{\sqrt{22}}\begin{pmatrix} 0 & 3d & 3d \\ 3d & 0 & -2d \\ 3d & -2d & 0 \end{pmatrix} \qquad (1)$$

TO_x and TO_y modes correspond to atomic displacement along directions X $[01\bar{1}]$ and Y $[\bar{2}33]$, (across and along the facets on (311)A surface). According to symmetry selection rules, in $Z(YY)\bar{Z}$ geometry, *Raman scattering intensity* for TO_y modes is $\left(\frac{54}{11\sqrt{22}}\right)^2 d^2$ ($\approx 1.09d^2$), but for TO_x modes the scattering is forbidden; whereas in $Z(YX)\bar{Z}$ geometry, Raman scattering intensity for TO_x modes is $\frac{2}{11}d^2$ ($\approx 0.18d^2$), but for TO_y modes the scattering is forbidden [4]. So, using these symmetry rules, one can resolve these modes even its splitting is very small. It should be noted that in the case of ultra thin corrugated layers, when lateral phonon localization occurs, one need to take into account intermixing of TO_X and TO_Y modes.

The Raman spectra of (311)A corrugated superlattices (CSLs) in various polarization geometries are shown in Figure 2. From this figure, one can clearly see the splitting of confined TO modes of first order - $TO1_y$ and $TO1_x$. The intensities of spectra registered in $Z(YX)\overline{Z}$ geometry were multiplied by 2.5 - 5 times, the intensity of the Raman spectrum of $GaAs_2/AlAs_8$ SL in $Z(YY)\overline{Z}$ geometry was divided by 2, for convenience of perception. The increase in *Raman intensity* for this sample must be as a result of near-resonant scattering condition. Comparing the spectra of samples, one can observe that the changes in spectra registered in different polarizations. Although the picture is complex, and, as one can see, the Raman spectra contain several peaks, the general feature is "splitting" of TO_x and TO_x modes. As one can see, when the thickness of GaAs layers decrease, the width of the TO peaks increases, which can be caused by the fact that the AlAs/GaAs interfaces are not ideal.

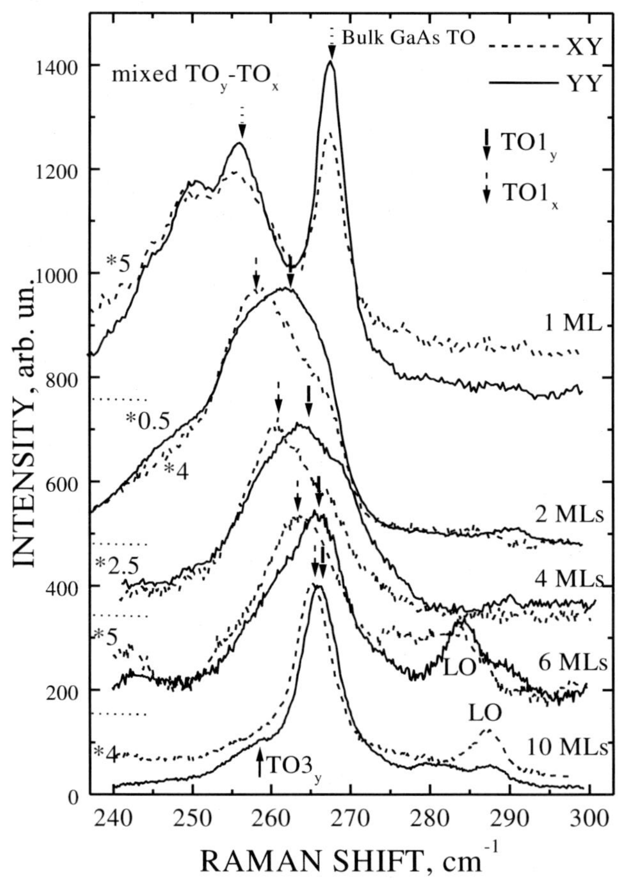

Figure 2. Raman spectra of (311)A GaAs/AlAs CSLs with various thickness of GaAs layers in YY (solid lines) and XY (dashed lines) polarizations.

It should be noted that for the satellite set of SLs grown on the GaAs (311)B surface the Raman spectra didn't reveal the splitting, in various polarizations Raman spectra were very similar. So, one can conclude that the effect of "splitting" is caused just by the GaAs (311)A surface corrugation, not by low symmetry of (311) direction and consequent strong intermixing of TO and LO modes [4].

The positions of all peaks corresponding to Raman scattering on $TO1_y$ and $TO1_x$ modes were determined using the approximation of the experimental spectra by *Lorenz curves* with minimizing of deviation. The effect of "splitting" increases with decreasing of the average thickness of GaAs layers for (311)A CSLs down to 2 MLs. For these samples the frequency of $TO1_y$ mode (atoms displace along to the facets) is higher than the frequency of $TO1_x$ mode (atoms displace normal to the facets).

The splitting increases sharply when the average thickness of GaAs layers is equal or less than the height of facets in one of the model of (311)A surface reconstruction (6 monolayers). This fact can be an indirect evidence of the QWW-like growth model. As mentioned above, the strong intermixing of TO_x, TO_y and LO modes can take place in the case of lowering symmetry caused by interface corrugation and other factors. One of these factors can be finite size of the QWWs in Y direction.

This effect will be essential for very thin and short QWWs. Indeed, for CSL $GaAs_1/AlAs_8$, Raman spectra in $Z(YX)\overline{Z}$ and $Z(YY)\overline{Z}$ geometries look like all modes became Raman active in all polarizations. So, one can observe no "splitting" in this sample. The width of the peaks increased, but one can still resolve two modes in Raman spectra of (311) SLs caused by scattering on LO-like and TO-like modes. The nature of the triplet features of this spectrum can be caused by lateral localization of GaAs-type optical phonons in very small 0-D clusters [4]. It should be noted that for the case of SLs grown on not-corrugated [5] surface (311)B, the splitting is negligible.

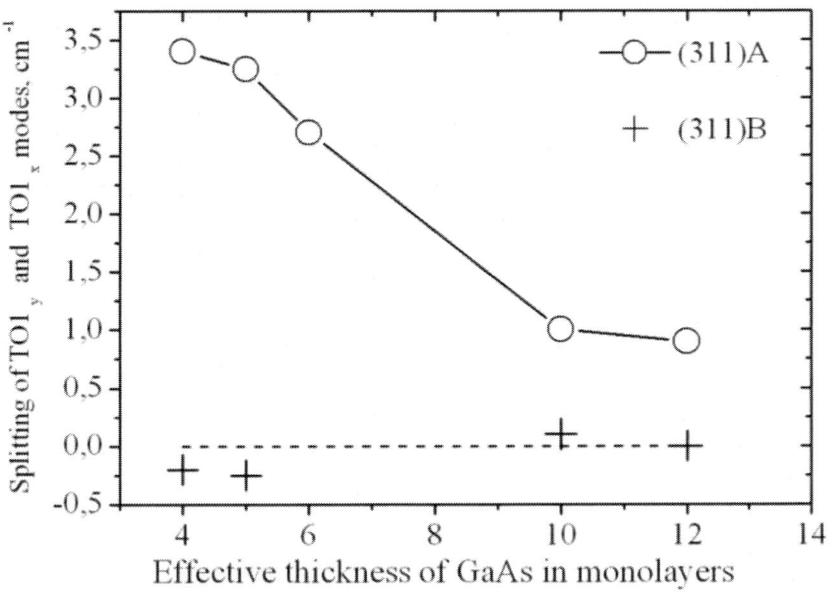

Figure 3. Splitting of TOx and TOy modes in (311) A and (311)B SLs versus effective thickness of GaAs in monolayers. One monolayer thickness for (311) direction is 1.7 angstrom.

3.2. Not Doped SLs – Angle Anisotropy of Phonons

It is also interesting to study not only polarization anisotropy, but also angle anisotropy, in other words anisotropy of phonon dispersion in SLs and CSLs.

Raman tensors for back-scattering geometry from (311) surface for LO and TO phonons were shown above (Figure 3). Raman tensors for back-scattering geometry from (100) surface for LO and TO phonons are given as follows:

$$\text{LO-}\begin{pmatrix} 0 & 0 & 0 \\ 0 & 0 & d \\ 0 & d & 0 \end{pmatrix};\ \text{TO}_1\text{-}\begin{pmatrix} 0 & d & 0 \\ d & 0 & 0 \\ 0 & 0 & 0 \end{pmatrix};\ \text{TO}_2\text{ -}\begin{pmatrix} 0 & 0 & d \\ 0 & 0 & 0 \\ d & 0 & 0 \end{pmatrix} \qquad (2)$$

Here, atomic displacements for LO phonons are along direction (100), atomic displacements for TO_1 and TO_2 phonons are along directions (001) and (010) respectively. Because light is transverse electromagnetic wave, in backscattering geometry from (100) surface only LO phonons are active in Raman scattering.

When one uses backscattering from edge of the (100) sample, Raman tensors for back-scattering geometry from (011) surface for LO and TO phonons are as follows:

$$\text{LO-}\ \frac{1}{\sqrt{2}}\begin{pmatrix} 0 & 0 & d \\ 0 & 0 & d \\ d & d & 0 \end{pmatrix};\ \text{TO}_{100}\text{-}\begin{pmatrix} 0 & d & 0 \\ d & 0 & 0 \\ 0 & 0 & 0 \end{pmatrix};\ \text{TO}_{01\text{-}1}\text{ -}\ \frac{1}{\sqrt{2}}\begin{pmatrix} 0 & 0 & -d \\ 0 & 0 & d \\ -d & d & 0 \end{pmatrix} \qquad (3)$$

In this equation, atomic displacements for LO phonons are along direction (011), atomic displacements for TO_{100} and $TO_{01\text{-}1}$ phonons are along directions (100) and (01-1) respectively. One can easily calculate that only TO phonons are active in this geometry.

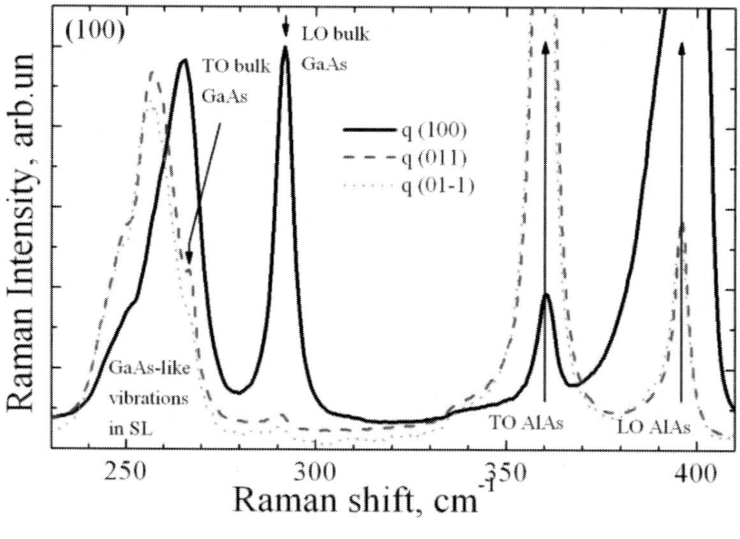

Figure 4. Not-polarized Raman spectra of (100) SL (sample 2, Table 1) registered for various direction of wave vector of incident light.

Figure 4 - Figure 6 show Raman spectra of samples grown on surfaces with different orientations. In the case of sample 2 ((100) orientation) if wave vector q is parallel to (100) direction, only longitudinal phonons are active in Raman scattering. If wave vector q is parallel to (011) or (01-1) directions, only transverse phonons are active in Raman scattering. But (011) and (01-1) directions are physically equivalent and one can see no difference in the Raman spectra (dashed and dotted lines in Figure 4).

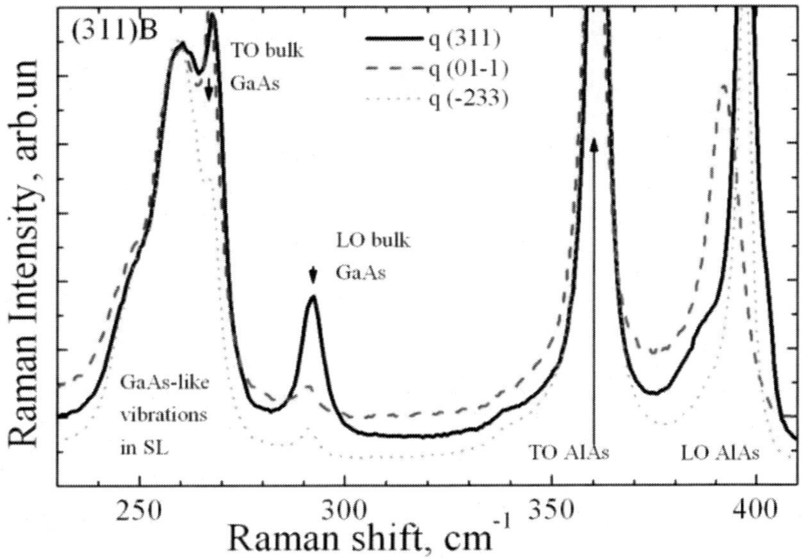

Figure 5. Not-polarized Raman spectra of (311)B SL (sample 2, Table 1) registered for various direction of wave vector of incident light.

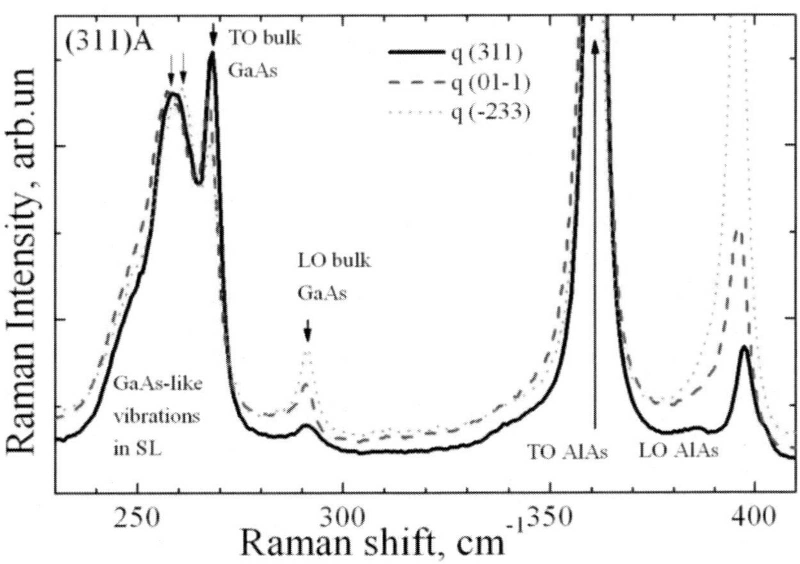

Figure 6. Not-polarized Raman spectra of (311)A CSL (sample 2, Table 1) registered for various direction of wave vector of incident light.

For the case of (311) SLs the directions (-233) and (01-1) directions are not physically equivalent and one can see some difference in Raman spectra (Figure 5). The *angle anisotropy* is more pronounced for the case of (311)A CSLs (Figure 6). Splitting of frequencies of optical phonons with different directions of wave vector is shown by arrows (Figure 6). It should be remarked that wave vectors are in lateral directions (-233) and (01-1), but anyway, one can see anisotropy. The observed anisotropy is supposed to be as a result of structural anisotropy of corrugated (311)A CSLs (see model of structure in figure 9). Optical phonons with wave vector along (01-1) directions propagate perpendicular to nano-facets, and due to localization should have low frequencies, what one can see in Figure 6.

3.3. Doped SLs – Phonon-Plasmon Coupled Modes Anisotropy

It is known that in polar semiconductors the *phonon-plasmon interaction* takes place. It leads to coupling of phonon-plasmon modes and to shift and broadening of Raman peaks [1]. Using micro-Raman system allows focusing laser beam into SL edge in sport about 1 micrometer (Figure 1). The Raman spectra of (100) GaAs/AlAs SLs for "normal" geometry are shown in Figure 7. One can see the shift of LO^+ phonon-plasmon modes versus variation of GaAs layer thickness. The peaks at 240-280 cm^{-1} are due to scattering on localized optic phonons in GaAs layers. One can see also longitudinal optic (LO) bulk modes from GaAs substrate. It should be noted that scattering on transverse optical (TO) phonons is forbidden in this geometry, but its appearance is supposed due to resonance phenomena.

In Figure 8 one can see Raman spectra of (100) SLs registered in Z(YY)-Z geometry, wave-vector of initial light was parallel to Z axis, wave-vector of scattered light was anti-parallel to Z axis, vector of polarization of light was parallel to SL layers (see Figure 1). In the case of (100) SLs axis X is along (100), axis Y is along (01-1), and axis Z is along (011) directions.

The spectra in this "in-plane" geometry practically does not depend on polarization of light (the configurations Z(XX)-Z and Z(XY)-Z were also used). If one compares the spectra of the same (100) SLs but in various geometry in Figure 7 and Figure 8, the anisotropy of phonon-plasmon modes is clearly seen where the wave-vector in the former is perpendicular to SL layers, while it is parallel to SL layers in the latter case (i.e. in Figure 8).

It should be noted that all SLs with the used thickness ratio of GaAs and AlAs must be of type II. The mixing of electron states of the Γ and *X* valleys occurs upon the transition of the GaAs/AlAs superlattice from type I to type II. The electron effective mass in the *X* valleys of both GaAs and AlAs is anisotropic, and the artificial symmetry of the superlattice period contributes to the mass anisotropy. Additional lateral anisotropy can be due to lateral nanofaceting of surface during SL growth.

In Figure 9 one can see an example of *lateral anisotropy* of (311)A SLs where a model of growth of QWW structure is also shown. In some growth condition in (311)A surface GaAs form quantum well wire (QWW) like structure [6]. The (311)A surface is nano-corrugated with lateral period 3.2 nm along (01-1) direction. So, it is possible to form structure with QWWs elongated along (-233) direction. In the case of (311) SLs axis X' is along (01-1), axis Y' is along (-233), and axis Z' is along (311) directions. The solid line in Figure 5 corresponds to Z'(X'X')-Z' geometry, dashed line corresponds to Y'(X'X')-Y' geometry, dot line corresponds to X'(Z'Z')-X' geometry. It should be noted, that axes X', Y', and Z' do not correspond to axes X, Y, and Z shown in Figure 1.

Anisotropy for "in-plane" geometry is shown in Figure 10. The wave vector can be parallel (dashed line) or perpendicular to QWWs (dot line). In the last case the spectrum is more like on spectrum in "normal" geometry (wave vector is also perpendicular to QWWs - solid line). Similar to (311)A SL, optical anisotropy of light absorbance (dichroizm) is observed [7].

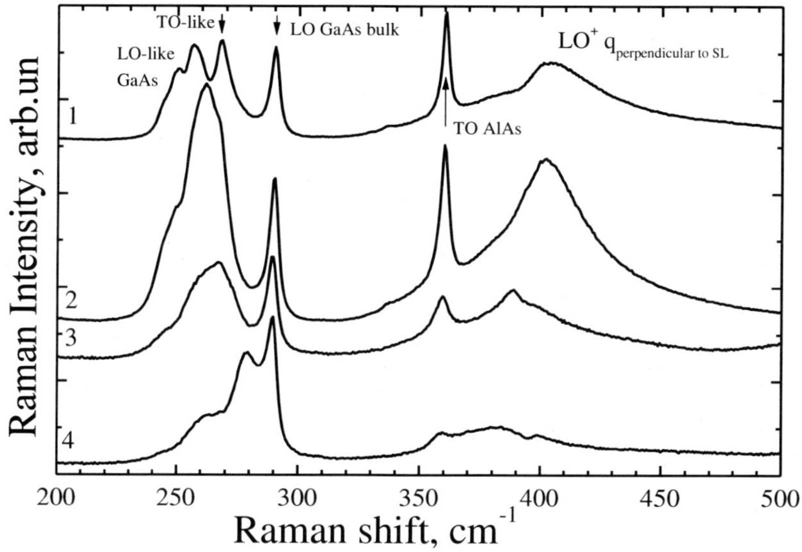

Figure 7. Raman spectra in "normal" geometry for (100) SLs, curves 1-4 correspond to samples 1-4 in Table 1.

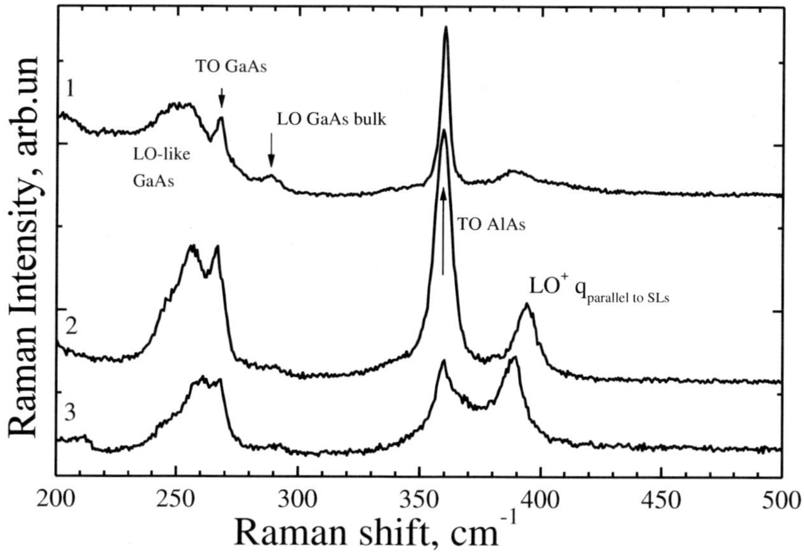

Figure 8. Raman spectra in "in-plane" geometry for (100) SLs 1-3.

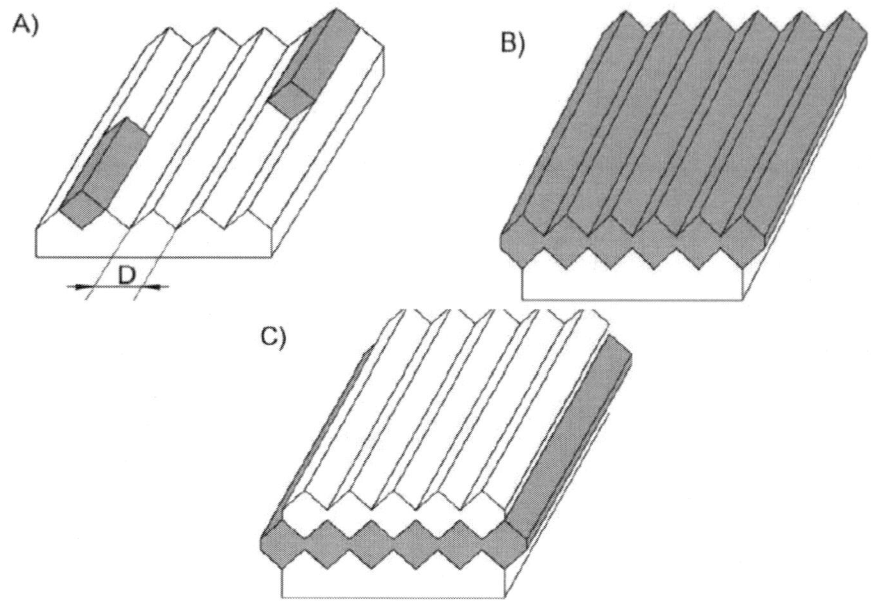

Figure 9. Model of growth of anisotropic quantum-well-wire like SLs on a corrugated surface. A – partial covering; B – full covering; C – covering by other material.

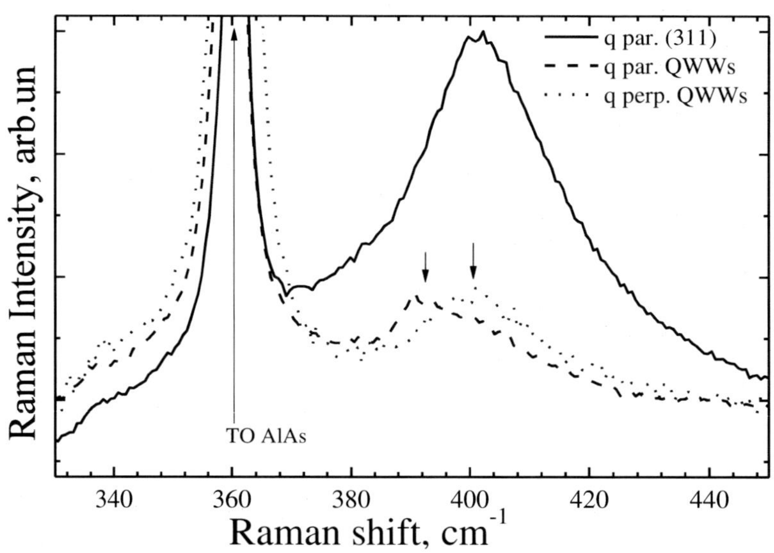

Figure 10. Raman spectra in "normal" and "in-plane" geometries for 311A SL (sample 2A – see Table 1). Arrows show the lateral anisotropy effect.

It has been already mentioned above that structural lateral anisotropy is not so pronounced in the case of (311)B SLs [5]. The Raman experiment was also carried out for 2B SLs (see Table 1). The solid line in Figure 11 corresponds to Z'(X'X')-Z' geometry, while the dashed line corresponds to Y'(X'X')-Y' geometry. One can also see anisotropy for "normal" and "in-plane" geometries. But for lateral effects, SLs (311)B were more like on (100) SLs without pronounced lateral anisotropy.

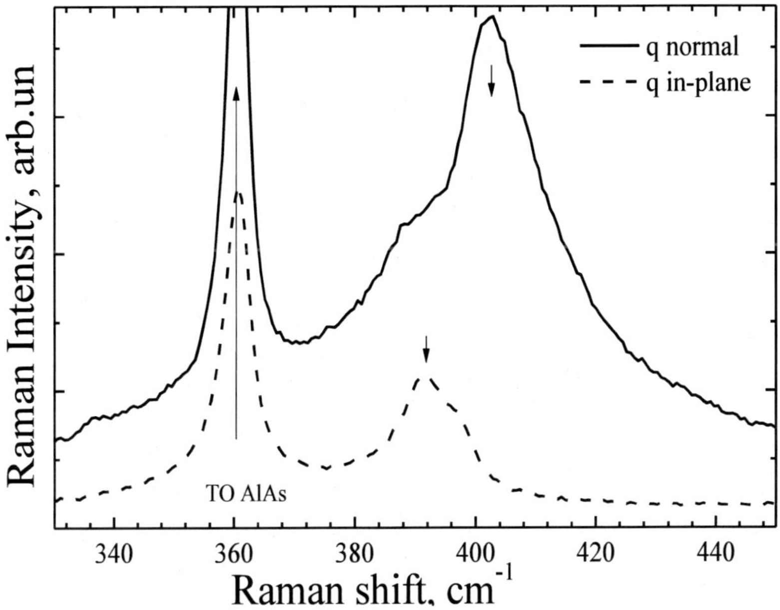

Figure 11. Raman spectra in "normal" and "in-plane" geometries for 311B SL (sample 2B – see Table 1). Arrows show the anisotropy effect.

CONCLUSION

In this chapter polarization and angle anisotropy of optical phonons were observed for corrugated GaAs/AlAs superlattices. As it was predicted in calculations [2, 3], the anisotropy in dispersion of phonon-plasmon modes, supposed due to effective mass anisotropy, has been observed. Further, the lateral anisotropy for (311)A SLs grown on nano-corrugated surfaces has been observed.

ACKNOWLEDGMENTS

The author is grateful to Drs. Preobrazhenskiy and Semyagin for growth of all SLs.

REFERENCES

[1] M. Cardona and G. Günterodt, *Light Scattering in Solids IV.Electronic Scattering, Spin Effects, SERC, and Morphic Effects*, Springer-Verlag, Berlin (1982).

[2] V.A. Volodin, M.D.Efremov, V.A.Sachkov, N.N.Ledentsov. Phys. Low-Dim. Strut. 5/6, 109 (2003).

[3] K. S. Gehar, *Nanophysics, Nanoclusters and Nanodevices: Phonon-plasmon coupling modes in tunneling thin GaAs/AlAs (311) and (001) SLs: Raman studies and modeling*, Nova Science Publishers Inc., New York, 2006.

[4] E. V. Dirote, *Trends in Nanotechnology Research: Interface reconstruction in GaAs/AlAs ultrathin superlattices grown on (311) and (001) surfaces*, Nova Science Publishers Inc., New York, 2004.
[5] N.N. Ledentsov, D.Litvinov, A. Rosenauer, D. Gerthsen, I.P. Soshnikov, V.A. Shchukin, V.M. Ustinov, A.Yu. Egorov, A.E.Zukov, V.A.Volodin, M.D. Efremov, V.V. Preobrazhenskii, B.P. Semyagin, D. Bimberg and Zh.I. Alferov, *Journal of Electronic Materials* 30, 464 (2001).
[6] R. Nötzel, N.N. Ledentsov, L.A. Däweritz, M. Hohenstein and K. Ploog, *Phys. Rev. Lett.* 67, 3812 (1991).
[7] V.A. Volodin, M.D. Efremov, R.S. Matvienko, V.V. Preobrazhenskii, B.R. Semyagin, N.N. Ledentsov, I.R. Soshnikov, D. Litvinov, A. Rosenauer, and D. Gerthsen. *Physics of the Solid State,* 47, 366 (2005).

In: Anisotropy Research
Editor: Hirpa Lemu
ISBN 978-1-62081-977-7

Chapter 7

MONITORING PROTEIN-DNA INTERACTIONS AND ENZYMATIC PROCESSES BY FLUORESCENCE ANISOTROPY

Xu-Guang Xi[1,*] and Eric Deprez[2]
[1]Laboratoire de Stress Génotoxiques et Cancer,
Institut Curie-Section de Recherche, Centre Universitaire, Orsay, France
[2]Laboratoire de Biologie et Pharmacologie Appliquées (LBPA),
Ecole Normale Supérieure Cachan, Institut d'Alembert, Cachan, France

ABSTRACT

Fluorescence anisotropy measurements are based on the principle of photoselective excitation of a fluorophore by a polarized light, providing information about rotational motions of fluorophores or fluorescently labeled molecules between photon absorption and emission. Some events such as overall rotational diffusion or flexibility are major causes of light depolarization. A high level of fluorescence anisotropy is generally associated with large molecules or complexes characterized by slow rotational diffusion or low flexibility level.

In this chapter, we describe several applications of steady-state fluorescence anisotropy for measuring biological enzymatic processes, and we focus our attention on two distinct examples of enzymatic assays based on the use of an extrinsic fluorophore covalently linked to DNA. These assays can simultaneously monitor DNA-binding and subsequent catalytic steps, using the same sample: (i) The first one is related to the unwinding activity of helicases. DNA helicases are motor proteins that use nucleotide triphosphate binding and hydrolysis to unwind duplex DNA. The fluorescence anisotropy assay is a rapid and sensitive fluorescence assay to measure helicase-mediated DNA unwinding. This assay takes the advantage of the substantial change in fluorescence polarization (or anisotropy) which occurs upon helicase binding to DNA and during the subsequent DNA unwinding reaction. The extent of light depolarization depends on the rate of tumbling of the fluorescently labeled DNA molecule, which decreases with increasing size. Thus, the fluorescence anisotropy-based assay can simultaneously

* Corresponding author: Xu-Guang.Xi@curie.fr.

monitor the DNA-binding of helicase as well as the unwinding reaction in real time. (ii) The second one is related to the HIV integrase activity. HIV integrase is responsible for the 3'-processing reaction of the extremities of the viral DNA. This reaction is the first of two sequential reactions catalyzed by integrase, leading to the covalent integration of viral DNA into the host chromosomic DNA. We have developed a fluorescence assay based on steady-state fluorescence anisotropy for monitoring the release of the terminal dinucleotide product – which occurs during the first reaction - in real time. As the initial anisotropy value obtained after DNA-binding and before catalysis depends on the fractional saturation of DNA sites, this assay can be also used to study simultaneously the DNA-binding step and the subsequent catalytic reaction in the same assay.

Finally, we present complementary analysis related to time-resolved fluorescence anisotropy for studying the oligomeric status of these enzymes either free in solution or bound to the DNA substrate.

1. Introduction to the Fluorescence Anisotropy Parameter

Fluorescence anisotropy (FA) measurements are based on the principle of photoselective excitation of a fluorophore by a polarized light, providing information about rotational motions of the fluorophore or a fluorescently labeled molecule between the time of photon absorption and the time of photon emission. Starting from a randomly oriented population, the photoselective excitation leads to a non-randomly oriented subpopulation of excited molecules. This subpopulation only is further observed during the emission process. Thus, FA is a powerful tool for measuring molecular orientation and rotational motions as well as processes that modulate them. In particular, FA can be used to measure complex formation (or dissociation), as well as enzymatic activities when a substantial difference in size exists between the substrate and the reaction product. Moreover, a precise determination of the molecular size is also accessible when necessary, according to the determination of the rotational correlation time.

The physical basis of fluorescence anisotropy is that, according to the photoselection principle, the fluorescent molecules with their absorption transition vectors aligned parallel to the electric vector of the linearly polarized light are selectively excited. This leads to an anisotropic subpopulation of non-randomly oriented excited molecules. Some events such as overall rotational diffusion or flexibility are major causes of light depolarization during the emission process. High levels of anisotropy are generally associated with large molecules or complexes characterized by slow rotational diffusion or low flexibility level. For small fluorescent molecules, rotating rapidly in solution, the initial preferential orientation of the photoselected subpopulation becomes randomized prior to emission, resulting in low fluorescence polarization or low fluorescence anisotropy.

In this context, binding of a small fluorescent probe to a large slowly rotating molecule results in an increase of fluorescence anisotropy, allowing quantification of the complex formation. Such quantification can be directly performed by a measurement of the fluorescence anisotropy in a solution containing a mixture of free and bound molecular species, without the need of physical separation of these molecular species, for instance using gel electrophoresis method. Moreover, the FA approach allows true equilibrium analysis with concentration of trace molecule as low as subnanomolar range.

Thus, every process (*e.g.* molecular interaction, enzymatic process), leading to a change in size of the fluorescent moiety, can be easily monitored and quantified, due to the additivity law of anisotropy (see below). After a brief overview and description of the instrumentation allowing the determination of the fluorescence anisotropy, we will see in this chapter, through two examples of enzymatic systems, *i.e.* the HIV-1 integrase and the *Escherichia coli* RecQ helicase, that steady-state fluorescence anisotropy is suitable for both equilibrium (protein-DNA interaction) and kinetic (enzymatic assay) studies. The determination of the corresponding time-resolved parameter, the rotational correlation time (θ), which can be directly calculated from the time-dependent decay of fluorescence anisotropy, requires a more sophisticated instrumentation. This approach allows to discriminate between fast and slow rotational motions for a given fluorophore (for instance, when a fluorescent specie is characterized by both an overall tumbling motion and local dynamics) and is, then, indispensable for determining the absolute molecular size while the steady-state approach allows to measure a relative change only. The last section of this chapter will present examples of time-resolved fluorescence anisotropy studies, again related to HIV-1 integrase and *E. coli* RecQ helicase.

2. Theory and Instrumentation

2.1. Steady-State Fluorescence Anisotropy

Fluorescence anisotropy (or *polarization*) has become one of the most popular techniques for the study of *molecular interactions* [1-3] in particular to study protein-DNA interactions [4-9]. Any equipment with a continuous excitation light source that allows determination of polarization can be used to measure the DNA-binding step and the subsequent catalytic activity based on the steady-state fluorescence *anisotropy parameter* (r). These equipments include spectrofluorimeters equipped with polarizers or commercial dedicated fluorescence polarization systems (such as the Beacon 2000 instrument, PanVera Corporation, Madison, USA). One advantage of the dedicated systems for polarization measurements is that these instruments correspond to compact, benchtop instruments that also provide subnanomolar or nanomolar sensitivity. The use of filters (instead of monochromators) ensures a high fluorescence intensity signal. Some of these systems (such as the Victor V, Perkin Elmer) are compatible with the 96-well plate format, for pharmacological applications.

All the above-mentioned instruments allow the determination of the parallel and the perpendicular components of the fluorescence emission intensity. Assuming a vertically polarized excitation, the steady state fluorescence anisotropy is defined by:

$$r = \frac{(I_V - I_H)}{(I_V + 2\,I_H)} \tag{1}$$

where I_v and I_h correspond to the fluorescence intensity parallel (vertical) and the fluorescence intensity perpendicular (horizontal) to the excitation plane, respectively.

As above-mentioned, the r parameter is suitable for monitoring equilibrium or kinetic parameters due to the additivity law of anisotropy.

Indeed, in the case of a mixture of molecular species (for instance, a mixture of fluorescent molecules which can be present either in a free or a bound form, during a titration process), each one is characterized by a proper r value, the resulting fluorescence anisotropy of the mixture is calculated according to:

$$r_{average} = \sum_i f_i x\, r_i \tag{2}$$

where r_i and f_i and indicate the anisotropies of the individual species and their fractional contribution in terms of population, respectively. Importantly, in the above-mentioned example, this relationship is correct only if the different fluorescent species (free or bound) have the same intensity. In this case, the calculation of the fraction of saturation (or the fraction of complexes, F_c) is direct:

$$F_C = \frac{r - r_f}{r_c - r_f} \tag{3}$$

where *r* is the measured *steady-state anisotropy* value, r_f and r_c are the r values corresponding to the free form (at the beginning of the titration curve) and the complexed form (at saturation) of the fluorescent specie. If the intensity value changes along the titration curve, i.e. if $I_f \neq I_C$, it is then necessary to correct Eq. 3 to take into account the ratio of the intensities of the bound and free forms (if not, F_c will be not anymore representative of the fractional populations, but instead, of the fractional intensities, leading to a misinterpretation of the *binding isotherms*):

$$F_C = \frac{r - r_f}{(r - r_f) + (\frac{I_C}{I_f})x(r_C - r)} \tag{4}$$

Note that, when $I_f = I_C$, Eq. (4) is identical to Eq. (3). Another way to apply the correction is to calculate the corrected anisotropy value (A) from the measured r value, according to:

$$A = \frac{\left[\left(\frac{r - r_f}{r_C - r}\right) x \left(\frac{I_f}{I_C}\right) x\, r_C\right] + r_f}{1 + \left[\left(\frac{r - r_f}{r_C - r}\right) x \left(\frac{I_f}{I_C}\right)\right]} \tag{4'}$$

2.2. Fluorescence Anisotropy Decays

In most cases, the size of the rotating unit cannot be estimated directly from the steady-state fluorescence anisotropy parameter. Indeed, for macromolecules, using either *intrinsic fluorescence* (tryptophan residues in proteins for example) or *extrinsic covalent labeling* (a small organic fluorophore attached to a terminal DNA end, for example), the fluorescence anisotropy decay is usually multiexponential, accounting for the presence of several types of rotational motions. This can be due first to the shape of the rotating unit which is not purely spherical. Second, even for a spherical molecule, the *anisotropy decay* can be complex because of the simultaneous presence of fast rotational motions (for instance, segmental motions of tryptophan residues in proteins or flexibilities of covalently linked extrinsic fluorophores) and slower rotational motions related to the overall tumbling motion of the whole fluorescently labeled species. Only the last rotational motion is related to the molecular size. In this context, the r value represents an average value accounting for several types of rotational motions, and thus is not necessarily directly related to the molecular size. The analysis of the fluorescence anisotropy decay allows discriminating between the different types of motions. The time-dependent anisotropy decay corresponds to:

$$r(t) = \sum_j r_{0,j} x(\exp(-\frac{t}{\theta_j})) \text{ with } r(0) = \sum_j r_{0,j} \quad (5)$$

r(0) represents the maximal anisotropy value (at t = 0). This value is experimental and cannot be higher than the fundamental anisotropy value r_0 (r_0 itself cannot be higher than 0.4 for 1-photon excitation, a consequence of the photoselection principle). θ_j represents the individual rotational correlation time. The long rotational correlation time (long) can be attributed to the tumbling motion of the whole macromolecule. For a spherical macromolecule, the hydrated molecular volume (V) is related to θ_{long} by the *Stokes-Einstein equation*:

$$\theta_{long} = \frac{\eta V}{kT} \quad (6)$$

where η is the viscosity, T the absolute temperature and k the Boltzmann constant.

Two main families of dedicated instrumentations for measuring the time-dependent anisotropy decay are available, operating either in the frequency or the time domain. Regarding the frequency domain approach, the sample is excited by an amplitude modulated excitation light source (see [10] for more details) while the sample needs to be excited by a pulsed excitation light source in the time domain method. Because the time domain method has been used for the two examples of time-resolved fluorescence anisotropy studies presented in Section 7.4 (*RecQ helicase* and *HIV-1 integrase*, respectively), only the corresponding instrumental set-up is presented below.

The time-resolved fluorescence anisotropy was obtained from the two polarized emission decays $I_V(t)$ and $I_H(t)$ using the time-correlated single-photon counting technique [10]. I_V corresponds to the emission intensity when both the excitation and emission polarizers are vertical while I_H corresponds to the emission intensity when *excitation polarizer* is vertical and *emission polarizer* is horizontal. The excitation light pulse source was a 100-fs pulse, 80-MHz mode-locked Mai Tai Ti:Sapphire tunable laser (Spectra Physics, CA, USA) associated with a second or a third harmonic generator for fluorescein (excitation wavelength, 490 nm) or tryptophan (excitation wavelength, 299 nm) fluorescence, respectively. The repetition of the laser was set down to 8 MHz (using a pulse picker), and to ensure the single-photon counting condition, the final counting rate never exceeded 80 kHz. The excitation light pulse was triggered by a Hamamatsu photodiode (S4753). The fluorescence emission was detected through a monochromator (SpectraPro 150, ARC) set at the appropriate wavelength (emission wavelength was 350 or 530 nm for tryptophan or fluorescein, respectively; $\Delta\lambda = 15$ nm). A time-correlated single-photon counting SPC-430 card (Becker-Hickl GmbH) was used for the acquisition. The time scaling was 11 ps per channel and 4096 channels were used. The two polarized components of the emission decay were collected until the total count of the Iv component reached 15-20 million. The microcell (volume 50 μL) was thermostated with a Haake type-F3 circulating bath. The data were analyzed by the Quantified Maximum Entropy Method (MEM) [11] in order to determine the distributions of lifetimes (i) or rotational correlation times (θ_j), according to Eqs (7) and (8):

$$I_V(t) = \frac{1}{3} x[(\sum_i \alpha_i \exp(-\frac{t}{\tau_i}))(1 + 2\sum_j r_{0,j} \exp(-\frac{t}{\theta_j}))] \quad (7)$$

$$I_H(t) = \frac{1}{3} x[(\sum_i \alpha_i \exp(-\frac{t}{\tau_i}))(1 - \sum_j r_{0,j} \exp(-\frac{t}{\theta_j}))] \quad (8)$$

with $\sum_i \alpha_i = 1$ and $\sum_j r_{0,j} = r(o)$

where α_i represents the population of fluorophores with lifetime τ_i and $r_{0,\,j}$ represents the apparent fundamental anisotropy of the population characterized by the rotational correlation time θ_j. The anisotropy decay r(t) is then calculated from both polarized components by:

$$r(t) = \frac{I_V(t) - G\,I_H(t)}{I_V(t) + 2\,I_H(t)} = \sum_j r_{0,j} x(\exp(-\frac{t}{\theta_j})) \quad (9)$$

where G is a correction factor accounting for the difference in the transmission efficiencies between the horizontally and vertically polarized intensities (mainly due to the monochromator).

3. Simultaneous Monotoring of the DNA-Binding Step and Enzymatic Activity by Steady-State Fluorescence Anisotropy

3.1. HIV Integrase Activity

HIV-1 integrase catalyzes the integration of viral DNA into the host genome ensuring its perpetuation in the host cell [12, 13]. The integration process requires two catalytic steps. During the first step, titled 3'-processing (3'-P), integrase specifically removes the 3'-terminal GT dinucleotide from each viral DNA end. Integrase then transfers both ends into target DNA by a one-step transesterification reaction, resulting in full-site integration. Integrase is necessary and sufficient for catalysis of the 3'-P reaction in vitro, using short-length double-stranded oligodeoxynucleotides (ODN) that mimic one viral long terminal repeat [12]. In the presence of the metallic cofactor Mg^{2+}, the 3'-P reaction is highly specific, in particular the CA sequence preceding the GT dinucleotide cleaved by integrase is strictly required [14].

We have developed an assay based on steady-state fluorescence anisotropy, for monitoring DNA-binding and subsequent 3'-P using the same sample as in [15]. This assay makes possible the separation of binding and catalytic parameters and the study of real-time kinetics.

Figure 1 summarizes the principle underlying the *anisotropy-based assay*. The fraction of released dinucleotides (F_{dinu}) is related to either the direct measurement of Δr during the reaction (real-time assay; Eq. (12)) or the measurement of Δr after disruption of complexes by SDS (fixed-time assay; Eq. (11)). Using a fluorescent probe covalently linked to the GT dinucleotide, this makes it possible to follow DNA binding and dinucleotide release, as both steps strongly influence the molecular size of the fluorescent moiety.

The DNA susbtrate is a 21-mer double-stranded ODN mimicking one end of the HIV-1 DNA (5'-G TGT GGA AAA TCT CTA GCA GT-*F). The GT dinucleotide which is removed by integrase is in bold and F designates the terminal fluorescein (Figure 2A).

For monitoring integrase activity, the ODN is labeled with fluorescein at the 3'-end of the CAGT sequence. The r value is expected to significantly increase upon integrase binding to the ODN (monitored at 25°C). Moreover, this assay allows quantification of the subsequent cleavage reaction (monitored at 37°C) resulting in physical separation of the processed 19/21-mer DNA (first reaction product) and the fluorescently labeled dinucleotide (second reaction product), leading to a decrease in the r value.

Unlike most standard 3'-P assays, which are based on quantification of the processed strand (first reaction product), the anisotropy-based assay monitors the released dinucleotide (second reaction product).

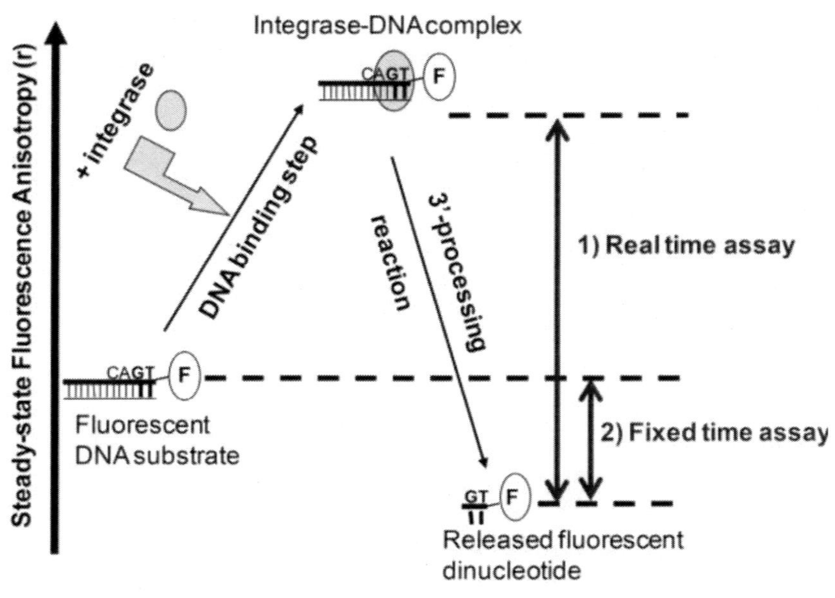

Figure 1. Principle of the fluorescence anisotropy-based integrase assay.

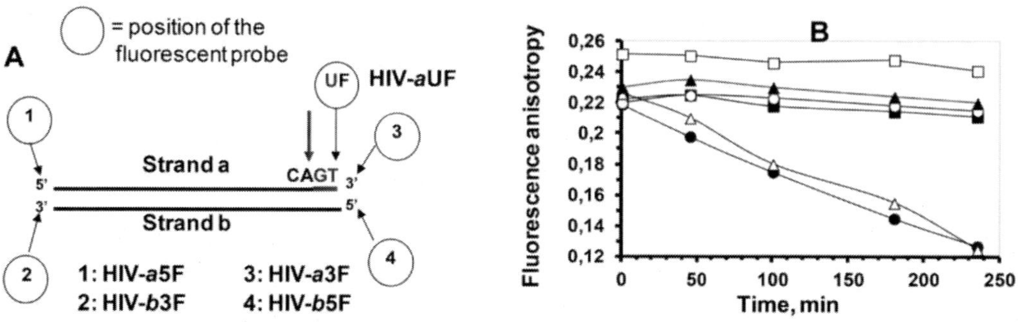

Figure 2. The dependence of the decrease in the r value during 3'-P on the position of fluorescein (F) on the DNA substrate. *(A)* Position of the fluorophore on the various ODNs. F was attached to the 3' or 5' extremity of one DNA strand. 3'-P involves the cleavage of the terminal GT dinucleotide, as indicated by the arrow. *(B)* After the DNA-binding step, the temperature was shifted from 25 to 37°C and r values were recorded. White squares: HIV-a3F, no Mg^{2+}. Black circles: HIV-a3F, 5 mM Mg^{2+}. White circles: HIV-b5F, 5 mM Mg^{2+}. HIV-1 integrase was purified under native conditions, as previously described[16].

We then studied the formation of integrase/DNA complexes at non permissive temperature for catalytic activity (25°C) by incubating the fluorescein-labeled DNA substrate with integrase in a final volume of 200 μl. The fractional saturation function (F_c) can be calculated as follows:

$$F_C = \frac{r - r_{ODN}}{r_{max} - r_{ODN}} \qquad (10)$$

where r_{max} and r_{ODN} correspond to the anisotropies of integrase-bound and free ODN, respectively (no significant change in fluorescence intensity was observed along the titration process).

Following the DNA-binding step, the temperature is raised from 25 to 37°C (permissive temperature for 3'-P) for monitoring the catalytic process. Two independent methods can then be used for quantification of the 3'-P activity. The fraction of dinucleotides released can be determined either after the disruption of all integrase/ODN complexes by the addition of SDS (r_{SDS} in Figure 1 corresponds to the difference in fluorescence anisotropies between unprocessed ODN and free dinucleotide) or from the real-time decrease in r observed during the reaction ($r_{real\text{-}time}$ in Figure 1 corresponds to the difference in fluorescence anisotropies between integrase/ODN complex and free dinucleotide). Typical integrase and DNA concentrations were 120 and 4 nM, respectively. The calculation of the 3'-P activity is then performed as follow:

(i) In fixed-time experiments, the reaction is stopped by adding SDS (0.25% final), disrupting all the integrase/ODN complexes in the sample. In this case, the solution contains two molecular species (covalently attached to the terminal fluorescein): the non-processed ODN and the dinucleotide released by the cleavage reaction. The fraction of dinucleotides released is then given by:

$$F_{dinu} = \frac{[GT]}{[ODN]_{total}} = \frac{r_{NP} - r}{r_{NP} - r_{dinu}} \tag{11}$$

where r_{NP} and r_{dinu} are the anisotropy values corresponding to non-processed 21-mer double-stranded ODN and GT dinucleotide, respectively (fluorescence intensity did not change significantly during the reaction). The r_{dinu} value was obtained independently using the 5'-GT-3'F dinucleotide.

(ii) An additional fluorescent population corresponding to the integrase/unprocessed ODN complex, is present in the sample real-time conditions. In this case, F_{dinu} is calculated using:

$$F_{dinu} = \frac{r_{t=0} - r}{r_{max} - r_{dinu}} \tag{12}$$

where r_{max} is the characteristic r value for optimal activity and $r_{t=0}$ is the r value obtained at the end of the DNA-binding step. The 3'-P activity calculated by using Eq. (11) and (12) is referred to as $activity_{SDS}$ and $activity_{real\text{-}time}$, respectively.

Regarding the DNA-binding step, the anisotropy-based assay was extensively used for characterizing the integrase/DNA complex formation (affinity, cooperativity...) [7, 17-18] as well as in pharmacological studies in order to measure IC50 values (inhibition concentration 50%) characterizing competitive inhibitors [8, 19]. The DNA-binding assay alone works, regardless of the position of fluorescein on the DNA (5' or 3' extremity of the *a* or *b* strand). By contrast, this position plays a crucial role for monitoring the 3'-P activity.

We tested several 21-mer DNA substrates labeled with fluorescein at one end, in all possible combinations (5' or 3' extremity of the *a* or *b* strand), or labeled on the 2'-amino group of a 3'-terminal 2'-aminouridine (Figure 2A). The free ODNs were characterized by r values of 0.045 - 0.105 at 37°C, depending on the sequence and location of fluorescein (r = 0.02 for the fluorescein-labeled dinucleotide). The binding of integrase to the ODN led to a significant increase in r value (>0.2).

The reaction was started by shifting the sample to a temperature of 37°C, and the initial r values obtained therefore correspond to $r_{t=0}$. In the presence of the divalent cationic cofactor Mg^{2+}, all tested ODNs gave similar $r_{t=0}$ values (0.22), indicating that the fluorescein position had no effect on DNA binding.

Experiments carried out in the absence of Mg^{2+} gave larger r values (0.25), most likely due to protein aggregation which is favored by an absence of Mg^{2+} [20]. The time-dependency of r under real-time conditions at 37°C shows that only the ODNs HIV-a3F and HIV-aUF, with the fluorescein directly attached to the small GT reaction product, displayed significant Mg-dependent decreases in r (Figure 2B).

In contrast, HIV-a5F, HIV-b5F and HIV-b3F displayed no change in r value. This result indicates that r is not sensitive enough to differentiate between the DNA substrate (ODN 21/21) and the first reaction product (ODN 19/21). This result also suggests that, under real-time conditions, the processed DNA product remains tightly bound to integrase after 3'-P while the terminal GT dinucleotide is normally released from integrase/ODN complexes after 3'-P. Fluorescence anisotropy is then reliable for monitoring the 3'-P reaction, but only if fluorescein is linked to the terminal GT.

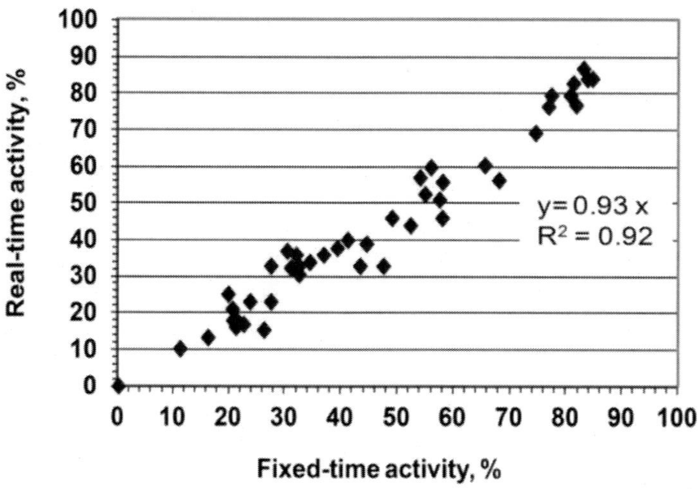

Figure 3. Correlation between $activity_{real\text{-}time}$ and $activity_{SDS}$ values.

The 3'-P reaction is strongly sequence-dependent [14]. The endonucleolytic site includes the conserved CA dinucleotide immediately preceding the GT dinucleotide. We assessed sequence specificity, using the anisotropy-based assay in terms of DNA-binding and 3'-P activity. We used several ODNs with identical lengths (21-mer): HIV-a3F corresponds to the cognate sequence; HIV-GTGT-a3F is a variant in which the 3'-terminal CAGT sequence is replaced by GTGT. We also tested three non-specific sequences, NS-TTCC-a3F, NS-CACC-a3F and NS-CAGT-a3F.

The DNA-binding step was not influenced by sequence since the final r values obtained after DNA binding were similar for all ODNs, irrespective of the sequence context. In contrast, sequence had a major effect on r value in catalysis conditions.

Only HIV-a3F gave a large decrease in r (not shown). This decrease was abolished by replacement of the 3'-terminal sequence CAGT by GTGT or the use of non-specific sequences, even with the CAGT sequence at the 3' end. Altogether, our data show that the decrease in r during incubation at 37°C is related to specific 3'-P activity.

Moreover, $activity_{real\text{-}time}$ (calculated from Eq. (12)) and $activity_{SDS}$ (calculated from Eq. (11)) were well correlated (Figure 3) (the r value was used to calculate the fraction of released dinucleotides under real-time conditions or after adding SDS, giving $activity_{real\text{-}time}$ or $activity_{SDS}$, respectively), demonstrating that both approaches are reliable for quantification and that the decrease in r under real-time conditions is due exclusively to the release of GT from the integrase/ODN complex, rather than another phenomenon, such as dissociation of the enzyme from the DNA substrate or product. Finally, the comparison of the 3'-P activity of non-fluorescent and fluorescein-labeled ODNs by gel electrophoresis confirmed that fluorescein has no influence on activity (not shown in the figure). This result sharply contrasts with that obtained using a TAMRA-labeled ODN for which the 3'-P activity was significantly affected [15].

In conclusion, the anisotropy-based assay is reliable as r decreased only in the presence of Mg^{2+} with the cognate HIV-1 sequence and only when fluorescein was attached to the GT 3'-end. The additivity law of anisotropy renders the quantification of activity easy and both real-time and fixed-time studies can be carried out. Moreover, it has been shown that the anisotropy-based method is also reliable for inhibitor characterization [7-8, 17-19].

3.2. DNA Unwinding Activity of Escherichia Coli RecQ Helicase

Helicases are molecular motor enzymes that unwind and translocate nucleic acids in an ATP hydrolysis-dependent manner [21-22]. They play essential roles in many important biological processes such as DNA replication, repair, recombination, transcription, splicing, and translation. *RecQ DNA helicases* constitute a ubiquitous family of helicases playing a key role in maintaining genome stability in a wide range of organisms, from bacteria to higher eukaryotes and are involved in many of the processes of DNA metabolism [23-25].

In humans, defects in RecQ family helicases give rise to several syndromes (Bloom, Werner, Rothmund-Thomson), characterized by genomic instability and susceptibility to cancer [26]. The E. coli RecQ helicase is the prototype enzyme of this family and was used to develop the *anisotropy-based helicase assay* [27-28].

Despite extensive biochemical and structural studies of helicases, the mechanism of DNA unwinding remains obscure and several enzymatic features of the RecQ family helicases are not clearly understood. A full description of the detailed mechanism by which helicase catalyzes duplex DNA unwinding requires a deeper understanding of the enzyme. The gold-standard gel assay has provided fundamental insights in the past into the mechanism of action of several helicases.

However, several mechanistic questions remain obscure and standard gel-electrophoresis approaches are limited, in particular for kinetic studies, because these approaches correspond to fixed-time methods. For this reason, several continuous fluorometric assays working in real time have been developed to measure the unwinding of duplex nucleic acids by helicase [30-33]. Recently, we have developed a fluorescence anisotropy-based assay for monitoring simultaneously DNA-binding and DNA unwinding activity of helicase in the same sample and in real time [28]. As shown above for HIV-1 integrase, this assay makes possible the separation of binding and catalytic parameters of RecQ helicases and the study of real-time kinetics.

Using a fluorescent probe (such as fluorescein) covalently linked to the released DNA strand, which is concomitant to DNA unwinding, it is possible to measure DNA-binding and subsequent unwinding, as both steps strongly influence the molecular size of the fluorescent moiety. The design of the DNA substrates depends on the nature and specific enzymatic properties of the helicase to be studied.

In the case of E. coli RecQ helicase, even though blunt DNA substrates may be unwound at high protein concentration, it is preferable that the DNA substrate harbors at least one single-stranded (ss) 3'-tail to improve the unwinding activity [34]. For studying E. coli RecQ helicase, we then used a 13-66 base pairs duplex ODN with a 3'-ss DNA tail varying between 7 and 30 nucleotides (nt). The short strand is labeled by fluorescein at the 3'-end (Figure 4A) for two reasons:

(i) E. coli RecQ helicase preferentially binds to the ss/ds DNA junction with the help of the 3'ss-tail and translocates along the ODN with the 3'->5' polarity [35]. At the end of the unwinding reaction, the short strand is released from the helicase/ODN complex while the helicase is still bound to the long strand. Importantly, when the long strand is fluorescein-labeled, no change in the anisotropy level can be detected. Moreover, the size of the short strand must be less than 20 nt. Indeed, the decrease in anisotropy accompanying unwinding activity may be very small if helicase has the possibility to re-bind the fluorescently labeled strand, complicating analysis (see below). In the case of E. coli RecQ helicase, this particular problem can be circumvented by using short fluorescently labeled strands (< 20-mer) which display very low affinities for helicase.

(ii) It is important to avoid the possibility that the fluorescein group may spatially hinder DNA-binding by helicase at the ss/ds DNA junction and, consequently, may inhibit unwinding initiation. Thus, the fluorescein labeling at the 3'-end is preferred over the 5'-end.

Figure 4A shows a schematic representation of the fluorescently labeled DNA strand (non-loading strand) used to monitor helicase-mediated DNA unwinding. The fluorescein-labeled double-stranded ODN obtained by hybridization of this strand with a complementary unlabeled DNA strand (loading strand) is characterized by a low r value, typically 0.065 (Figure 4B). This value significantly increased upon addition of helicase, indicating helicase/ODN complex formation (r = 0.17). This high r value indicates that the binding of helicase to DNA can be observed and quantified independently, prior to DNA unwinding. Typical helicase and DNA concentrations used in this assays are 30 and 5 nM, respectively.

The unwinding process can be initiated by addition of ATP (1mM). The progress of the reaction was clearly evidenced by a significant decrease in the r value, indicating unwinding of the duplex DNA and concomitant release of the fluorescently labeled strand from helicase. The r value of the released ss ODN (0.035) was found to be even lower than the duplex DNA substrate. The lower limit r value corresponds exactly to the value found for a ss ODN of similar size in the absence of any protein, thus indicative of the complete DNA unwinding.

From a quantification point of view, the determination of the fractional saturation function (related to the DNA-binding step) as well as the calculation of the unwinding activity are based on similar equations than those used for HIV-1 integrase in the previous section (Eq. (10) and (12)).

Again, due to the additivity law of anisotropy, the r value can be directly used to measure the binding of helicase to DNA or the unwinding rate, only if the emission intensity of the fluorescent probe does not change significantly before and after the binding of helicase to the ODN or during the unwinding process. This means that the fluorescence of the probe remains unchanged (i) when comparing the free DNA and the complex and (ii) when comparing the double- and the single-stranded labeled DNA. In the case of E. coli RecQ helicase, the fluorescence intensity remained constant along the binding and the unwinding processes. If the intensity varies (either in the case of another helicase or using another fluorophore) during these processes, the r value must be corrected according to Eq. (4) or (4').

Importantly, helicase displays a greater affinity for long ODNs (typically longer than 20 nt) [27]. The fluorescence anisotropy-based assay fails to monitor DNA unwinding when the fluorophore is linked to a strand longer than 20 nt.

As shown in Figure 5A, a modest decrease or no decrease occurs when the fluorescently labeled short strand is 36 or 40 nt long, respectively. This result shows that helicase (present in excess over DNA concentration in this assay) can bind efficiently to the released ODN and, consequently, underestimates both the anisotropy decrease and the unwinding activity. In contrast, using a 14 nt long fluorescently labeled short strand, the re-binding phenomenon cannot occur (ODNs shorter than 20-mer cannot re-bind to helicase) and the whole unwinding process can be easily monitored by recording the anisotropy decrease (Figure 5A).

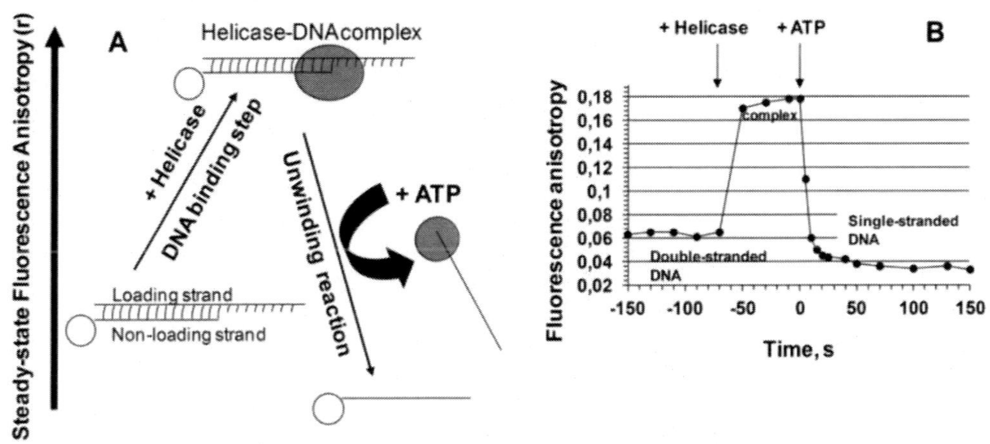

Figure 4. Principle of helicase activity assay by anisotropy. *(A)* Schematic illustration of the fluorescence anisotropy-based helicase assay using a fuorescein (open circle)-labelled ODN substrate. This ODN alone leads to a low r value which further increases upon the binding of helicase (closed circle) to the ODN. During unwinding, the fluorescent strand is released from the helicase/ODN complex, decreasing the r value. *(B)* DNA unwinding activity monitored by fluorescence anisotropy.

When using a long DNA substrate containing a fluorescein-labeled strand of 36 nt, it is possible to increase the amplitude of the reaction by decreasing the amount of helicases in the assay or by adding single-stranded DNA trap (Figure 5B). These two procedures may be used to minimize the re-binding process to the released fluorescent strand that originates in the excess of free proteins over the DNA substrate in this assay.

In conclusion, it is strongly recommended to optimize the FA study of helicase activity by using double-stranded DNA substrates containing a long and a short strand with the fluorescent probe attached to the short strand. Typically, the fluorescently labeled ODN must be shorter than 20 nt. Under such optimized conditions, FA is particularly suitable for monitoring helicase activity as it does not require any separation of the ss DNA product from the ds DNA substrate by gel electrophoresis and allow real-time kinetic studies.

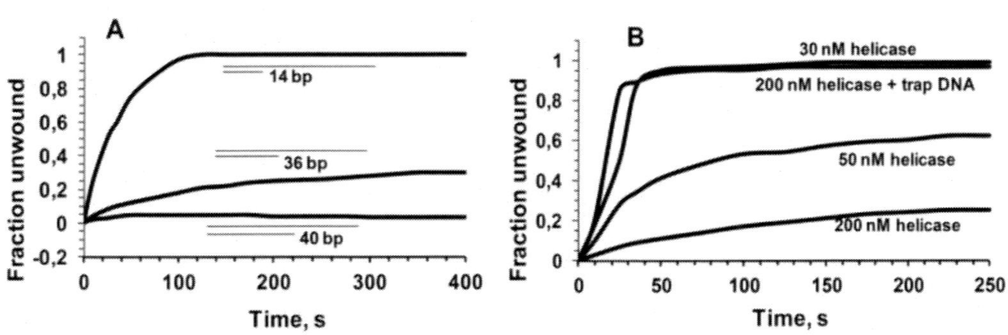

Figure 5. Influence of the length of the fuorescein-labeled DNA strand on fluorescence anisotropy (adapted from [28]). *(A)* DNA unwinding of substrates formed with fluorescein-labeled ss ODN of varying lengths. [Helicase] = 40 nM; [DNA] = 5 nM. *(B)* A DNA substrate containing a 36-bp duplex region was incubated with 30, 50 or 200 nM helicase. The unwinding reaction was initiated by adding either 1 mM ATP alone or 1 mM ATP + 2 mM ss DNA trap.

The FA approach is compatible with single-turnover conditions (ensuring initial saturation of the DNA substrate) and it is particularly easy to get information on the DNA-binding process. The FA method presented here has been validated using mainly the E.coli RecQ helicase.

It can probably be used for studying any other helicase although it is crucial to adapt the unwinding assay for each helicase (nature of the buffer, nature of the DNA substrate (size, blunt or with a 3'/5' ss DNA tail...)). This FA assay is suitable for studying the mechanism by which helicase unwinds duplex DNA (for instance regarding the polarity of the reaction, the identity of the strand which remains associated to the enzyme during unwinding...) as well as it could be suitable for screening helicase inhibitors [36-37].

4. TIME-RESOLVED FLUORESCENCE ANISOTROPY EXPERIMENTS

More precise details about the size of molecules require *time-resolved fluorescence anisotropy* experiments. For a protein alone (free in solution), the time-dependent fluorescence anisotropy decay can be studied via the fluorescence of tryptophan (trp) residues as exemplified below with E. coli RecQ helicase. Through the example of HIV- integrase, we will see that the oligomeric status of the DNA-bound protein can be studied using the fluorescence of a small organic fluorescent probe linked to the DNA substrate.

4.1. Escherichia Coli RecQ Helicase

The fluorescence anisotropy decay of E. coli RecQ helicase (free in solution) was studied using intrinsic trp fluorescence [29]. E. coli RecQ helicase contains 11 trp residues distributed along the sequence of the entire protein. The analysis of the time-resolved fluorescence anisotropy data by MEM displays the distribution of rotational correlation times (θ) related to the *hydrodynamic volume* (Eq. (6)). An example of such a distribution (corresponding to [helicase] = 1 μM) is shown in Figure 6. The short rotational correlation times, typically related to local dynamics and segmental flexibilities of trp (below 1 ns), and the lifetime distribution (also calculated by the MEM analysis: $\tau_1 = 0.37 \pm 0.04$ ns (39%); $\tau_2 = 1.6 \pm 0.1$ ns (33%); $\tau_3 = 4.05 \pm 0.2$ ns (21%); $\tau_4 = 7.35 \pm 0.25$ ns (7%); $\tau_{average} = 2.05$ ns) are not represented here.

A 1 μM-RecQ helicase preparation displayed a long correlation time of about 50 ns at 20°C, consistent with a lower-order of *oligomerization*, mainly a monomeric form. Indeed, serum albumin, characterized by a similar mass (66-kDa) than E. coli RecQ helicase, has a correlation time in the same range (46 ns) [10], and the dimeric form of γ-interferon (34 kDa), half of the mass of E. coli RecQ helicase, has a correlation time divided by 2 (27 ns) [38]. Another peak was detected at about 4.9 ns. This peak corresponds most likely to a domain motion that is independent of the global tumbling of the whole protein. E. coli RecQ helicase displayed a similar long rotational correlation time between 0.5 and 20 μM indicating that the monomeric state is stable and concentration-dependent oligomerization does not occur in this concentration range. The monomer is also stable regarding the temperature. From 4 to 37°C, the long correlation time decreased from 90 ± 13 to 38 ± 5 ns.

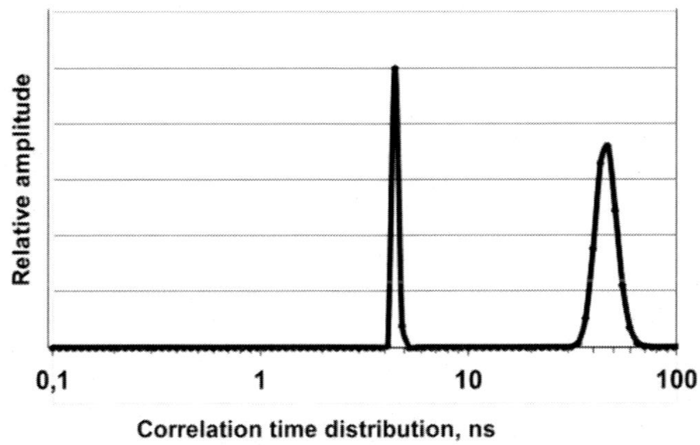

Figure 6. Rotational correlation time distribution of DNA-free E. coli RecQ helicase.

To take into account the modulation of correlation times by external factors such as the temperature and viscosity, all the data were normalized at 20°C according to the *Stokes-Einstein equation* (Eq. (6)). After normalization, the rotational correlation times did not change significantly as a function of temperature, showing that E. coli RecQ helicase is mainly monomeric in the 4 to 37°C temperature range. In conclusion, time-resolved fluorescence anisotropy experiments established that E. coli RecQ helicase exists as a monomer up to a concentration of at least 20 μM and that this monomer is stable from 4 to 37 °C. In subsequent functional studies, it was shown that E. coli RecQ helicase functions as a monomeric form despite the fact that large helicase/DNA complexes (larger than expected assuming a monomer/DNA complex) can be observed by mixing helicase and DNA [33-34, 39]. This is due to the ability of multiple monomers to functionally cooperate for unwinding (several monomers - the exact number depends on the length of the DNA substrate - may align on the same DNA and function in a cooperative manner).

4.2. HIV Integrase

The correlation time distribution of HIV-integrase, as studied using fluorescence anisotropy decay of trp, is more complex than RecQ helicase. HIV-1 integrase is characterized by a mixture of monomeric, dimeric, tetrameric and aggregative forms, depending on several factor such as the protein concentration, temperature, the purification protocol [20, 40-43]. The behaviour of another *retroviral integrase* (from foamy virus (FV)) is easier to undestand as it is more soluble than HIV-1 integrase [17, 44]. This integrase is characterized by a monomer:dimer equilibrium in the micromolar range [17]. The oligomeric status of integrase responsible for the 3'-P reaction was studied using fluorescein-labeled ODNs (similar to those presented in Section 7.3.1). The results are more consistent when comparing the self-association properties of HIV-1 and FV integrases bound to their respective DNA substrate, regardless of the initial oligomeric state (corresponding to DNA-free proteins). The results presented below were obtained with HIV-1 integrase [15], but similar results were obtained with FV integrase [17].

As underlined in previous sections, under site saturation conditions, the steady-state r value only reflects the molecular size of the complexes in a purely indicative manner.

However, this value is not directly related to the molecular size. Indeed, steady-state fluorescence anisotropy corresponds to an average parameter accounting for a mixture of fast and slow rotational motions. This statement is well illustrated by the steady-state r values characterizing free ODNs which are comprised between 0.045 and 0.105 (see Section 7.3.1), even though all the tested double-stranded ODNs have a length of 21 base pairs. This apparent discrepancy appears to be related to the location of fluorescein (5' versus 3') and the sequence adjacent to fluorescein. Indeed, all the fluorescein-labeled ODNs on the 5' DNA end displayed a low r value (0.045). Among the fluorescein-labeled ODNs on the 3' DNA end, only those with a 3'-terminal GT were characterized by a high r value (0.105). The time-resolved fluorescence anisotropy study indicates that all the 21-mer double-stranded ODNs displayed similar correlation time values (in particular, the long correlation times are comprised between 8.6 and 9.6 ns) but with different $r_{0,j}/r_0$ ratio. The ODNs with a 3'-terminal GT displayed a higher r_{slow}/r_0 ratio (0.5) than other ODNs of similar length (r_{slow}/r_0 = 0.125), indicating that the fractional amplitude of fast motions is lower (flexibility of the fluorophore is lower) in the presence of a 3'-terminal GT, accounting for a higher r value despite similarities in the rotational motion of the whole molecule when comparing the different 21-mer ODNs.

The molecular sizes of active integrase/DNA complexes were studied in greater detail by studying representative samples corresponding to different r values by *time-resolved fluorescence anisotropy*. This approach allows the determination of the long rotational correlation times (θ_{long}). The 21-mer ODN was the most suitable substrate for studies of integrase/DNA complexes as longer free ODNs are characterized by large θ_{long} values which could complicate the further analysis of complexes. The time-dependent anisotropy decays of different samples corresponding to various protein:DNA ratio were then studied by using the fluorescence of fluorescein. The 3'-P activity plotted as a function of the protein:DNA ratio (or as a function of the r value measured after integrase/DNA complex formation) displays a bell-shaped behavior (Figure 7), with the increasing phase corresponding to DNA site saturation.

In order to study the molecular size of the active complexes, we restricted our analysis to samples corresponding to the peak and to the decreasing phase of the curve. The distribution in the short θ domain was not dependent on the protein:DNA ratio. In contrast, the characteristic θ_{long} of 8.6 ns found for free DNA was replaced by a $\theta_{long} \geq 37.8$ ns (values obtained at 25°C), depending on the ratio (Figure 7). The sample corresponding to the peak of the bell-shaped curve was rather mono-disperse, as suggested by the small difference between the maximum and the center of gravity of the peak in the θ distribution. The complexes are characterized by a θ_{long} value of 38 ns, compatible with hydrodynamic properties of a dimeric form of integrase on the DNA [15]. The shift of the center of gravity toward longer θ values probably results from the presence of small amounts of aggregated forms of integrase on DNA. Further increases in the protein: DNA ratio displace the distribution toward longer θ values, suggesting that the less active samples - corresponding to the highest r value (0.272) - consist essentially of aggregated forms of integrase on DNA, as shown by the broad distribution above 100 ns, while the catalytically competent complexes correspond to a dimeric protein bound to its DNA substrate.

It is important to note that the determination (and its accuracy) of the absolute molecular size via the measurement of rotational correlation times is limited for large macromolecules (or complexes). This limitation is related to the difference in the values of the lifetime and the correlation time. The accuracy in the determination of the long correlation time is itself related to the possibility to discriminate between $I_V(t)$ and $I_H(t)$ at a time which can be much longer compared to the fluorescence lifetime and then is related to the photon counting statistics. For instance, using trp fluorescence (characterized in HIV-1 integrase by an average lifetime value of 2.3 ns), simulation analysis shows that the recovery of a value (up to 80 ns) is satisfying with a reasonable accuracy with a "high" counting statistics (*i.e.* 12×10^6 and 16×10^6 counts for $I_V(t)$ and $I_H(t)$, respectively) [20]. This limit is θ = 40 ns with a "low" counting statistics (12×10^5 and 16×10^5 counts for $I_V(t)$ and $I_H(t)$, respectively). This means that, for example in the case of HIV-1 integrase (32-kDa), it is possible to discriminate between monomeric, dimeric and tetrameric forms (characterized by θ_{long} = 20, 40 and 80 ns, respectively) but higher values (> 80 ns) cannot differentiate higher-order oligomeric forms (such as hexameric or octameric forms for instance) and aggregates.

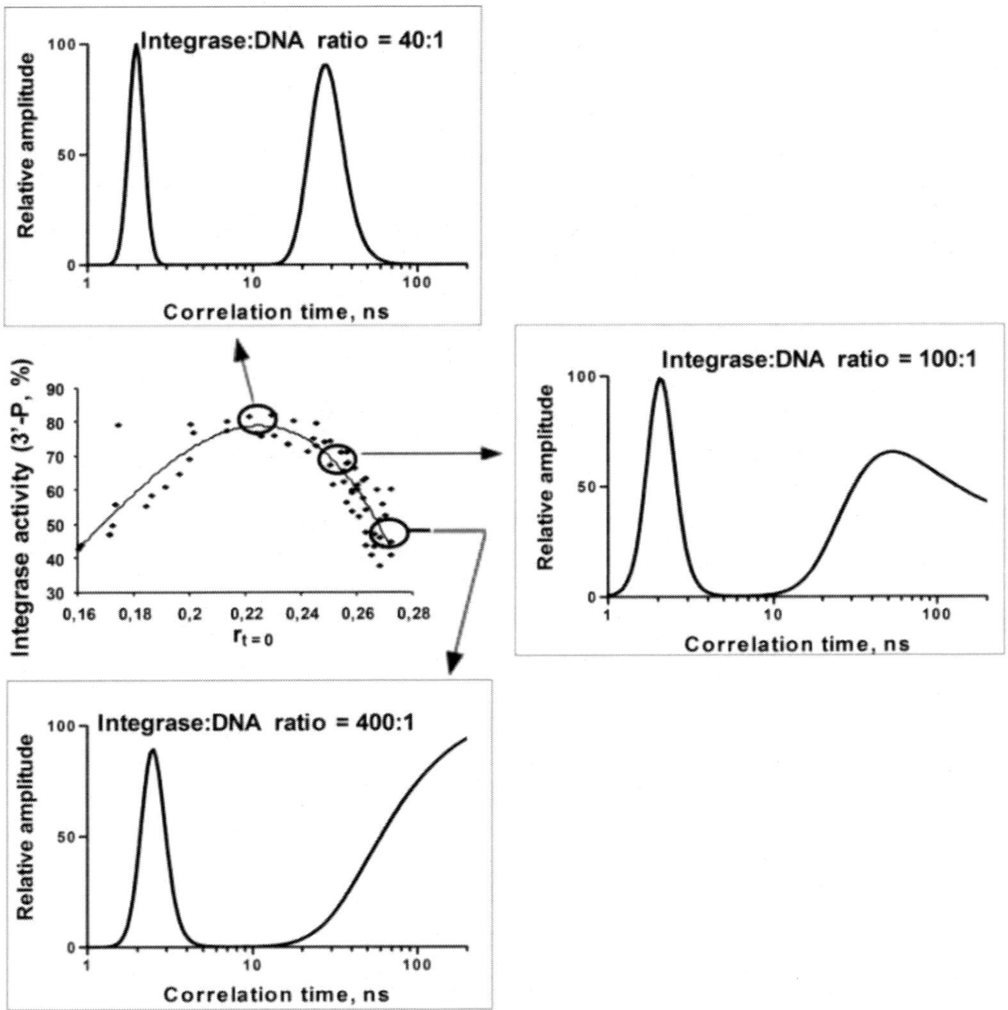

Figure 7. Correlation time distributions of representative samples of integrase/DNA complexes obtained at different protein:DNA ratio. 3'-P was plotted as a function of the r value measured after the DNA-binding step.

This aspect represents the main limitation of the fluorescence anisotropy approach for the determination of the absolute molecular size when the associated correlation time is much longer compared to the lifetime value. Such a limitation does not exist when using fluorescence correlation spectroscopy (FCS) which is based on the determination of the translational diffusion [17, 45]. However, it is important to note that the *rotational diffusion* is more sensitive to the relative change in size compared to the *translational diffusion* and therefore is more adapted for studying the interactions between molecules, especially when the two molecular partners have similar size. This is because the rotational diffusion is related to the third power of the hydrodynamic radius while the translational diffusion is directly related to the first power of the radius [10].

CONCLUSION

Through two enzymes using DNA as a substrate, HIV-1 integrase and RecQ helicase, we have shown that steady-state fluorescence anisotropy is well suitable for measuring both protein/DNA interactions and the subsequent catalytic activities, 3'-processing and DNA unwinding, respectively. Based on a relative change in size, the fluorescence anisotropy approach allows equilibrium and kinetic studies from a quantitative point of view, due to the additivity law of anisotropy, at very low molecule concentrations - in the low nanomolar range - due to the sensitivity of fluorescence.

For the two above-mentioned proteins, it was possible to monitor, first, their binding to DNA and, then, to perform real-time measurements of their enzymatic activities, using the same sample. This approach could be applied to many enzymatic systems. However, for an accurate measurement of the catalytic activity, the steady-state fluorescence anisotropy approach requires a significant change in the size of the fluorescent moiety during catalysis. It is important to note that, in most cases, no information regarding the stoichiometry can be directly obtained from steady-state fluorescence anisotropy measurements.

This is due to the inherent flexibilities of macromolecules accounting for the superimposition of several depolarization modes such as those related to fast rotational motions (local flexibilities, segmental motions) and those related to slow tumbling motions. More precise information can be obtained by performing time-resolved fluorescence anisotropy measurements.

These studies allow the determination of the long correlation time characterizing the hydrodynamic volume. This correlation time, not the steady-state anisotropy parameter which is a composite parameter, is directly related to the size. The oligomeric status of proteins can be then studied by the time-resolved approach. Combined together, the steady-state and the time-resolved fluorescence anisotropy approaches provide deep insights into the functional macromolecules and fundamental mechanisms by which they perform their functions.

REFERENCES

[1] S. Debnath, S. Chatterjee, M. Arif, T. K. Kundu, and S. Roy, *J. Biol. Chem.* 286, 25076 (2011).

[2] B. T. Glaser, J. P. Malerich, S. J. Duellman, J. Fong, C. Hutson, R. M. Fine, B. Keblansky, M. J. Tanga, and P.B. Madrid, *J. Biomol. Screening*, 16, 230 (2011).

[3] J. van Dieck, T. Brandt, D. P. Teufel, D. B. Veprintsev, A. C. Joerger, and A. R. Fersht, *Oncogene* 29, 2024 (2010).

[4] J. S. Yang, J. Xi, Z. H. Zhuang, and S. J. Benkovic, *J. Biol. Chem.* 280, 25416 (2005).

[5] R. L. Weinberg, D. B. Veprintsev, and A.R. Fersht, *J. Mol. Biol.* 341, 1145 (2004).

[6] O. Grillo, M. P. Brown, and C. A. Royer, *J. Mol. Biol.* 287, 539 (1999).

[7] K. Carayon, H. Leh, E. Henry, F. Simon, JF. Mouscadet, and E. Deprez, *Nucleic Acids Research.* 38, 3692 (2010).

[8] Levin, H. Benyamini, Z. Hayouka, A. Friedler, and A. Loyter, *Febs Journal* 278, 316 (2011).

[9] R. P. Ghosh, R. A. Horowitz-Scherer, T. Nikitina, L. S. Shlyakhtenko, and C. L. Woodcock, *Mol. Cell. Biol.* 30, 4656 (2010).

[10] J. R. Lakowics, in Principles of Fluorescence Spectroscopy, 3rd Ed., *Springer-Verlag New York Inc. New York* (2006).

[11] JC. Brochon, *Methods Enzymol.* 240, 262 (1994).

[12] O. Delelis, K. Carayon, A. Saib, E. Deprez, and JF. Mouscadet, *Retrovirology*, 5, 114 (2008).

[13] N. Neamati, in *HIV-1 Integrase, Mechanisms and Inhibitor Design*, *John Wiley and Sons, Inc. publication* (2011)

[14] D. Esposito and R. Craigie, *EMBO J.* 17, 5832 (1998).

[15] E. Guiot, K. Carayon, O. Delelis, F. Simon, P. Tauc, E. Zubin, M. Gottikh, JF. Mouscadet, JC. Brochon, and E. Deprez, *J. Biol. Chem.* 281, 22707 (2006).

[16] H. Leh, P. Brodin, J. Bischerour, E. Deprez, P. Tauc, JC. Brochon, E. LeCam, D. Coulaud, C. Auclair, and JF. Mouscadet, *Biochemistry.* 39, 9285 (2000).

[17] O. Delelis, K. Carayon, E. Guiot, H. Leh, P. Tauc, JC. Brochon, JF. Mouscadet, and E. Deprez, *J. Biol. Chem.* 283, 27838 (2008).

[18] M. Pinskaya, E. Romanova, E. Volkov, E. Deprez, H. Leh, JC. Brochon, JF. Mouscadet, and M.Gottikh, *Biochemistry.* 43, 8735 (2004).

[19] E. Deprez, S. Barbe, M. Kolaski, H. Leh, F. Zouhiri, C. Auclair, JC. Brochon, M. Le Bret, and JF. Mouscadet, *Mol.Pharmacol.*65, 85 (2004).

[20] E. Deprez, P. Tauc, H. Leh, JF. Mouscadet, C. Auclair, and JC. Brochon, *Biochemistry* 39, 9275 (2000).

[21] S. S. Patel and I. Donmez, *J. Biol. Chem.*281, 18265 (2006).

[22] M. R. Singleton, M. S. Dillingham, and D. B. Wigley, *Annu.Rev.Biochem.* 76, 23 (2007).

[23] M. P. Killoran and J. L. Keck, *Nucleic Acids Res.*34, 4098 (2006).

[24] R. M. Jr Brosh and V. A. Bohr, *Nucleic Acids Res.*35, 7527 (2007).

[25] S. Sharma and R. M. Jr. Brosh, *Cell Cycle.*7, 989 (2008).

[26] R. J. Monnat, *Seminars in Cancer Biology* 20, 329 (2010).

[27] S. X. Dou, P. Y. Wang, H. Q. Xu, and X. G. Xi, *J. Biol. Chem.* 279, 6354 (2004).

[28] H. Q. Xu, A. H. Zhang, C. Auclair, and X. G. Xi, *Nucleic Acids Res.*31, e70(2003).
[29] H. Q. Xu, E. Deprez, A. H. Zhang, P. Tauc, M. M. Ladjimi, JC. Brochon, C. Auclair, and X. G. Xi, *J. Biol. Chem.*, 278, 34925 (2003).
[30] K. P. Bjornson, M. Amaratunga, K. J. M. Moore, and T. M. Lohman, *Biochemistry*.33, 14306 (1994).
[31] K. D. Raney, L. C. Sowers, D. P. Millar, and S. J. Benkovic, *Proc. Natl. Acad. Sci. U.S.A.* 91, 6644 (1994).
[32] P. Houston and T. Kodadek, *Proc. Natl. Acad. Sci. U.S.A.*91, 5471 (1994).
[33] X. D. Zhang, S. X. Dou, P. Xie, J. S. Hu, P. Y. Wang, and X. G. Xi, *J.Biol.Chem.*281, 12655 (2006).
[34] N. Li, E. Henry, E. Guiot, P. Rigolet, JC. Brochon, X.G. Xi, and E. Deprez, *J. Biol. Chem.* 285, 6922 (2010).
[35] K. Umezu, K. Nakayama, and H. Nakayama, *Proc. Natl. Acad. Sci. U.S.A.* 87, 5363 (1990).
[36] M. Aggarwal, J. A. Sommers, R. H. Shoemaker, and R. M. Brosh, *Proc. Natl. Acad. Sci. U.S.A.* 108, 1525 (2011).
[37] H. Tani, N. Akimitsu, O. Fujita, Y. Matsuda, R. Miyata, S. Tsuneda, M. Igarashi, Y. Sekiguchi, and N. Noda, *Biochemical and Biophysical Research Communications*. 379, 1054 (2009).
[38] J. C. Brochon, P. Tauc, F. Merola, and B. M. Schoot, *Analytical Chemistry* 65, 1028- (1993).
[39] D. A. Bernstein, M. C. Zittel, and J. L. Keck, *EMBO J.* 22, 4910 (2003).
[40] E. Deprez, P. Tauc, H. Leh, F. Mouscadet, C. Auclair, M. E. Hawkins, and JC. Brochon, *Proc. Natl. Acad. Sci. U.S.A.* 98, 10090 (2001).
[41] Faure, C. Calmels, C. Desjobert, M. Castroviejo, A. Caumont-Sarcos, L. Tarrago-Litvak, S. Litvak, and V. Parissi, *Nucleic Acids Res.* 33, 977 (2005).
[42] S. Baranova, F. V. Tuzikov, O. D. Zakharova, N. A. Tuzikova, C. Calmels, S. Litvak, L. Tarrago-Litvak, V. Parissi, and G. A. Nevinsky, *Nucleic Acids Res.*35, 975 (2007).
[43] S. P. Lee, J. Xiao, J. R. Knutson, M. S. Lewis, and M. K. Han, *Biochemistry* 36, 173 (1997).
[44] S. Hare, S. S. Gupta, E. Valkov, A. Engelman, and P. Cherepanov, *Nature*. 464, 232 (2010).
[45] E. Haustein and P. Schwille, *Annu. Rev. Biophys. Biomol. Struct.* 36, 151 (2007).

In: Anisotropy Research
Editor: Hirpa Lemu

ISBN 978-1-62081-977-7

Chapter 8

SURFACE MAGNETIC ANISOTROPY IN MAGNETIC SUBSTANCES INDUCED BY REVERSIBLE ADSORPTION

*V. E. Zubov**

Faculty of Physics, M.V. Lomonosov Moscow State University,
Leninskie Gory, Moscow, Russia

ABSTRACT

A review of recent investigations of the influence of molecules adsorbed through the mechanism of the formation of hydrogen bonds on magnetic properties of ferromagnetic alloys and ferrodielectrics is presented in this chapter. It has been found that domain structures of amorphous magnetically soft iron-based ferromagnetic ribbons in the atmosphere of water vapor and methyl alcohol are significantly different. In the first case, several domains separated by domain walls oriented at an angle of $\mathbf{30^0}$-$\mathbf{40^0}$ to the longer side of the sample are observed. In the second case, two domains separated by one wall located in the middle of the sample parallel with the longer side are observed. While the normal component of the magnetization on the sample surface has been detected, in the former case, using the polar magnetooptical Kerr effect, this component is almost absent in the latter case. The normal component of the magnetization is induced by desorption of water (methyl alcohol) molecules, which are adsorbed through the mechanism of the formation of hydrogen bonds, **(!)** from the sample surface. According to the performed estimate, the value of the effective field of the perpendicular magnetic anisotropy reaches 1.6 kA/m. It has been found, using the Faraday-effect magnetooptic method that the magnetic domain width of the labyrinth domain structure in bismuth-containing iron garnet films with perpendicular anisotropy considerably changes after the adsorption of water (methyl alcohol) molecules. A maximum change in domain width of 40% has been observed in methanol saturated vapor, and this change in water vapor is 15%. A decrease in domain width under adsorption has been attributed to the decrease in the effective perpendicular magnetic anisotropy constant of the film caused by the adsorption of water (methyl alcohol) molecules. These observed effects are reversible. A conclusion has been

* E-mail address: zubov@magn.ru.

drawn that the mechanisms of influence of weak adsorption on magnetic properties of ferromagnetic alloys and ferrodielectrics are similar. Presence of atoms with partly filled electron 3d-shell is in both cases important, and electrical conductivity of the samples is not essential for the observed effects.

1. INTRODUCTION

The correlation between *adsorption-desorption processes*, catalysis properties of transition elements and magnetic properties of ferromagnets was mentioned in the literature [1-3]. Adsorption, catalysis and magnetic properties of *transition elements* and their compounds are determined by electrons on partly filled 3d- and 4f-shells. Investigation of catalysis activity of alloys shows that filling 3d-shells by electrons (for example, Ni-Cu and Pd-Ag alloys) leads to sharp decrease of *catalysis activity* [4].

It has been possible to conclude, from analysis of experimental data, that adsorption and catalysis activity of transition elements (in metals) and compounds of transition elements have the same reason. Big difference in electrical conductivity of metals and their oxides weakly affects the adsorption and catalysis properties. Adsorption and catalysis properties of transition elements and their oxides are determined by presence of localized electron 3d-states of the elements [4].

The surface is one of the main defects of solids because absence of nearest neighbors at one side of surface atoms leads to a change of surrounding symmetry of the atoms. Unsaturated chemical bonds appear that are ready to form new chemical bonds. According to the binding energy of the adsorbed molecules, adsorption is classified into two types: *physical* and *chemical*. While physical adsorption is caused by the *van der Waals interaction*, chemical adsorption takes place by coordination, covalent and ion bonds. There exists also intermediate adsorption type, i.e., hydrogen bonds (H-bonds), whose energy is in the range 0.1 - 0.3 eV. *H-bond* is closer to chemical bond as it is characterized by saturation and directivity, in contrast to *physical bond*. The main types of adsorbed molecule bonds and their energy levels are presented in Table 1.

Real surface of solids is of special interest. It represents the surface that comes in contact with the atmosphere. All of the active centers on the real surface are occupied by chemisorbed molecules. For this reason new chemical bonds with air molecules are not formed. Film of its own oxide exists usually on the metal real surface. The hydrate cover forms on the oxide film where adsorbed coordination-bond water molecules $(H_2O)_c$ and OH groups take place on the oxide film [4].

OH groups are connected with 3d- atoms by covalent bonds. In case of coordination bond, electron cloud transfer of oxygen electron unshared pair of the coordination-bond water molecule $(H_2O)_c$ to the 3d-shell of atom-acceptor takes place. Weakening of valence bonds in an adsorbed molecule $(H_2O)_c$ happens in this case, and protonization of adsorbed water molecule takes place. According to chemical classification adsorbed molecule $(H_2O)_c$ becomes proton-donor center [5].

In addition to water molecules, alcohol molecules and other molecules can form coordination bonds with active centers on the real surface [1]. After filling all active centers on the surface by coordination-bond water molecules and OH groups, water molecules are adsorbed via H-bond mechanism.

Table 1. Energy of different types of bond

Types of bond	Energy of bond [kJ/mole]	[eV]
Van der Waals	0.1-0.5	0.001-0.05
H-bond	10-30	0.1-0.3
Chemical bond	≥100	≥1

Bond energy of coordination-bond water molecules with oxides is in the range 0.7 - 0.85 eV (70 - 85 kJ/mole) [5]. This is by 3 - 4 times higher than the energy of H-bond. The degree of protonization of the $(H_2O)_c$ molecules is more than that of protonization of OH groups. Thus, the $(H_2O)_c$ molecules are more active adsorption centers than OH groups. It has been found by the NMR spectroscopy at the silicon surface that adsorption of water molecules takes place first on the $(H_2O)_c$ molecules [1]. When the mean number of adsorbed water molecules corresponds to one monolayer, the adsorbed phase becomes clusters of adsorbed water, with thickness 2 - 3 molecule diameters and fastened on the coordination-bond water molecules. Presence of OH groups on the surface stimulates a growth of the clusters through the surface. If a mean number of adsorbed water molecules correspond to several monolayers, a junction of clusters takes place [5]. Adsorption of water molecules leads to the surface deformation where the surface stretches due to adsorption of the $(H_2O)_c$ molecules, and the surface is contracted due to adsorption of H-bond molecules [1]. Study of adsorption isotherms in gaze phase on different oxides has shown that the quantity of adsorbed molecules weakly depends on oxide composition [1]. That is the reason to estimate the quantity of adsorbed molecules using the isotherm of adsorption of the same adsorbates on other adsorbents. It is natural assumption that similar situation happens on the hydrated surfaces of ferromagnetic metals and ferrodielectrics.

The influence of the adsorption of water molecules on magnetic properties is analogous to a break of translation symmetry on the surface of ferromagnets. Change of the environment of the near-surface magnetic atoms leads to the change of near-surface magnetic properties of the ferromagnets. The influence of the surface on magnetic properties is multiform. Cause of change of the surface magnetic properties can be *surface magnetic anisotropy*, a change of exchange interaction in the near-surface region and other reasons (for example, as given [6] and references in it). The surface magnetic anisotropy was discussed for the first time by Neel [7] who received an estimate of the surface magnetic anisotropy energy to be $(0.1\text{-}1)\cdot 10^{-3}$ J/m^2. As given in [8], surface magnetic anisotropy was observed for the first time in hematite, and it is stated that the value of the surface magnetic anisotropy energy obtained in this study ([8]) is in line with the estimate given by Neel [7]. A number of works are devoted to the investigation of the influence of different gaze's chemical adsorption on the surface properties of ferromagnets (for example, in [9-15]).

The binding energy of molecules with the surface in chemical adsorption is *huge*. The return to the initial state (desorption) may be realized by powerful influence on the surface such as by etching, thermal desorption at high temperature, etc. Such influence changes the structural and magnetic characteristics of materials. Weak physical adsorption caused by the van der Waals interaction does not change surface properties of a solid, while H-bond type adsorption is often reversible at room temperature in contrast to chemical adsorption. That means, it is possible to return to the initial state of the sample surface and to receive reversible experimental results at the same surface of the sample under investigation. Adsorption occurring through mechanism of the formation of hydrogen bonds is very frequent, where molecules of water, ammonia, alcohols, etc., are adsorbed through the formation of hydrogen bonds. Some of these molecules exist in the atmosphere and create contact with ferromagnetic surfaces. However, there are practically no published investigations of the influence of weak adsorption on ferromagnetic properties. Investigations conducted at the Faculty of Physics of M. V. Lomonosov Moscow State University are exceptional. In these investigations, the influence of reversible adsorption on dynamic properties of domain walls, magnetic susceptibility and domain structure of ferromagnetic metals and magnetic dielectrics have been observed. It has been shown that reversible *adsorption-desorption processes* proceeding through the formation of hydrogen bonds induce *surface magnetic anisotropy*.

This chapter presents the investigation results of the surface magnetic anisotropy in ferromagnetic metals and magnetic dielectrics. The mechanisms of surface magnetic anisotropy appearance and the influence of surface anisotropy on magnetic susceptibility and domain structure of the samples are well covered.

2. Experimental Procedure

The experimental investigation was carried out using a magnetooptical micromagnetometer built on an MIM-8 metallographic reflecting microscope. This micromagnetometer operates by measuring changes in the intensity of light reflected from a microregion of a ferromagnet in the image plane of the microscope when its magnetization is reversed. The operations required to measure any particular magnetooptic effect were described in detail in [16]. Magnetization components lying in the magnetic mirror plane were determined using the linear (in the magnetization) equatorial and *meridional Kerr effects*. The component perpendicular to the surface of a sample was measured using the *polar Kerr effect*. The sensitivity of measurements of the relative change in the intensity of light was equal to $\sim 10^{-5}$ in the studies of part of ferromagnet surface of $\sim 1\ \mu m^2$ area at frequencies from 20 Hz to 100 kHz. For investigation of domain structure of samples *equatorial Kerr effect* was measured, which was caused by change in magnetization of illuminated part of the sample's surface through the domain wall (DW) motion in alternating magnetic field. The slit of the photoelectron multiplier was scanned along the axis perpendicular to the DW direction, and reconstruction of magnetic domain structure of the sample was carried out by detection of DW location on the sample's surface.

In order to investigate the influence of adsorption on magnetic samples, the magnetooptical micromagnetometer was supplemented with a vacuum cell in which a sample was placed and with a system of controlled introduction of various gases. The pressure of the gas in the cell could vary from atmospheric pressure to 10^{-3} Pa.

In addition, investigation of transparent ferrite-garnet thin films was conducted with the help of magnetooptical micromagnetometer at room temperature. The sample under investigation was placed in the vacuum cell and fastened with the help of a special holder. It was connected to a system with inflow and pumping of various gases. Observation of the domain structure was conducted by use of the *Faraday rotation effect*. Linear polarized light from an objective lens was passed through ferrite-garnet film, reflected back from its second surface and retuned to the objective lens. Faraday rotation of light polarization plane takes place in the domains.

Since sign of rotation depends on the direction of magnetization in the domains, placing the analyzer on the path of reflected light allows us to observe domain structure of the sample, which is registered by digital camera DCM-500 and the signal from the camera goes to the computer. This enables us to see the domain structure under study on the screen of the computer and the results in the computer memory. Alternating magnetic field perpendicular to the film surface was created by the coil with alternating current. The domain walls coercive field (H_c) is the minimum value of the field, for which DW oscillations start. To eliminate the effect of the attachment of domain walls to defects in the process of their displacement, a perpendicular alternating magnetic field, which exceeds H_c, was applied to the sample. In the presence of this field, the domain structure rearrangement becomes gradual without jumps.

3. EFFECTS INDUCED BY REVERSIBLE ADSORPTION IN MAGNETIC ALLOYS AND DIELECTRICS

Amorphous ferromagnetic ribbons: The influence of weak adsorption of water and methyl alcohol molecules on domain structure and on the value of normal component of magnetization on the surface of ferromagnetic ribbons of amorphous iron-based alloy has been investigated in [17]. In this study, the samples of composition $Fe_{76.5}Cu_1Nb_3Si_{13.5}B_6$ were obtained by the spinning technique. Before *amorphization*, the melt was subjected to special time-temperature treatment to obtain a uniform distribution of doping elements and, as a result, the soft magnetic properties of the alloy have improved [18]. The amorphous structure of the ribbons was established with the help of X-ray structural analysis. The samples were 25-30 μm thick, 0.55 mm wide and 15-20 mm long.

Hereinafter, the results of the experiments were described by using the coordinate system with *x* and *y* axis being directed along the sample width and length respectively, and the direction of the *z* axis corresponds to the shortest dimension. The initial magnetic structure of the samples consisted of two domains, which were magnetized parallel to the *y* axis and separated by one 180^0 domain wall. The width of the DW and coercive field of the DW measured by the magnetooptical method was ~10 μm and ~1 A/m, respectively. Saturation induction is $B_s = \mu_0 I_s = 1.2$ T (where I_s = *saturation magnetization*).

The *equatorial Kerr effect* was measured during the investigation of the domain structure. The slit of the photoelectron multiplier was scanned along the *x* axis perpendicular to the longer side of the sample. The magnetization reversal of the sample was performed due to the oscillation of the domain wall in the alternating magnetic field directed along the *y* axis. The frequency of the magnetic field was 900 Hz. To study the normal component of magnetization on the sample surface, we used the *polar Kerr effect*, which is proportional to this magnetization component.

The measurement of the polar Kerr effect at a setup with sensitivity of $\sim 10^{-5}$ was unsuccessful. The estimates showed that the expected polar Kerr effect was $\sim 10^{-6}$ and, to measure it, the sensitivity should be increased by one or two orders of magnitude. To increase the sensitivity of the magnetooptical measurements, it was found necessary to increase the intensity of light reflected from the sample region under investigation. This problem was resolved by the removal of a homal from the optical scheme of the microscope (homal is a lens that is situated after objective lens, and is responsible for the formation of the sample surface image). In this case, light reflected from the sample almost entirely enters the photoelectric multiplier. The size of the illuminated surface region is controlled by a special field diaphragm of the microscope. The diameter of the illuminated region can vary from several micrometers to several tens of micrometers. The improvements increased the sensitivity of the measurements to $\sim 10^{-7}$. The measurements of the polar Kerr effect were carried out at the normal incidence of light in the alternating magnetic field parallel to the y axis.

A change in the intensity of linearly polarized light passing through the analyzer was observed. This change is caused by the rotation of the polarization plane of a light beam reflected from the sample and is proportional to the normal magnetization component on the sample surface. The polar Kerr effect is observed due to the oscillations of the DW when the normal magnetization component in neighboring domains has different signs.

The magnetooptical investigation of the domain structure was performed in the magnetic field with amplitude of 50 A/m. The samples were magnetized almost to saturation in the field of $H_y = 600$ A/m. The equilibrium position of the DW is most accurately determined in the small-amplitude field, when the amplitude of the wall oscillations is comparable with its effective width. The domain structure of the samples changed significantly in vacuum. Instead of one 180^0 DW in the initial state, which was oriented along the y axis, several domain walls oriented at an angle in the range 30^0-40^0 to the longer side of the sample appeared.

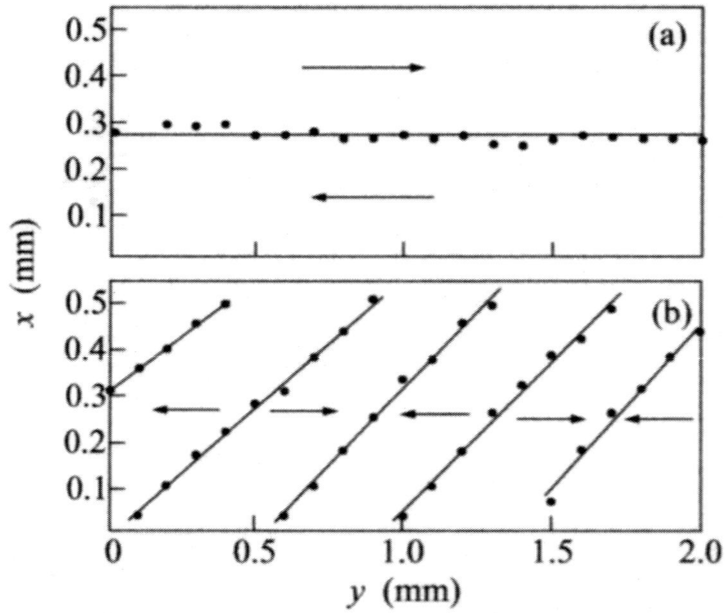

Figure 1. Domain structure in the amorphous ferromagnet sample as reconstructed using the magnetooptical Kerr effect in the field of H = 50 A/m [17]. Note: The points in the figure indicate the position of the domain wall determined with the Kerr effect. The arrows show the magnetization direction in the domains. (a) The structure in the initial state and in atmospheres of water vapor or methyl alcohol at a pressure of 0.5 p_s. The domains are separated by one 180^0 domain wall parallel to the longer side of the sample. (b) The structure of the same sample region in a vacuum.

Figure 1 shows the domain structure of the sample fragment with width of 0.55 mm and length of 2 mm reconstructed by means of measurement of the *equatorial Kerr effect* in the initial state (Figure 1(a)) and after aging in vacuum with a residual gas pressure of about 1 Pa (Figure 1(b)). The introduction of methyl alcohol vapor into a preliminary evacuated cell with the sample at the pressure of 7 kPa (this corresponds to the pressure ~$0.5p_s$, where $p_s \approx 13$ kPa is the pressure of a saturated methyl alcohol vapor at room temperature) leads to the recovery of the initial domain structure; i.e. the structure shown in the Figure 1(b) is transformed into the structure shown in Figure 1(a). The complete recovery was observed after about one day from the introduction of methyl alcohol vapor. The effect was reversible, i.e., the subsequent evacuation of the cell with the sample led to the appearance of the domain structure shown in Figure 1(b).

The introduction of water vapor into a preliminarily evacuated cell with the sample at the pressure of 1 kPa (~$0.5p_s$, where $p_s \approx 1.8$ kPa is the pressure of the saturated water vapor at room temperature) induced a change in the domain structure. The structure shown in Figure 1(b) was then transformed into the structure shown in Figure 1(a). As in the case of methyl alcohol, the total time of transformation of the structure was about one day. Thus, the effect of the reversible adsorption of water and methyl alcohol molecules on the domain structure of amorphous ferromagnet is similar.

The effect of the adsorption of oxygen, nitrogen and argon (which is an inert gas) molecules on the domain structure of the amorphous ferromagnet sample was also investigated. These molecules are adsorbed on the real surface of the ferromagnet at room temperature through the mechanism of the formation of very weak van der Waals bonds. The introduction of these gases into the cell with the sample up to pressure of 100 kPa was not accompanied by a change in the domain structure of the sample. The last results indicate that in contrast to the adsorption through the formation of hydrogen bonds, weaker *van der Waals* adsorption does not lead to a change in the magnetic properties of the amorphous ferromagnet.

The normal magnetization component on the sample surfaces was studied using the *polar Kerr effect*. The measurements of the polar Kerr effect were carried out in the alternating magnetic field with amplitude of 100-600 A/m [17]. The deviation of the light beam incident on the sample surface was no more than 0.5^0. The rotation of the polarization plane of light reflected from the sample was measured for the analyzer arrangement angles of $+45^0$ and -45^0 with respect to the polarizer. For both angles, the measured effect has the same magnitude but different signs.

For such measurements, possible errors associated with asymmetric incidence of light on the sample and with the appearance of other *magnetooptical Kerr effects* in the measurements of the polar Kerr effect are compensated. The rotation angle, α, of the polarization plane of light reflected from the sample is related to the measured magnetooptical signal δ as $\alpha = 2\delta$.

Figure 2 shows the *polar Kerr effect* measured in the field of H_y =600 A/m on the same fragment of the sample surface on which the domain structure was investigated (see also Figure 1). Figure 2(a) shows the polar Kerr effect as a function of the *y* coordinate in the initial state. The polar Kerr effect was measured at the center of the sample when a light spot was scanned along the *y* axis parallel to the longer side of the sample. The diameter of the illuminated section on the sample surface was ~10 μm. The polar Kerr effect increased linearly with the field and was saturated in a field of 600 A/m. For this reason, the measurements were carried out in the field in which the effect was maximal.

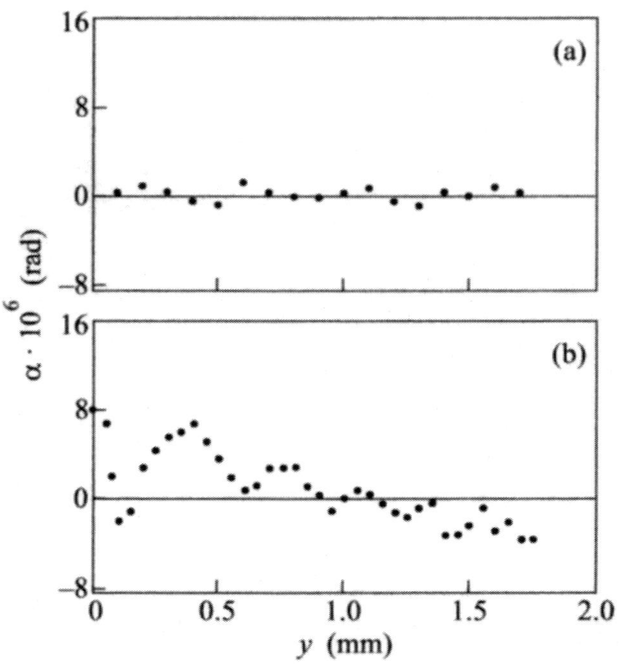

Figure 2. Polar Kerr effect in the amorphous ferromagnet in a field of H = 600 A/m in the (a) initial state and in atmospheres of water vapor or methyl alcohol at a pressure of 0.5 p_s and (b) in a vacuum [17].

Figure 2(b) shows the polar Kerr effect measured after the evacuation of the cell with the sample. According to this figure, $\alpha(y)$ dependence is oscillatory. The maximum α value in this figure is $8{\cdot}10^{-6}$. The oscillation period is about the period of the domain structure shown in Figure 1(b). After the introduction of methyl alcohol vapor at the pressure of 7 kPa ($\sim 0.5p_s$) into the evacuated cell with the sample, dependence of $\alpha(y)$ has the same form as in the initial state (see Figure 2(a)). The introduction of water vapor at the pressure of 1 kPa ($\sim 0.5p_s$) into the evacuated cell led to a similar result, i.e., to a strong decrease in the polar Kerr effect. A comparison of Figure 2(a) with (b) shows that the polar Kerr effect in the initial state and in vacuum differs by an order of significant magnitude. Similar to the case of the investigation of the domain structure (see Figure 1), the effect is reversible, which means subsequent evacuation of the cell with the sample is accompanied by the recovery of the dependence of the polar Kerr effect as shown in Figure 2(b).

The introduction of oxygen, nitrogen, and argon into the evacuated cell with the sample up to the pressure of 100 kPa did not change the polar effect. Thus, the molecules adsorbed through the formation of *van der Waals* bonds do not change the polar Kerr effect, and the effect of the adsorption of water and methyl alcohol molecules on the polar Kerr effect is similar. The results make it possible to estimate the normal magnetization component induced by the perpendicular *surface magnetic anisotropy*. The saturation induction of the amorphous ferromagnet is $B_s=\mu_o I_s=1.2$ T, while the saturated polar Kerr effect in pure iron is $\alpha_s \sim 0.01$. The value of α_s in the amorphous ferromagnet is smaller than in pure iron owing to the dilution of iron by nonmagnetic atoms. Taking $\alpha_s \sim 0.005$, we obtained the maximum value of the normal component of the saturation induction of the amorphous ferromagnet to be $B_n=\mu_o I_n=2$ mT (where I_n is the normal magnetization component).

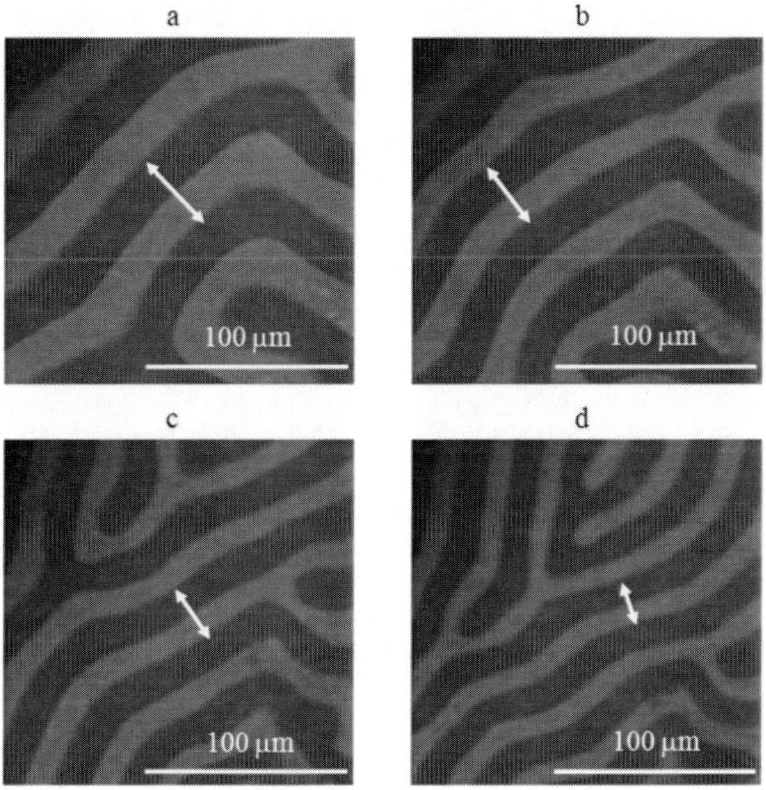

Figure 3. Domain structures of the sample fragment (a) in a vacuum and in (b) 1, (c) 3 and (d) 8 hrs after methanol input to the cell with the sample at pressure p = p_s [**19**].

Thus, the normal magnetization component induced by the reversible adsorption-desorption processes on the surface of the amorphous ferromagnet is ~0.2% of the saturation magnetization. The effective field of the *induced surface* magnetic anisotropy reaches H_a= I_n~1.6 kA/m.

Ferrite-garnet films: Investigation of the influence of the reversible adsorption on the ferrodielectric magnetic properties is of special interest. This is because concentration of magnetic atoms in ferrodielectrics is much less than in ferromagnetic metals, and ferro-dielectrics as usual have more complicated magnetic structure than ferromagnetic metals. Ferrite-garnet thin films of composition $(Bi,Lu)_3(Fe,Ga)_5O_{12}$ with perpendicular magnetic anisotropy were selected for the investigation. Films were prepared with the use of gadolinium-gallium garnet substrates with crystallographic orientation (111) by the method of *liquid phase epitaxy*. The following parameters were valid: thickness of films h = 4.5 μm, $B_s = \mu_0 I_s$ = 3 mT (where I_s = magnetization), DW coercive field $H_c \approx$ 16 A/m and width of domains $d \approx$ 20 μm. Relatively large domain width is determined by small value of I_s and small value of *perpendicular magnetic anisotropy* constant (K) of the films [**19**]. Small value of I_s is determined by dilution of iron sublattices by Ga ions. The content of Ga is 1.2 on formula unit. First, *perpendicular magnetic anisotropy* was induced by the mechanical stresses in the film due to distinction between dimensions of the unit cells of *ferrite-garnet film* and gadolinium-gallium garnet substrate (Δa). Value of Δa in the investigated films is not more than 0.003 Å. Another mechanism of perpendicular magnetic anisotropy appearance is growth anisotropy due to chemical composition of ferrite-garnet film and technological conditions of growth [20]. Perfect gadolinium-gallium garnet substrates were then used, and careful preparation of substrate surface was conducted. Technological conditions and small value of Δa were allowed making magnetic films with high crystallographic quality. The films are characterized by high constant of Faraday rotation due to presence of Bi (~1 on formula unit). Such films are used for magneto-optical visualization of magnetic field distribution [20].

The domain structure of ferrite-garnet films was studied in vacuum with the pressure of residual gases in the cell with the sample of about 1 Pa, and then at water and methyl alcohol vapor inflow under pressure from zero to p_s. The domain width was observed to decrease when water or methyl alcohol vapor was input into the cell with the sample.

The variation of average domain width in vacuum and at water vapor pressure of $p = p_s$ was about 15%, while the variation for methyl alcohol vapor was equal to 40%. Then a new equilibrium domain structure forms in about 3 hrs after water vapor input (i.e., 8 hrs after methyl alcohol vapor input).

Figure 3 shows photos of the domain structure of a sample fragment in vacuum and at different times after methanol vapor input to the cell with the sample at $p = p_s$. The arrows in the figure show, for example, the regions on the sample image where the domain width was measured. One can see that the domain structure of the sample changes and the width of the domain decreases with time.

The values of average domains width are presented in the outline for each photo in Figure 3. These values, calculated using 5 points on each photo, are (a) 21 μm, (b) 17.5 μm, (c) 15 μm, and (d) 13 μm (Figure 3). The points were chosen on the straight pieces of the domains. Investigation of influence of the water molecules adsorption on domain structure of ferrite-garnet films showed similar results.

As stated above, the investigated monocrystal ferrite-garnet films possess perfect crystal structure. Therefore, adsorbed water molecules cannot penetrate into the film volume. These molecules interact only with the near-surface layers of ferrite-garnet film. The photos show that adsorption changes the domain structure not only in the surface region, but also throughout the sample bulk because the Faraday-effect carries information on the bulk magnetization. So, the interaction of the adsorbed water and methyl alcohol molecules with the near-surface layers of ferrite-garnet film changes the domain structure in its volume.

4. Mechanisms of Influence of Weak Adsorption on Magnetic Properties of Magnetically Ordered Substances

As stated above, adsorption of molecules adsorbed through the mechanism of the formation of hydrogen bonds leads to the appearance of the mechanical stresses in the near-surface layer of the solid. These mechanical stresses lead to the existence of surface magnetic anisotropy because of magneto-elastic interaction. The effect is intensified due to the fact that the magneto-elastic energy constants for surface atoms are higher than for atoms in the volume because of lower surrounding symmetry of the surface atoms [21]. This conclusion may be used for the explanation of experimental results described above. The model which describes effects observed in amorphous ribbons is formulated below.

Amorphous ferromagnetic ribbons: As a result of staying of amorphous ribbons in the air, the oxide film and hydrated cover were formed on their surface. After that the surface was covered by a layer of water molecules adsorbed by the H-bond mechanism. The mechanical stresses existed in the near-surface layer due to adsorption of water molecules, and the stresses disappear with time owing to the diffusion of atoms; and thus the new equilibrium state is formed on the surface. The evacuation of the cell with the sample is accompanied by desorption, which leads to the transition of the surface region of the sample from the equilibrium state and to the appearance of the mechanical stress on its surface. Owing to the surface stress and surface magnetostriction, the perpendicular surface magnetic anisotropy appears and induces the normal magnetization component on the sample surface. As a result, additional stray magnetic fields appear on the surface. In this case, an increase in the number of the domain is energetically favorable, because the sign of the normal magnetization component on the surface alternates, and correspondingly the stray magnetic field decreases.

The proposed model explains why the evacuation of the cell with the amorphous sample is accompanied by the rearrangement of the domain structure from the two-domain state shown in Figure 1(a) to a multidomain state presented in Figure 1(b). The introduction of water vapor into the cell leads to filling of vacancies on the sample surface, the removal of the mechanical stress in the surface region, and as a result, the initial domain structure is recovered. Since the size of water and methyl alcohol molecules are close to each other, they can fill vacancies on the ferromagnetic surface almost identically. Assuming that the molecules are spherical, we estimated dimensions (d_0) of water and methyl alcohol molecules and received d_0 = 3.1 Å for the water molecule and d_0 = 4.1 Å for the methyl alcohol molecule. For this reason, the introduction of methyl alcohol vapor into the cell with the sample gives the same result as the introduction of water vapor. The appearance of the normal magnetization component on the surface after the evacuation of the cell with amorphous samples, as well as a nearly complete disappearance of the normal component after the introduction of water or methyl alcohol vapors, completely confirms the described model.

Ferrite-garnet films: Domain width in ferrite-garnet thin films with perpendicular magnetic anisotropy is determined by the competition of 180-degree domain wall energy and energy of magnetic stray field over a film surface due to a sudden change of the magnetization normal component on the film surface. Domain width can be estimated with the help of the expression: $d \sim \sqrt{\frac{\sigma h}{1,75 M_s^2}}$, where $\sigma = 4\sqrt{AK}$ is energy of the domain wall unit area and A is an exchange parameter [22]. These formulas show that density of DW energy σ and domain width decrease when *anisotropy constant* K decreases. The above-mentioned formulas give $K \sim 170$ J/m^3 and $A = 1.4 \cdot 10^{-12}$ J/m. The value of parameter A is obtained by taking into account the dilution of ferrite-garnet iron sublattices by gallium [23]. It is natural to explain that the domain width of the film decreased with reduction of effective magnetic anisotropy constant due to the water molecules adsorption. The sign of surface uniaxial magnetic anisotropy constant induced by adsorption must be opposite to the sign of magnetic anisotropy constant of ferrite-garnet film in initial state. The sum of initial magnetic anisotropy constant and surface magnetic anisotropy induced by water molecules adsorption leads to the reduction of effective perpendicular magnetic anisotropy constant of the film. It seems that the origin of *surface magnetic anisotropy* of the films induced by adsorption is analogous to the origin of the uniaxial magnetic anisotropy induced by the distinction between dimensions of the unit cells of ferrite-garnet film and gadolinium-gallium garnet substrate.

An elastic deformation arises in the ferrite-garnet film owing to the distinction between dimensions of the unit cells of the ferrite-garnet film and the substrate. The elastic deformation leads to the appearance of the uniaxial magnetic anisotropy of the film. The elastic deformation appears also on the free surface of the ferrite-garnet film due to the adsorption of water molecules. Addition of the initial magnetic anisotropy of the film and the magnetic anisotropy induced by the adsorption of water molecules leads to the reduction of the effective perpendicular magnetic anisotropy.

Energy of initial perpendicular magnetic anisotropy attributed to the film unit area ($K{\cdot}h$) is ~ $0.8{\cdot}10^{-3}$ J/m^2. This value is obtained by using the value of K presented above. Influence of the adsorption of water molecules on magnetic properties is analogous to the influence of a break of translation symmetry on the surface of the magnetics. Deformation of the surface caused by the water molecules adsorption leads to a displacement of iron ions in the near surface region from initial positions. This changes symmetry of the nearest neighbors to these ions. Hence, the water molecules adsorption leads to appearance of the surface magnetic anisotropy. The value of this anisotropy may be the same order of magnitude as *Neel surface magnetic anisotropy*.

Energy of Neel surface anisotropy is equal to $(0.1\text{-}1){\cdot}10^{-3}$ J/m^2 [7]. Thus, changing the width of film domains can be explained by the reduction of effective perpendicular magnetic anisotropy constant due to water molecules adsorption. Reduction of effective perpendicular magnetic anisotropy leads to decrease in σ and domains width. Similar arguments are true in the case of adsorption of the methyl alcohol molecules.

CONCLUSION

Investigation of the influence of weak adsorption occurring through the mechanism of the formation of hydrogen bonds on magnetic properties of magnetically ordered substance shows that 1) such influence exists; and 2) this influence can considerably change the magnetic properties of the above-mentioned substance. It is established on the example of some gases, which are adsorbed by the mechanism of formation of van der Waals bonds, that van der Waals adsorption does not change magnetic properties of ferromagnetic. Besides, it has been shown that there is no essential distinction of the influence of H-bond adsorption on magnetic metals and dielectrics.

It is important in both cases that there are atoms with partially filled 3d-shells on the real surface of the ferromagnetic metals and ferrodielectrics. Formation of hydrate covering the surface of the ferromagnetic metals and ferrodielectrics takes place with the participation of 3d-atoms.

Subsequent adsorption of water (methyl alcohol) molecules takes place through the mechanism of the formation of hydrogen bonds. This result is in agreement with the conclusion given in [4] that the process of adsorption of molecules, adsorbed on the surface of oxides and on the real surface of metals, takes place in similar way.

REFERENCES

[1] V. F. Kiselev, O.V. Krylov, *Adsorption processes on semiconductor and dielectric surface,* Springer-Verlag, Berlin, New York, Tokio (1984).
[2] V. E. Zubov, A. D. Kudakov, N. L. Levshin, V. V. Pilipenko, *J. Magn. Magn. Mater.* 140-144, 1895 (1995).
[3] M. Getzlaff, J. Bansmann, G. Schönhense, *J. Magn. Magn. Mater.* 192, 458 (1999).
[4] O.V. Krylov, V. F. Kiselev, *Adsorption and catalysis on the transition metals and their oxides*, Himiya, Moscow (1981) (in Russian).
[5] V. F. Kiselev, S. N. Kozlov, A. I. Zoteev, *Foundations of physics of the surface of solids*, Publ. House of Moscow Univ., Moscow (1999) (in Russian).
[6] T. Kaneyoshi, *J. Phys: Condens. Matter.* 3, 4497 (1991).
[7] L. Neel, *J. Phys. Rad.*, 5, 225 (1954).
[8] G.S. Krinchik, V.E. Zubov. *Sov. Phys. JETP* 42 359 (1975).
[9] P. Štěpán, D. Hugues, *J. Magn. Magn. Mat*er. 198 – 199, 312 (1999).
[10] P. Štěpán, D. Hugues, *Surf. Sci.* 394, 192 (1997).
[11] M., Getzlaff, G. Schönhense, *Surf. Sci. A* 331, 213 (1995).
[12] H. J, Elmers, U. Gradmann, *J. Appl. Phys.* 63, 3664 (1988).
[13] J. Bansmann, M. Getzlaff, C. Westphal, G. Schönhense, *J. Magn. Magn. Mater.* 117, 38 (1992).
[14] M. Donath, G. Schönhense, K. Ertl, V. Dose, *Appl. Phys. A.* 50, 49 (1990).
[15] F. Bodker, S. Morup, S. Linderoth, *J. Magn. Magn. Mater.* 140–144, 373 (1995).
[16] G. S. Krinchik, O. M. Benidze, *Sov. Phys. JETP* 40, 1081 (1975).
[17] V. E. Zubov, N. L. Levshin, *JETP Letters* 90, 268 (2009).
[18] Yu. N. Starodubtsev, L. D. Son, V. S. Tsepelev *et al.*, *Rasplavy* 4, 76 (1992) (in Russian).
[19] V. E. Zubov, A. D. Kudakov, N. L. Levsнin, M. Yu. Gusev, N. S. Neustroev, *JETP Letters* 94, 44 (2011).
[20] M.Yu. Gusev, Yu.F. Kozlov, N.S. Neustroev. *Elektronnaya promishlennost* №3, 93 (2003) (in Russian).
[21] R. Zuberec, H. Szymczak, R. Krishnan, *et al.*, *J. Magn. Magn. Mater.* 121, 510 (1993).
[22] G.S. Krinchik, *Physics of Magnetic Phenomena*, Publ. House of Moscow Univ., Moscow (1985) (in Russian).
[23] A.H. Eschenfelder, *Magnetic Bubble Technology*, Springer-Verlag, Berlin (1981).

In: Anisotropy Research
Editor: Hirpa Lemu

ISBN 978-1-62081-977-7

Chapter 9

EFFECTS OF LAYERED STRUCTURE AS ONE OF ANISOTROPY REALIZATIONS IN THERMOELECTRICITY

P. V. Gorskyi*

Chernivtsi National University, Chernivtsi, Ukraine

ABSTRACT

Conventionally, overwhelming majority of theoretical and experimental works is devoted to investigation of layered crystals with highly open Fermi surfaces. The effects of layered structure may be observed in magnetism, galvanomagnetic phenomena and thermoelectricity even if the Fermi surface of a layered crystal is closed. In this chapter the dependence of thermoelectromotive coefficient on quantizing magnetic field at helium temperatures is investigated. The calculation is performed for the case of both the quantizing magnetic field and the temperature gradient parallelism to superlattice axis of crystal. In the calculation, process several models of charge carriers scattering are considered. These models are: (1) the models of constant lifetime and constant free path for electrons scattering by charged impurities and (2) the models of lifetime proportional to total and longitudinal electron velocity on the Fermi surface for electrons scattering by acoustic phonons. The effects of layered structures for all scattering models manifest themselves in the decrease of thermoelectromotive coefficient value in the region of the quantum oscillations and phase retardation of quantum oscillations thermoelectromotive coefficient. In ultraquantum magnetic fields for all scattering models the layered structure effects manifest themselves in a faster decrease of thermoelectromotive coefficient, which tends to zero in very strong magnetic fields. For all scattering models, however, a peak of dependence of the thermoelectromotive coefficient on magnetic field is observed. The peak value of thermoelecromotive coefficient at selected crystal parameters and temperature is almost independent of the layered structure effects and scattering mechanisms of the charge carriers. This value is approximately equal to 77 μV/K. However, in general, the best conditions for manifestation of the layered structure effects for closed Fermi surfaces are observed, in case of the charge carriers scattering by acoustic phonons.

* E-mail address: gena.grim@gmail.com.

1. Introduction

The model of layered crystal's band spectrum was proposed by R.F Fivaz [1] in 1967. In this model, the motion of *charge carriers* in the planes of layers is described by *effective mass approximation*, and in the super lattice axis direction is described by strong coupling (cosinusoidal) approximation. Since then, many theoretical and experimental papers devoted to various physical characteristics of layered crystals were published. But the overwhelming majority of these works are related to the layered crystals, having the *open Fermi surfaces* (FS) [2-5]. It means that FS occupies the whole one-dimensional *Brillouin zone* and, when being periodically continued, it forms a connected surface. This case corresponds to the inequality $\zeta/\Delta \geq 2$, where ζ is *chemical potential,* which is counted out from mini-band bottom, and Δ is *half-width* of mini-band. The inequality sign corresponds to the transition from the *closed FS* to the open one. Usually only the case $\zeta/\Delta >> 2$ is investigated. Thus, the majority of researchers suppose that the layered structure does not manifest itself by anyway, if FS is closed. But the author of this chapter believes that the conventional effective mass approximation is valid only if $\zeta/\Delta << 1$.

The influence of the layered structure effects for the closed FS on electron gas diamagnetism is investigated in papers [6-9], while the influence on longitudinal conductivity of layered crystals in quantifying magnetic field is discussed in [10, 11]. The aim of this chapter is to demonstrate the influence of the layered structure on *thermoelectromotive coefficient* of the layered crystals with the closed FS. The calculation of thermolectromotive coefficient is performed in the wide range of quantizing magnetic field and for different scattering mechanisms.

2. The General Formula for Thermoelectromotive Coefficient

The most general form of the layered crystal band spectrum in quantizing magnetic field with induction B, which is parallel to super lattice axis (C-axis), is given as:

$$\varepsilon(n,x) = \mu^* B(2n+1) + W(x) \tag{1}$$

where $\mu^* = \mu_B m_0 / m^*$, μ_B –*Bohr magneton,* m_0–mass of free electron, m^*– effective electron mass in the plane of layers, n– Landau level number, $W(x)$ – dispersion law along C-axis, $x = ak_z$, a – *super lattice constant* and k_z– quasi-momentum component along C-axis.

In the *Fivaz model* we have:

$$W(x) = \Delta(1-\cos x) \tag{2}$$

where Δ is as defined above.

For *thermoelectromotive coefficient* calculation *Boltzmann equation* has been used. This approach gives the following formula for the volume component of thermoelectromotive coefficient α :

$$\alpha = -\left[\frac{\partial}{\partial T}\sum_{\gamma}\tau_{\gamma}g_{\gamma}v_{z\gamma}^{2}f^{0}(\varepsilon_{\gamma})\right]\left[e\frac{\partial}{\partial \zeta}\sum_{\gamma}\tau_{\gamma}g_{\gamma}v_{z\gamma}^{2}f^{0}(\varepsilon_{\gamma})\right]^{-1} \quad (3)$$

where, in this formula, T – absolute temperature, ζ – chemical potential of charge carriers system, e – elementary charge, g_{γ} – statistical weight of energy level with energy ε_{γ} and set $\gamma \equiv (n, x)$ of quantum numbers, $f^{0}(\varepsilon_{\gamma})$– Fermi-Dirac distribution, $v_{z\gamma}$ – longitudinal charge carrier velocity and τ_{γ} – relaxation time of charge carrier.

This formula is valid if the temperature gradient is parallel to C-axis. Summation over quantum numbers in (3) may be performed completely if we assume that relaxation time τ_{γ} depends only on longitudinal dimensionless quasi-momentum component x. For further transformation of formula (3) the equation for chemical potential definition at low temperatures is to be used [10]:

$$n_0 = \frac{4m^*}{ah^2}\int_{W(x)\le\zeta}[\zeta - W(x)]dx + \\ + \frac{8\pi m^* kT}{ah^2}\sum_{l=1}^{\infty}\frac{(-1)^l}{\sinh\left(\pi^2 lkT/\mu^* B\right)}\int_{W(x)\le\zeta}\sin\left[\pi l\frac{\zeta - W(x)}{\mu^* B}\right]dx \quad (4)$$

where n_0 – concentration of charge carriers and other symbols are conventional or as defined above. By substituting (4) into (3) one obtains:

$$\alpha = \alpha_0 A(B + C)^{-1} \quad (5)$$

where $\alpha_0 = k/e$ and A, B, C may be defined as:

$$A = \sum_{l=1}^{\infty}(-1)^{l-1} f_l^{th}\int_{W(x)\le\zeta}\tau(x)W'(x)^2\sin\left[\pi l\frac{\zeta - W(x)}{\mu^* B}\right]dx \quad (6)$$

$$B = 0.5\int_{W(x)\le\zeta}\tau(x)W'(x)^2 dx \quad (7)$$

$$C = \sum_{l=1}^{\infty}(-1)^{l} f_l^{\sigma}\int_{W(x)\le\zeta}\tau(x)W'(x)^2\cos\left[\pi l\frac{\zeta - W(x)}{\mu^* B}\right]dx. \quad (8)$$

In these formulae:

$$f_l^{th} = \left[\sinh\left(\pi^2 lkT/\mu^* B\right)\right]^{-1}\left[1-\left(\pi^2 lkT/\mu^* B\right)\coth\left(\pi^2 lkT/\mu^* B\right)\right] \quad (9)$$

$$f_l^{\sigma} = \left(\pi^2 lkT/\mu^* B\right)\left[\sinh\left(\pi^2 lkT/\mu^* B\right)\right]^{-1} \quad (10)$$

Thus, taking into account formulae (6-10), Eq. (5) completely defines the thermoelectromotive coefficient of the layered crystal in quantifying magnetic field at low temperatures.

3. THE DEFINITION OF CHEMICAL POTENTIAL

For the band spectrum (2), Eq. (4) has the following form:

$$\begin{aligned} n_0 &= \frac{4m^*\Delta}{ah^2}\left[(\gamma-1)C_0+\sqrt{2\gamma-\gamma^2}\right]+ \\ &+\frac{8m^*\pi kT}{ah^2}\sum_{l=1}^{\infty}\frac{(-1)^l}{\sinh\left(\pi^2 lkT/\mu^* B\right)}\left\{\sin\left(\pi l\frac{\zeta-\Delta}{\mu^* B}\right)\times\right. \\ &\times\left[C_0 J_0\left(\frac{\pi l\Delta}{\mu^* B}\right)+2\sum_{r=1}^{\infty}(-1)^r C_{2r}J_{2r}\left(\frac{\pi l\Delta}{\mu^* B}\right)\right]+ \\ &\left.+2\cos\left(\pi l\frac{\zeta-\Delta}{\mu^* B}\right)\sum_{r=0}^{\infty}(-1)^r C_{2r+1}J_{2r+1}\left(\frac{\pi l\Delta}{\mu^* B}\right)\right\} \end{aligned} \quad (11)$$

where $J_n(x)$ – *Bessel functions* of the first kind and real argument, $\gamma=\zeta/\Delta$, $C_0=\arccos(1-\zeta/\Delta)$, and $C_m=\sin mC_0/m(m\neq 0)$.

The left-hand side of (11) may be written as:

$$n_0 = \frac{4m^*\Delta}{ah^2}\left[(\gamma_0-1)\arccos(1-\gamma_0)+\sqrt{2\gamma_0-\gamma_0^2}\right] \quad (12)$$

where $\gamma_0=\zeta_0/\Delta$, and ζ_0 – Fermi energy of ideal electron gas in the layered crystal at $T=0$ and $B=0$.

For the open FS, when $\zeta_0/\Delta\geq 2$, we shall substitute $C_0=\pi, \sqrt{2\gamma_0-\gamma_0^2}=0$. For an effective mass approximation, Eq. (4) may be rewritten as:

$$\begin{aligned} n_0 &= \frac{8m^*\zeta}{3ah^2}\sqrt{\frac{2\zeta}{\Delta}}+\frac{8\pi m^* kT}{ah^2}\sqrt{\frac{\mu^* B}{\Delta}}\times \\ &\times\sum_{l=1}^{\infty}\frac{(-1)^l}{l^{1/2}\sinh\left(\pi^2 lkT/\mu^* B\right)}\left[\sin\left(\frac{\pi l\zeta}{\mu^* B}\right)C\left(\sqrt{\frac{2l\zeta}{\mu^* B}}\right)+\cos\left(\frac{\pi l\zeta}{\mu^* B}\right)S\left(\sqrt{\frac{2l\zeta}{\mu^* B}}\right)\right] \end{aligned} \quad (13)$$

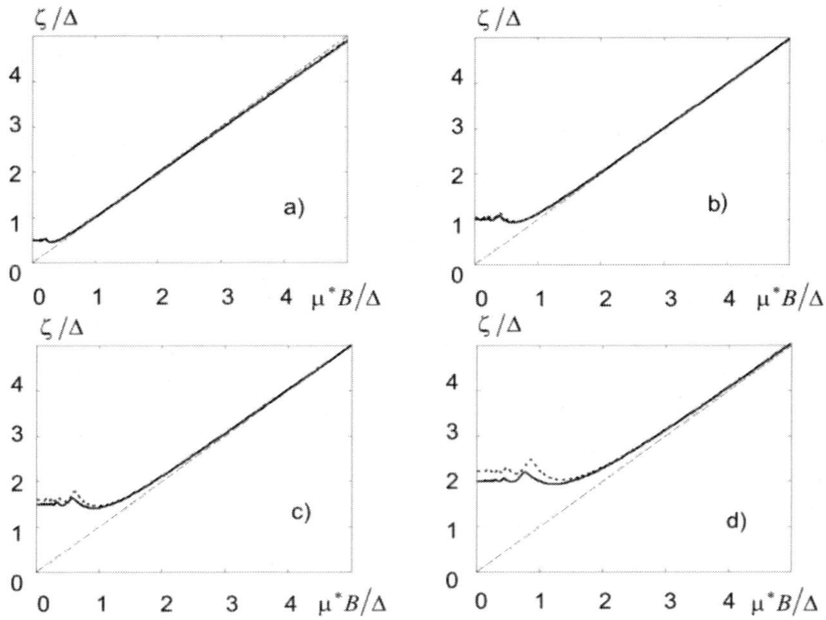

Figure 1. [1,2] Field dependence of chemical potential at $0 \le \mu^* B/\Delta \le 5$. The thin dashed lines represent asymptotic solution.

Concentration of charge carriers n_0 in left-hand side of Eq. (13) is defined according to formula (12), $C(x)$ and $S(x)$ are Fresnel cosine and sine integrals respectively.

Equations (11) and (13) may be solved numerically and the results of these numerical solutions are shown in Figure 1.

The figure demonstrates that at equal charge carriers' concentrations, the chemical potential of charge carriers for real layered crystal is smaller than that in effective mass approximation because the density of states for band spectrum (2) in energy region $0 < \varepsilon/\Delta \le 2$ is higher than that in effective mass approximation.

Moreover, the density of states for dispersion law (2) has two singularities: at $\varepsilon/\Delta = 0$ and $\varepsilon/\Delta = 2$, whereas the density of states in effective mass approximation has singularity only at $\varepsilon/\Delta = 0$. This is the reason why the difference between solid and dashed curves in Figure 1 increases, when the concentration of charge carriers grows, i.e., the ratio ζ_0/Δ grows. The case $\zeta_0/\Delta = 2$ corresponds to the transition from the closed FS to the open one.

However, we can see that at strong magnetic fields, i.e., when $\mu^* B/\Delta \ge 1$, the solid and dashed curves in Figure 1 are close to one another. The reason for such behavior of chemical potential lies in condensation of charge carriers on the bottom of the lowest *Landau band* with number $n = 0$. Equation (4) at strong magnetic field has the following asymptotic solution:

[1] Note 1: Unless stated otherwise, the plots in this chapter are drawn for the following conditions: $kT/\Delta = 0.03$ and plot a) $\zeta_0/\Delta = 0.5$, b) $\zeta_0/\Delta = 1$, c) $\zeta_0/\Delta = 1.5$, d) $\zeta_0/\Delta = 2$.

[2] Note 2: The following is also valid for all curves in this chapter: Solid curves are for real layered crystal, dashed curves are for effective mass approximation.

$$\zeta(B) = \mu^* B + \Delta\left[1 - \cos\left(\frac{f(\gamma_0)\Delta}{2\mu^* B}\right)\right] \tag{14}$$

For the closed FS we have:

$$f(\gamma_0) = (\gamma_0 - 1)\arccos(1 - \gamma_0) + \sqrt{2\gamma_0 - \gamma_0^2} \tag{15}$$

In effective mass approximation Eq. (14) has the form:

$$\zeta(B) = \mu^* B + \frac{\Delta^3 f^2(\gamma_0)}{8(\mu^* B)^2} \tag{16}$$

Formulae (14) and (16) mean that “Landau’s magnetic tube”, which represents the FS in this case, is squeezed in C-axis direction. This squeeze has to be taken into account, when we calculate the thermoelectromotive coefficient.

4. Calculation and Analysis of Thermoelectromotive Coefficient Dependence

This section focuses on the calculation and analysis of thermoelectromotive coefficient dependence on magnetic field for different scattering mechanisms.

4.1. The Model of Constant Relaxation Time

At first the variants of thermoelectromotive coefficient dependence on the magnetic field under conditions of charge carriers scattering by charged impurities is considered. This scattering mechanism is the leading one at helium temperatures.

Let's begin from the model of constant relaxation time [4], i.e., assume that $\tau(x) = \tau_0$. In the framework of this model after reducing fractions we obtain such dimensionless expressions for the quantities A, B, C:

$$A = \sum_{l=1}^{\infty}(-1)^{l-1} f_l^{th}\left\{\sin\left(\pi l\frac{\zeta-\Delta}{\mu^* B}\right)\left[(C_0 - C_2)J_0\left(\frac{\pi l\Delta}{\mu^* B}\right) + \sum_{r=1}^{\infty}(-1)^r \times \right.\right.$$
$$\left.\times(2C_{2r} - C_{2r+2} - C_{2r-2})J_{2r}\left(\frac{\pi l\Delta}{\mu^* B}\right)\right] + \cos\left(\pi l\frac{\zeta-\Delta}{\mu^* B}\right)\sum_{r=0}^{\infty}(-1)^r \times \tag{17}$$
$$\left.\times\left(2C_{2r+1} - C_{2r+3} - C_{|2r-1|}\right)J_{2r+1}\left(\frac{\pi l\Delta}{\mu^* B}\right)\right\}$$

$$B = 0.5(C_0 - C_2) \tag{18}$$

$$C = \sum_{l=1}^{\infty}(-1)^l f_l^\sigma \left\{ \cos\left(\pi l \frac{\zeta - \Delta}{\mu^* B}\right)\left[(C_0 - C_2)J_0\left(\frac{\pi l \Delta}{\mu^* B}\right) + \sum_{r=1}^{\infty}(-1)^r \times \right.\right.$$
$$\left. \times (2C_{2r} - C_{2r+2} - C_{2r-2})J_{2r}\left(\frac{\pi l \Delta}{\mu^* B}\right)\right] - \sin\left(\pi l \frac{\zeta - \Delta}{\mu^* B}\right)\sum_{r=0}^{\infty}(-1)^r \times \tag{19}$$
$$\left. \times \left(2C_{2r+1} - C_{2r+3} - C_{|2r-1|}\right)J_{2r+1}\left(\frac{\pi l \Delta}{\mu^* B}\right)\right\}$$

Here

$$C_0 = \arccos\left(1 - \frac{\zeta - \mu^* B}{\Delta}\right), C_m = \frac{\sin mC_0}{m} \tag{20}$$

In effective mass approximation, the above formulae may be rewritten as:

$$A = \sum_{l=1}^{\infty}(-1)^{l-1} f_l^{th}\left\{\frac{\mu^* B}{\pi l \Delta}\sqrt{\frac{2(\zeta - \mu^* B)}{\Delta}} - \frac{1}{\pi}\left(\frac{\mu^* B}{l\Delta}\right)^{3/2}\left[\cos\left(\frac{\pi l \zeta}{\mu^* B}\right)\times \right.\right.$$
$$\left.\left. C\left[\sqrt{2l\left(\frac{\zeta - \mu^* B}{\mu^* B}\right)}\right] + \sin\left(\frac{\pi l \zeta}{\mu^* B}\right)S\left[\sqrt{2l\left(\frac{\zeta - \mu^* B}{\mu^* B}\right)}\right]\right]\right\} \tag{21}$$

$$B = \frac{1}{6}\left[\frac{2(\zeta - \mu^* B)}{\Delta}\right]^{3/2} \tag{22}$$

$$C = \sum_{l=1}^{\infty}(-1)^l f_l^\sigma \frac{1}{\pi}\left(\frac{\mu^* B}{l\Delta}\right)^{3/2}\left\{\sin\left(\frac{\pi l \zeta}{\mu^* B}\right)C\left[\sqrt{2l\left(\frac{\zeta - \mu^* B}{\mu^* B}\right)}\right] - \right.$$
$$\left. \cos\left(\frac{\pi l \zeta}{\mu^* B}\right)S\left[\sqrt{2l\left(\frac{\zeta - \mu^* B}{\mu^* B}\right)}\right]\right\} \tag{23}$$

While subtracting the value of charge carriers minimal energy in the quantizing magnetic field from the value of chemical potential, in formulae (20)-(23) the FS squeeze is explicitly taken into account.

In quasi-classical approximation, i.e. at $\Delta/\mu^* B >> 1$, $\zeta/\mu^* B >> 1$, formulae (17) and (21) as well as (19) and (23) give the same results, namely:

$$A = \frac{1}{\sqrt{2}\pi}\left(\frac{\mu^* B}{\Delta}\right)^{3/2}\sum_{l=1}^{\infty}(-1)^l f_l^{th} l^{-3/2}\sin\left(\frac{\pi l \zeta}{\mu^* B} + \frac{\pi}{4}\right), \tag{24}$$

$$C = \frac{1}{\sqrt{2}\pi}\left(\frac{\mu^* B}{\Delta}\right)^{3/2}\sum_{l=1}^{\infty}(-1)^l f_l^{\sigma} l^{-3/2}\cos\left(\frac{\pi l\zeta}{\mu^* B}-\frac{\pi}{4}\right), \tag{25}$$

Therefore, in quasi-classical approximation the thermoelectromotive coefficient may be represented as:

$$\alpha = \frac{1.5\alpha_0}{\pi}\left(\frac{\mu^* B}{\zeta}\right)^{3/2}\sum_{l=1}^{\infty}(-1)^l f_l^{th} l^{-3/2}\cos\left(\frac{\pi l\zeta}{\mu^* B}-\frac{\pi}{4}\right), \tag{26}$$

It can be seen from formula (26) that in quasi-classical approximation the parameters that characterize the effects of the layered structure does not appear explicitly. However, the difference between the values of the chemical potential for real layered crystal and in effective mass approximation must be taken into account. Namely, under the condition of constant concentration of charge carriers the chemical potential of real layered crystal is smaller than that in effective mass approximation. Thus, even in quasi-classical approximation, the effects of the layered structure will manifest themselves in the decrease of oscillation frequencies, phase retardation and increase of oscillation amplitudes.

The calculation results of the thermoelectromotive coefficient in the model of constant relaxation time in the magnetic field range $0.04 \le \mu^* B/\Delta \le 0.1$ are shown in Figure 2. This figure demonstrates that the effects of the layered structure in weak quantizing magnetic fields manifest themselves in the phase retardation of the thermoelectromotive coefficient oscillations, the decrease of the total thermoelectromotive coefficient and the increase of relative contribution of thermoelectromotive coefficient oscillations into the total thermoelectromotive coefficient.

Yet it can be seen that if the ratio ζ_0/Δ increases and $\zeta_0/\Delta \le 2$, then oscillation frequency and retardation of oscillations phase increase and the total thermoelectromotive coefficient decreases.

The physical reason of this phenomenon consists in a slower dependence of the FS cross-section on longitudinal quasi-momentum for real layered crystal in comparison with such dependence in effective mass approximation. If $\zeta_0/\Delta = 2$, then the field dependences of thermoelectromotive coefficient for real layered crystal and those in effective mass approximation are quite different. The physical reason for such a difference lies in the singularity of the density of states for electron spectrum of real layered crystal at $\zeta/\Delta = 2$. As mentioned above, this singularity is not taken into account in effective mass approximation. This leads to overestimation of the contribution of low-frequency oscillations into the thermoelectromotive coefficient.

At $\mu^* B/\Delta = 0.1$ and $\zeta_0/\Delta = 0.5; 1; 1.5; 2$ the thermoelectromotive coefficients for real layered crystal are $3.278 \cdot 10^{-6}$, $2.156 \cdot 10^{-6}$, $1.984 \cdot 10^{-6}$, $7.514 \cdot 10^{-7}$ and $4.487 \cdot 10^{-7}$ V/K respectively. In effective mass approximation these thermoelectromotive coefficients are $3.191 \cdot 10^{-6}$, $1.294 \cdot 10^{-6}$ and $9.487 \cdot 10^{-7}$ V/K respectively.

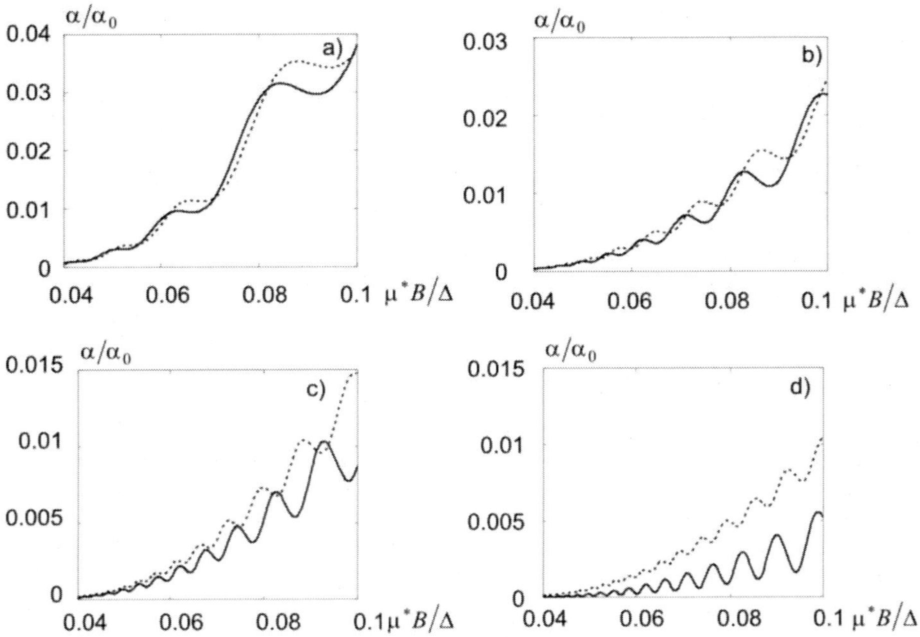

Figure 2. Field dependence of thermoelectromotive coefficient in the model of constant relaxation time at $0.04 \le \mu^* B/\Delta \le 0.1$.

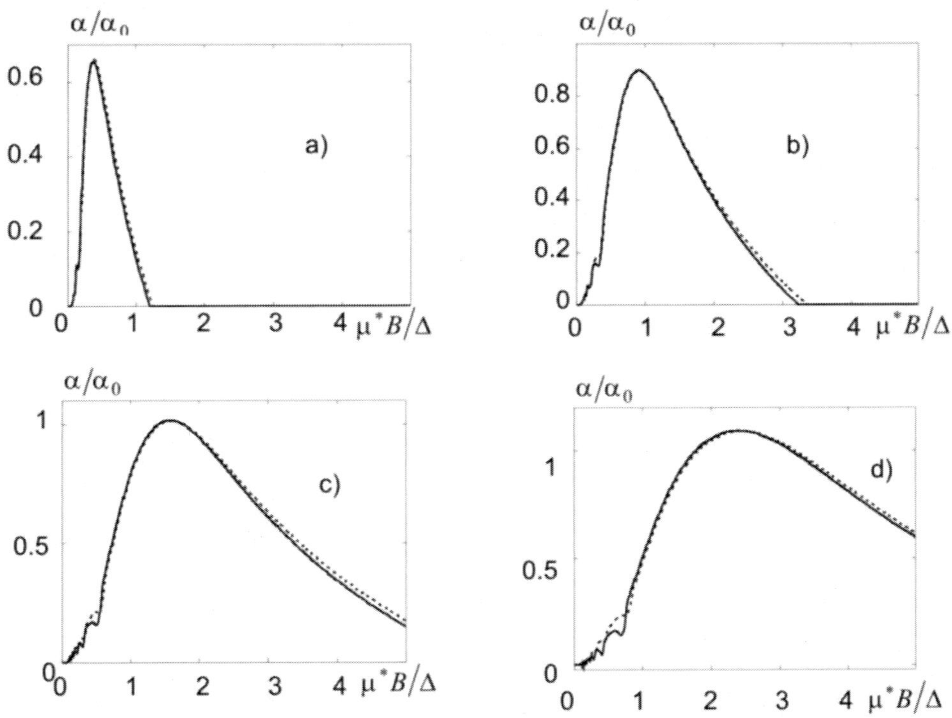

Figure 3. Field dependence of thermoelectromotive coefficient in model of constant relaxation time at $0 \le \mu^* B/\Delta \le 5$.

The results of calculations of thermoelectromotive coefficient at $0 \le \mu^* B/\Delta \le 5$ are shown in Figure 3. This figure demonstrates that layered structure effects become more sharply manifested when ratio ζ_0/Δ increases. The field dependence of thermoelectromotive coefficient for real layered crystal is characterized by not only the principal maximum, but also by many side maxima, which are absent in effective mass approximation. But the principal maxima for real layered crystal and in effective mass approximation are practically coincident. At $\zeta_0/\Delta = 0.5; 1; 1.5; 2$ these maximal values are $5.632 \cdot 10^{-5}$, $7.728 \cdot 10^{-5}$, $8.754 \cdot 10^{-5}$, $9.41 \cdot 10^{-5}$ eV/K respectively.

In effective mass approximation the maximal values of the thermoelectromotive coefficient are $5.692 \cdot 10^{-5}$, $7.737 \cdot 10^{-5}$, $8.772 \cdot 10^{-5}$, $9.393 \cdot 10^{-5}$ eV/K respectively. Such small differences may be explained by the FS squeeze in the strong quantizing magnetic field. This leads to the validity of effective mass approximation. While $0.5 \le \zeta_0/\Delta \le 1$, the principal maximum increases quickly, but if $1.5 \le \zeta_0/\Delta \le 2$, it increases slowly and achieves saturation. If ζ_0/Δ increases, the maximum of thermoelectromotive coefficient shifts to higher inductions of the magnetic field. The existence of the principal maximum may be explained in the following way.

We can see that if the induction of magnetic field becomes greater than zero, then the total thermoelectromotive coefficient increases. But if the quantizing magnetic field is strong enough, the thermoelectromotive coefficient must decrease due to the FS squeeze. Moreover, in certain sense the increase of the quantizing magnetic field induction is equivalent to the sample cooling. Therefore, the thermoelectromotive coefficient must have a maximum at certain induction of the magnetic field. If ratio ζ_0/Δ increases, then this induction also increases. After the thermoelectromotive coefficient maximum decreases, it tends to zero in a strong quantizing magnetic field according to certain asymptotic law. We can define this law as follows. Let us substitute the value of ζ in formulae (6) – (8) according to (16). Then the factor $(-1)^l$ will be compensated. Further, for small x we shall substitute $\sin x \approx x$ and $\cos x \approx 1$. Then f_l^{th} is represented as:

$$f_l^{th} = \frac{\left(\pi^2 lkT/\mu^* B\right)^2}{2\sinh\left(\pi^2 lkT/\mu^* B\right)} \tag{27}$$

Then we shall take into account the fact that the numerical analysis demonstrates the validity of such formulae:

$$\sum_{l=1}^{\infty} \frac{\delta l}{\sinh(\delta l)} = \frac{2.467}{\delta} \tag{28}$$

$$\sum_{l=1}^{\infty} \frac{\delta l^3}{\sinh(\delta l)} = \frac{12.176}{\delta^3} \tag{29}$$

Before integrating over x, $W(x)$ is expanded into a series up to quadratic term. After all transformations of formulae (6 – 8) by taking into account the formulae (16) and (26) the following asymptotic law of thermoelectromotive coefficient decrease in a strong quantizing magnetic field (in ultra-quantum limit) is obtained:

$$\alpha = 0.039\alpha_0 f^2(\gamma_0)\frac{\Delta^3}{(\mu^* B)^2 kT} \tag{30}$$

This asymptotic law is rather difficult to understand, because the thermoelectromotive coefficient does not tend to zero when $T = 0$. But this law does not lead to physically incorrect results at real finite temperatures. Really, for $\Delta = 0.01\text{eV}$, $m^* = 0.01m_0$, $B = 60\text{T}$, $T = 3\text{K}$, $\zeta_0/\Delta = 0.5;1;1.5;2$ the values of the thermoelectromotive coefficient α are 12.6, 107.4, 393.2 and 1060nV/K respectively. These values are much less than the maximal thermoelectromotive coefficient values at corresponding ζ_0/Δ. Yet we can see that while $\zeta_0/\Delta < 2$, the thermoelectromotive coefficient maximum in effective mass approximation is higher than that for a real layered crystal. But if $\zeta_0/\Delta = 2$, then the maximum of thermoelectromotive coefficient in effective mass approximation is lower than that for a real layered crystal. This may be explained, if we shall take into account that at strong magnetic fields, the dependence of the thermoelectromotive coefficient on the ratio ζ_0/Δ is determined only by singularity of the density of charge carriers' states at $\zeta_0/\Delta = 0$. On the contrary, the dependence of thermoelectromotive coefficient on the ratio ζ_0/Δ in the region of oscillations is determined by two singularities of density of charge carriers' states. Therefore, for every concentration of charge carriers there exists certain "optimal" interval of the quantizing magnetic fields where the effects of the layered structure in the field dependence of the thermoelectromotive coefficient are observed best.

4.2. Model of "Constant Free Path" with Relaxation Time, Independent of Magnetic Field

Now let us consider the model of "*constant free path*" of charge carriers. In this model, the following expression for relaxation time is typical [12]:

$$\tau(x) = l/v_f(x), \tag{31}$$

where l – free path, caused by scattering of charge carriers by charged impurities, $v_f(x)$ – total Fermi-velocity of current carrier with dimensionless longitudinal quasi-momentum x.

In the framework of this model we may obtain such expressions for quantities A, B, C:

$$A = \sum_{l=1}^{\infty}(-1)^{l-1} f_l^{th} \int_0^{(\zeta-\mu^* B)/\Delta} \sqrt{\frac{2y - y^2}{A_m(\zeta\Delta^{-1} - y) + 2y - y^2}} \sin\left[\frac{\pi l \Delta}{\mu^* B}(\zeta\Delta^{-1} - y)\right] dy, \tag{32}$$

$$B = 0.5 \int_0^{(\zeta-\mu^* B)/\Delta} \sqrt{\frac{2y - y^2}{A_m(\zeta\Delta^{-1} - y) + 2y - y^2}} dy, \tag{33}$$

$$C = \sum_{l=1}^{\infty} (-1)^l f_l^\sigma \int_0^{(\zeta-\mu^* B)/\Delta} \sqrt{\frac{2y - y^2}{A_m(\zeta\Delta^{-1} - y) + 2y - y^2}} \cos\left[\frac{\pi l\Delta}{\mu^* B}(\zeta\Delta^{-1} - y)\right] dy. \tag{34}$$

In these formulae we have denoted $A_m = 2m_l^*/m^*$, where m_l^* – effective mass of current carrier along C-axis direction. In effective mass approximation formulae (31) – (33) may be rewritten as:

$$A = \sum_{l=1}^{\infty} (-1)^{l-1} f_l^{th} \int_0^{(\zeta-\mu^* B)/\Delta} \sqrt{\frac{2y}{A_m(\zeta\Delta^{-1} - y) + 2y}} \sin\left[\frac{\pi l\Delta}{\mu^* B}(\zeta\Delta^{-1} - y)\right] dy, \tag{35}$$

$$B = 0.5 \int_0^{(\zeta-\mu^* B)/\Delta} \sqrt{\frac{2y}{A_m(\zeta\Delta^{-1} - y) + 2y}} dy, \tag{35}$$

$$C = \sum_{l=1}^{\infty} (-1)^l f_l^\sigma \int_0^{(\zeta-\mu^* B)/\Delta} \sqrt{\frac{2y}{A_m(\zeta\Delta^{-1} - y) + 2y}} \cos\left[\frac{\pi l\Delta}{\mu^* B}(\zeta\Delta^{-1} - y)\right] dy. \tag{36}$$

Further we shall note that if we do not consider the influence of the magnetic field on the scattering process, then we must take into account the dependence of chemical potential on the magnetic field only in arguments of trigonometric functions and in integration limits of formulae (31)-(36), whereas under the square radicals we must substitute the chemical potential at zero magnetic field. However, if we take into consideration the influence of the magnetic field on the scattering process, then we must take into account the dependence of chemical potential on magnetic field everywhere in these formulae.

The results of the thermoelectromotive coefficient calculations in “constant free path” model with relaxation time independent of the magnetic field at $A_m = 5$ in the magnetic field range $0.04 \le \mu^* B/\Delta \le 0.1$ for ratios $\zeta_0/\Delta = 0.5; 1; 1.5; 2$ are shown in Figure 4.

The plots in Figure 4 demonstrate that the effects of the layered structure are observed in phase retardation and in the increase of amplitude of the thermoelectromotive coefficient and (or) its derivative oscillations. We can see that at $\zeta_0/\Delta = 0.5$ the thermoelectromotive coefficient represents a monotonously increasing function of the magnetic field, while its derivative is an oscillating function. At $\zeta_0/\Delta \ge 1$, however, the thermoelectromotive coefficient also becomes an oscillating function of the magnetic field. The difference between the solid and dashed curves is best observed when $\zeta_0/\Delta = 2$. This may be explained by singularity of the density of states for the real layered crystal at $\zeta_0/\Delta = 2$ and by the absence of planes tangent to the FS and perpendicular to the magnetic field direction in this case.

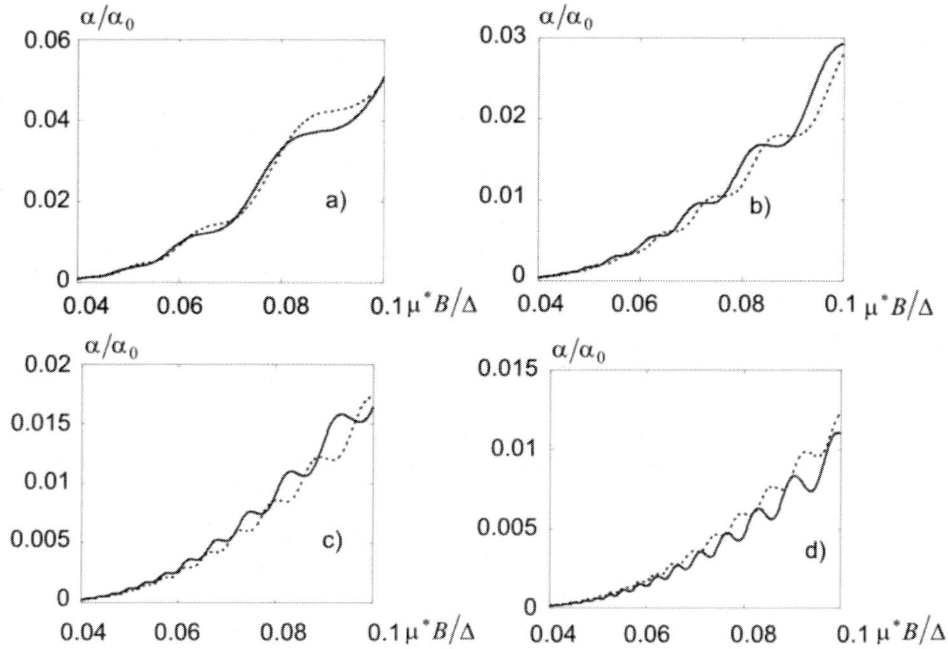

Figure 4. Field dependence of thermoelectromotive coefficient in the model of "constant free path" with relaxation time, independent of magnetic field at $0.04 \le \mu^* B/\Delta \le 0.1$, $A_m = 5$.

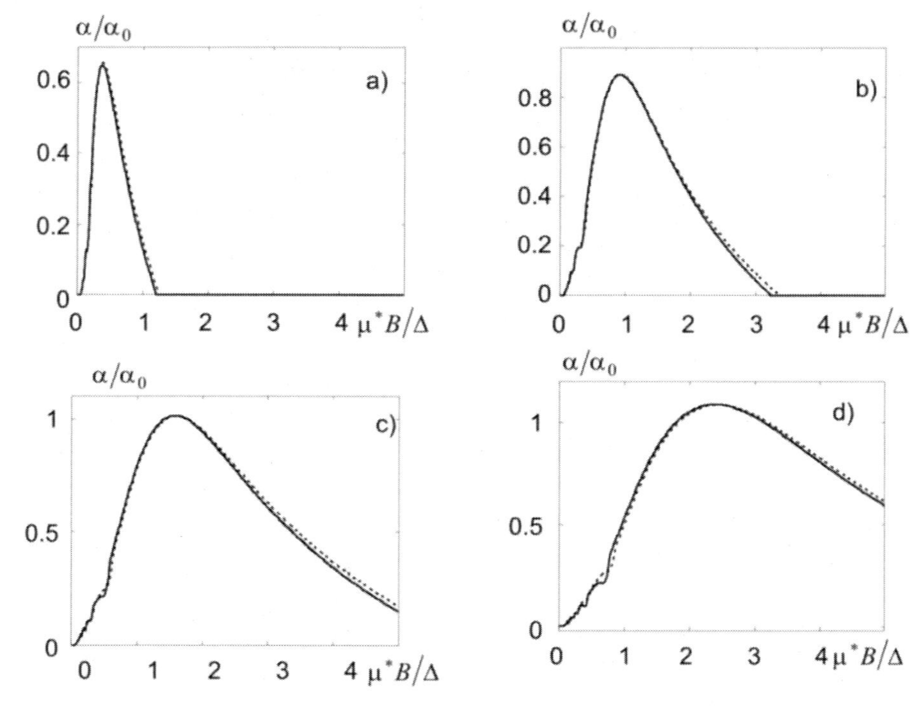

Figure 5. Field dependence of thermoelectromotive coefficient in the model of "constant free path" with relaxation time, independent of the magnetic field at $0 \le \mu^* B/\Delta \le 5$, $A_m = 5$.

At $\mu^* B/\Delta = 0.1$ and $\zeta_0/\Delta = 0.5; 1; 1.5; 2$ for the real layered crystal the thermoelectromotive coefficients are $4.313 \cdot 10^{-6}$, $2.501 \cdot 10^{-6}, 1.380 \cdot 10^{-6}$ and $9.487 \cdot 10^{-7}$ V/K respectively. In effective mass approximation these thermoelectromotive coefficients are $4.054 \cdot 10^{-6}$, $2.415 \cdot 10^{-6}$, $1.466 \cdot 10^{-6}$ and $1.035 \cdot 10^{-6}$ V/K respectively.

The results of thermoelectromotive coefficient calculations in "constant free path" model with relaxation time independent of the magnetic field at $A_m = 5$ in magnetic field range $0 \le \mu^* B/\Delta \le 5$ for ratios $\zeta_0/\Delta = 0.5; 1; 1.5; 2$ are shown in Figure 5.

Comparing Figure 5 with Figure 3, one finds that the model of relaxation time influences the difference between the dashed and solid curves. Namely, the effects of the layered structure are better observed, if the relaxation time is constant. Maxima of the thermoelectromotive coefficient in the "constant free path" model with the relaxation time independent of the magnetic field for the real layered crystal at $\zeta_0/\Delta = 0.5;1;1.5;2$ are $5.580 \cdot 10^{-5}, 7.693 \cdot 10^{-5}, 8.729 \cdot 10^{-5}$ and $9.375 \cdot 10^{-5}$ V/K respectively.

These maxima are somewhat smaller than those in the model of constant relaxation time. In effective mass approximation these maxima are $5.649 \cdot 10^{-5}, 7.693 \cdot 10^{-5}, 8.72 \cdot 10^{-5}, 9.367 \cdot 10^{-5}$ V/K respectively. Such small differences occur due to validity of the effective mass approximation in the super strong quantizing magnetic fields. These maxima are smaller than the corresponding maxima in the model of constant relaxation time.

The results of thermoelectromotive coefficient calculations in the "constant free path" model with relaxation time independent of the magnetic field at $A_m = 15$ in magnetic field range $0.04 \le \mu^* B/\Delta \le 0.1$ for ratios $\zeta_0/\Delta = 0.5; 1; 1.5; 2$ are shown in Figure 6. As mentioned before, the effects of layered structure are manifested in phase retardation and in the increase of the amplitude of the thermoelectromotive coefficient and (or) its derivative oscillations.

Now Figure 6 demonstrates that at $A_m = 15$ and $\zeta_0/\Delta = 0.5; 1$ only the derivative of the thermoelectromotive coefficient is an oscillating function of the magnetic field. But at $A_m = 15$ and $\zeta_0/\Delta = 1.5; 2$ the thermoelectromotive coefficient also becomes an oscillating function of the magnetic field. At $\mu^* B/\Delta = 0.1$, $A_m = 15$ and $\zeta_0/\Delta = 0.5; 1; 1.5; 2$ the thermoelectromotive coefficients for the real layered crystal are $4.916 \cdot 10^{-6}, 2.933 \cdot 10^{-6}, 1.725 \cdot 10^{-6}, 9.487 \cdot 10^{-7}$ V/K respectively. In effective mass approximation, these thermoelectromotive coefficients are $4.830 \cdot 10^{-6}$, $2.933 \cdot 10^{-6}, 1.897 \cdot 10^{-6}$ and $1.380 \cdot 10^{-6}$ V/K respectively.

As depicted in Figure 4, the *anisotropy parameter* A_m does not influence the frequencies of thermoelectromotive coefficient oscillations. On the other hand, the form of oscillations and the difference between the dashed and solid lines on each figure are determined by this parameter, A_m.

The results of thermoelectromotive coefficient calculations in the "constant free path" model with the relaxation time independent of the magnetic field at $A_m = 15$ in magnetic field range $0 \le \mu^* B/\Delta \le 5$ for ratios $\zeta_0/\Delta = 0.5; 1; 1.5; 2$ are shown in Figure 7. At $A_m = 15$ and $\zeta_0/\Delta = 0.5; 1$, the effects of the layered structure are observed better after the principal maximum of the thermoelectromotive coefficient than before this maximum.

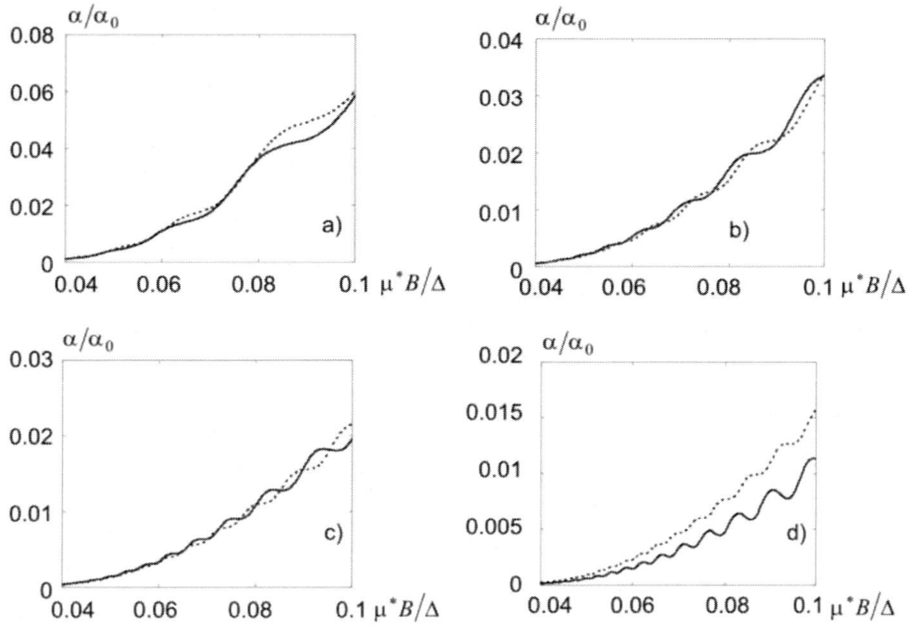

Figure 6. Field dependence of thermoelectromotive coefficient in the model of "constant free path" with relaxation time, independent of the magnetic field at $0.04 \leq \mu^* B/\Delta \leq 0.1$, $A_m = 5$.

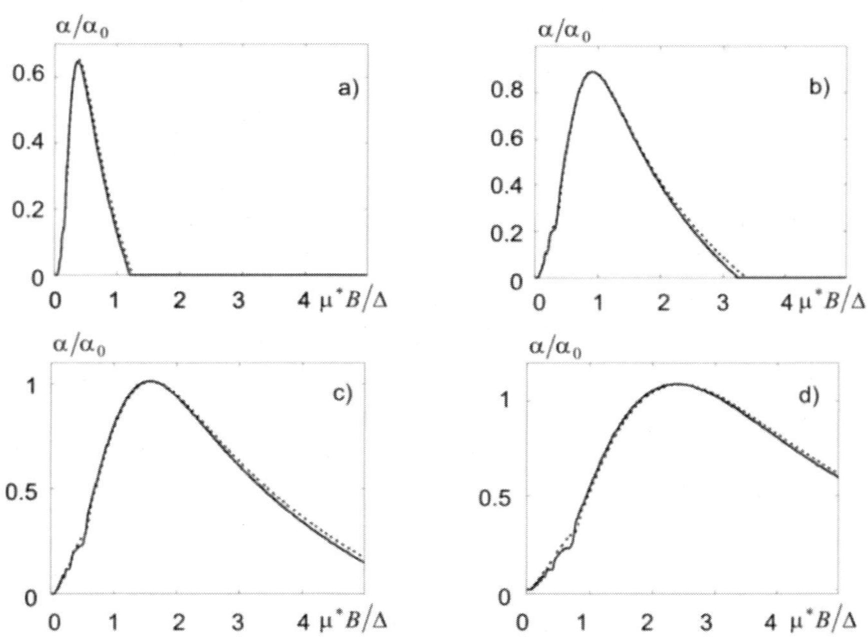

Figure 7. Field dependence of thermoelectromotive coefficient in the model of "constant free path" with the relaxation time, independent of the magnetic field at $0 \leq \mu^* B/\Delta \leq 5$, $A_m = 15$.

But at $A_m = 15$ and $\zeta_0/\Delta = 1.5;2$ the effects of the layered structure become well observed before principal maximum of the thermoelectromotive coefficient. Such behavior of the thermoelectromotive coefficient may be explained by the above mentioned peculiarities of the thermoelectromotive coefficient oscillations and its derivative, as well as by the difference between the density of charge carriers' states for the real layered crystal and that in effective mass approximation.

The maxima of the thermoelectromotive coefficient for the real layered crystal at $A_m = 15$ and $\zeta_0/\Delta = 0.5;1;1.5;2$ are $5.563\cdot10^{-5}, 7.676\cdot10^{-5}, 8.720\cdot10^{-5}, 9.367\cdot10^{-5}$ V/K respectively. In effective mass approximation these maxima of the thermoelectromotive coefficient are $5.623\cdot10^{-5}, 7.668\cdot10^{-5}, 8.711\cdot10^{-5}, 9.349\cdot10^{-5}$ V/K respectively.

The results of thermoelectromotive coefficient calculations in the "constant free path" model with the relaxation time independent of the magnetic field at $A_m = 25$ in magnetic field range $0.04 \le \mu^* B/\Delta \le 0.1$ for ratios $\zeta_0/\Delta = 0.5;1;1.5;2$ are shown in Figure 8.As mentioned before, the effects of the layered structure are manifested in phase retardation of and in the increase of amplitude of thermoelectromotive coefficient and (or) its derivative oscillations. Now Figure 8 demonstrates that at $A_m = 25$ and $\zeta_0/\Delta = 0.5;1$ for the real layered crystal as well as in effective mass approximation only the derivative of the thermoelectromotive coefficient represents an oscillating function of the magnetic field.

But at $A_m = 25$ and $\zeta_0/\Delta = 1.5;2$ for the real layered crystal the thermoelectromotive coefficient also becomes an oscillating function of the magnetic field. And vice versa, at $A_m = 25$ and $\zeta_0/\Delta = 1.5;2$ in effective mass approximation only the derivative of the thermoelectromotive coefficient is an oscillating function of the magnetic field, while the actual thermoelectromotive coefficient remains a monotonously increasing function of the magnetic field.

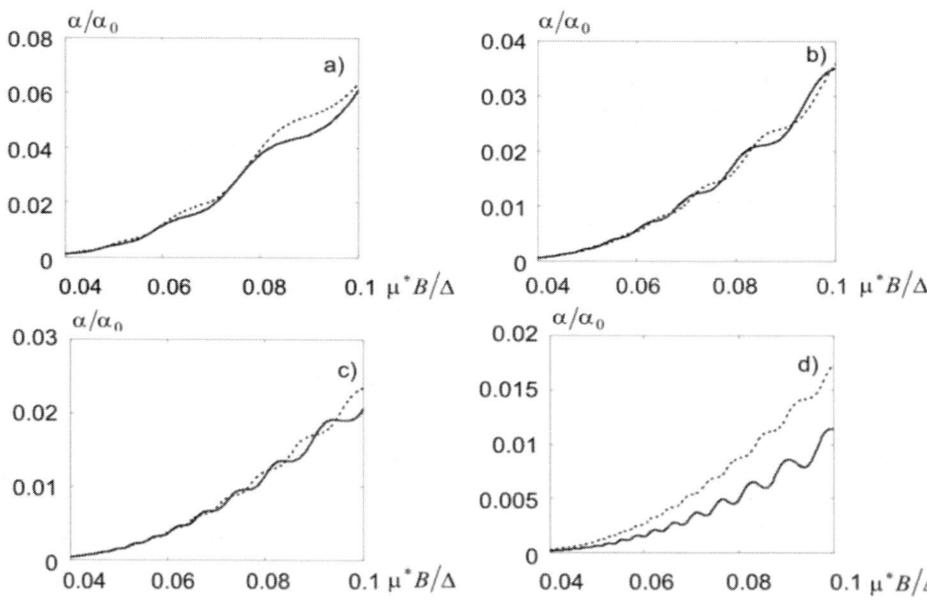

Figure 8. Field dependence of thermoelectromotive coefficient in model of "constant free path" with relaxation time, independent on magnetic field at $0.04 \le \mu^* B/\Delta \le 0.1$, $A_m = 25$.

At $\mu^* B/\Delta = 0.1$, $A_m = 25$ and $\zeta_0/\Delta = 0.5; 1; 1.5; 2$ the thermoelectromotive coefficients for the real layered crystal are $5.089 \cdot 10^{-6}, 3.019 \cdot 10^{-6}, 1.811 \cdot 10^{-6}, 9.487 \cdot 10^{-7}$ V/K respectively. In effective mass approximation these thermoelectromotive coefficients are $5.089 \cdot 10^{-6}$, $3.105 \cdot 10^{-6}$, $2.070 \cdot 10^{-6}$, $1.466 \cdot 10^{-6}$ V/K respectively. Comparing Figure 8 with Figure 4, one observes, as before, that *anisotropy parameter* A_m has no influence on the frequencies of thermoelectromotive coefficient oscillations. But the form of oscillations and the difference between the dashed and the solid line on each figure are determined by A_m.

The results of thermoelectromotive coefficient calculations in the "constant free path" model with the relaxation time independent of the magnetic field at $A_m = 25$ in magnetic field range $0 \le \mu^* B/\Delta \le 5$ for ratios $\zeta_0/\Delta = 0.5; 1; 1.5; 2$ are shown in Figure 9.

Similar to earlier ones, the plots in this figure demonstrate that at $A_m = 25$ and $\zeta_0/\Delta = 0.5;1$ the effects of the layered structure after the principal maximum of thermoelectromotive coefficient are better observed than those before this maximum. But at $A_m = 25$ and $\zeta_0/\Delta = 1.5;2$ the effects of the layered structure become well observed before the principal maximum of the thermoelectromotive coefficient. Such behavior of the thermoelectromotive coefficient, as seen in the case of $A_m = 5$ and 15, may be explained by the above mentioned peculiarities of the thermoelectromotive coefficient oscillations and of its derivative as well as by the difference between the density of states of charge carriers for the real layered crystal and in effective mass approximation.

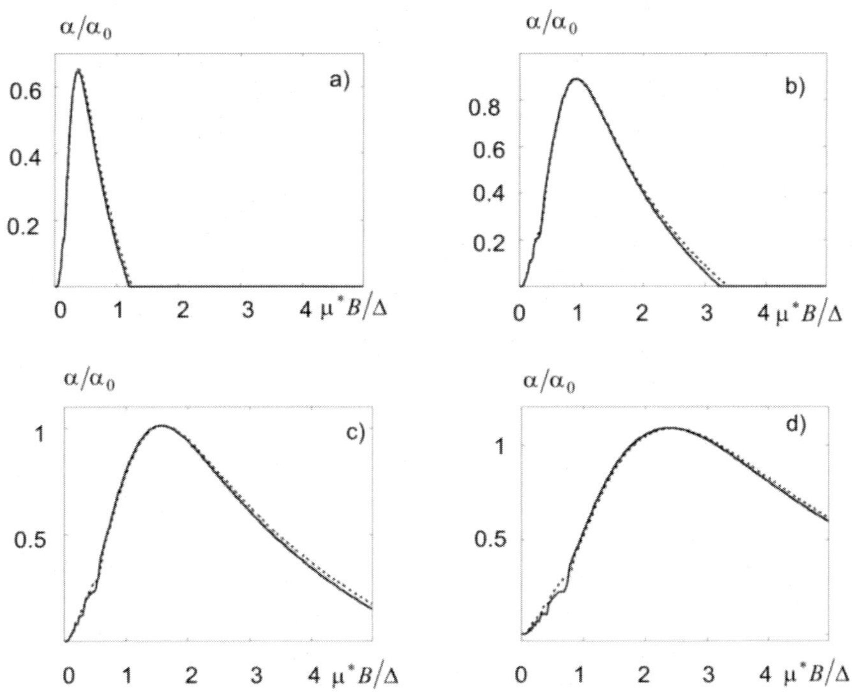

Figure 9. Field dependence of thermoelectromotive coefficient in the model of "constant free path" with the relaxation time, independent of the magnetic field at $0 \le \mu^* B/\Delta \le 5$, $A_m = 25$.

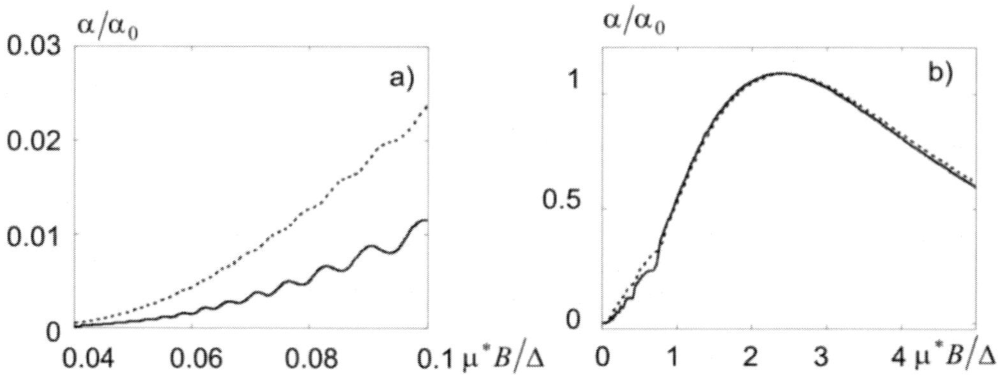

Figure 10. Field dependence of thermoelectromotive coefficient in the model of "constant free path" with the relaxation time, independent of the magnetic field at $\zeta_0/\Delta = 2,\ A_m = 10^5$ and a) $0.04 \le \mu^* B/\Delta \le 0/1$, b) $0 \le \mu^* B/\Delta \le 5$.

The maxima of the thermoelectromotive coefficient for the real layered crystal at $A_m = 25$ and $\zeta_0/\Delta = 0.5;\ 1;\ 1.5;\ 2$ are $5.554 \cdot 10^{-5}$, $7.668 \cdot 10^{-5}$, $8.711 \cdot 10^{-5}$ and $9.367 \cdot 10^{-5}$ V/K respectively. In effective mass approximation, these maxima of the thermoelectromotive coefficient are $5.615 \cdot 10^{-5}$, $7.668 \cdot 10^{-5}$, $8.703 \cdot 10^{-5}$ and $9.349 \cdot 10^{-5}$ V/K respectively.

Now, for instance, we consider the special case $A_m = 10^5, \zeta_0/\Delta = 2$. This case approximately corresponds to dichalcogenides of transition metals or intercalated graphite. The field dependence of the thermoelectromotive coefficient for this case is shown in Figure 10. This figure also shows that even at such A_m the thermoelectromotive coefficient and its derivative for the real layered crystal remain the oscillating functions of the magnetic field. In effective mass approximation for such A_m only the derivative of the thermoelectromotive coefficient represents an oscillating function of the magnetic field. As before, such difference between the dashed and solid curves may be explained by the singularity of the density of states for the real layered crystal at $\zeta_0/\Delta = 2$ and by the absence of planes, tangent to the FS and perpendicular to the magnetic field direction in this case. For the real layered crystal at $A_m = 10^5, \zeta_0/\Delta = 2, \mu^* B/\Delta = 0.1$ the thermoelectromotive coefficient is $1.035 \cdot 10^{-6}$ eV/K. In effective mass approximation the thermoelectromotive coefficient is $2.070 \cdot 10^{-6}$ eV/K.

At $A_m = 10^5$ and $\zeta_0/\Delta = 2$, the effects of the layered structure before the principal maximum of the thermoelectromotive coefficient are well observed. As before, such behavior of the thermoelectromotive coefficient may be explained by the above mentioned peculiarities of thermoelectromotive coefficient oscillations and of its derivative. These peculiarities may be explained by the singularity of the density of states for the real layered crystal at $\zeta_0/\Delta = 2$ and by the absence of planes, tangent to the FS and perpendicular to the magnetic field direction in this case. The maximum of the thermoelectromotive coefficient for the real layered crystal at $A_m = 10^5$ and $\zeta_0/\Delta = 2$ is $9.358 \cdot 10^{-5}$ V/K. In effective mass approximation this maximum is $9.349 \cdot 10^{-5}$ V/K. It means that as $A_m \to \infty$ the field dependence of the thermoelectromotive coefficient tends to some limiting law. But at high magnetic field, i.e.

when $\mu^* B/\Delta >> 1$ the thermoelectromotive coefficient is completely determined by formula (30).

4.3. Model of "Constant Free Path" with Relaxation Time, Dependent on Magnetic Field

In this section, the case of the magnetic field influence on the scattering process due to the field dependence of chemical potential is considered. It means that we take into account the field dependence of the total velocity of charge carriers, i.e. we assume that $\tau(x,B) = l/v_f(x,B)$.

The results of thermoelectromotive coefficient calculations in the "constant free path" model with the relaxation time, dependent on the magnetic field, at $A_m = 5$ in magnetic field range $0.04 \le \mu^* B/\Delta \le 0.1$ for ratios $\zeta_0/\Delta = 0.5;1;1.5;2$ are shown in Figure 11.

Similar to the model with $\tau(x) = l/v_f(x)$, this figure demonstrates that the effects of the layered structure are manifested in phase retardation and in the increase of the thermoelectromotive coefficient amplitude and (or) its derivative oscillations.

It can be seen that at $\zeta_0/\Delta = 0.5$ the thermoelectromotive coefficient is a monotonously increasing function of the magnetic field, but its derivative represents an oscillating function of the magnetic field. However at $\zeta_0/\Delta \ge 1$ the thermoelectromotive coefficient also becomes an oscillating function of the magnetic field.

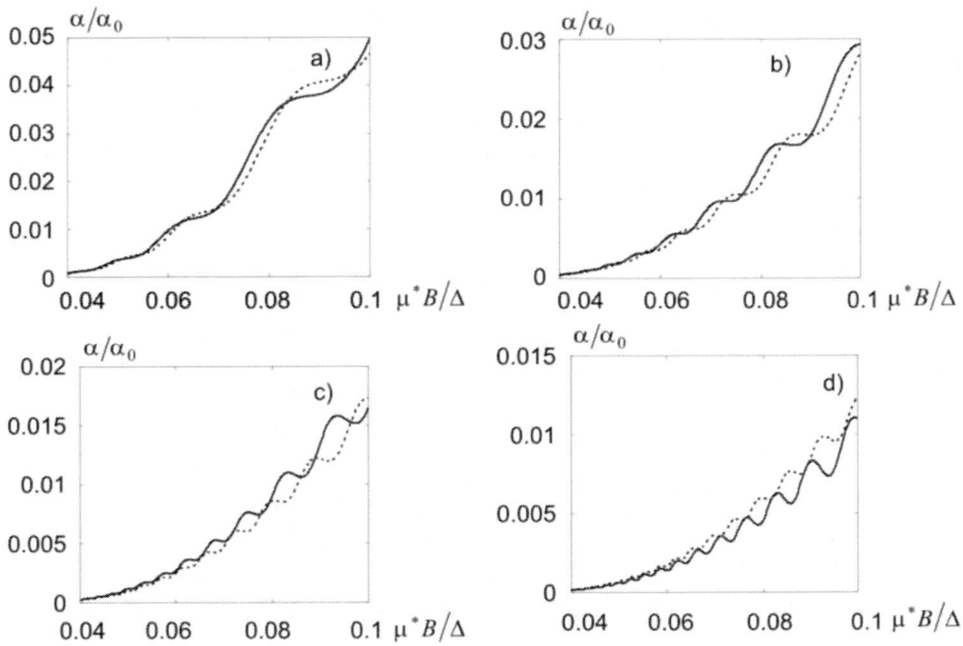

Figure 11. Field dependence of thermoelectromotive coefficient in the model of "constant free path" with the relaxation time, dependent on magnetic field at $0.04 \le \mu^* B/\Delta \le 0.1$, $A_m = 5$.

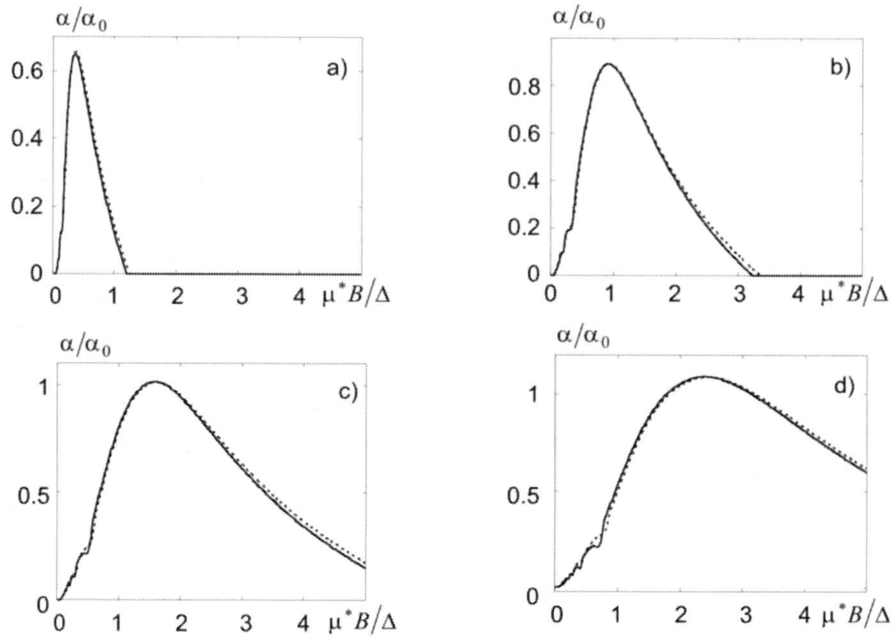

Figure 12. Field dependence of thermoelectromotive coefficient in the model of "constant free path" with the relaxation time, dependent on the magnetic field a $0 \le \mu^* B/\Delta \le 5$, $kT/\Delta = 0.03$, $A_m = 5$.

The difference between the solid and dashed curves is best observed when $\zeta_0/\Delta = 2$. It may be explained by the singularity of the density of states for the real layered crystal at $\zeta_0/\Delta = 2$ and by the absence of planes tangent to the FS and perpendicular to the magnetic field direction in this case. At $\mu^* B/\Delta = 0.1$ and $\zeta_0/\Delta = 0.5;1;1.5;2$ for the real layered crystal the thermoelectromotive coefficients are $4.313 \cdot 10^{-6}$, $2.501 \cdot 10^{-6}$, $1.380 \cdot 10^{-6}$ and $9.487 \cdot 10^{-7}$ V/K respectively. In effective mass approximation these thermoelectromotive coefficients are $4.054 \cdot 10^{-6}$, $2.415 \cdot 10^{-6}$, $1.466 \cdot 10^{-6}$ and $1.035 \cdot 10^{-6}$ V/K respectively.

The results of thermoelectromotive coefficient calculations in the "constant free path" model with the relaxation time, dependent on the magnetic field, at $A_m = 5$ in magnetic field range $0 \le \mu^* B/\Delta \le 5$ for ratios $\zeta_0/\Delta = 0.5;1;1.5;2$ are shown in Figure 12.

The maxima of the thermoelectromotive coefficient in "constant free path" model with the relaxation time, dependent on the magnetic field, for the real layered crystal at $\zeta_0/\Delta = 0.5;1;1.5;2$ are $5.580 \cdot 10^{-5}$, $7.693 \cdot 10^{-5}$, $8.737 \cdot 10^{-5}$ and $9.384 \cdot 10^{-5}$ V/K respectively.

In effective mass approximation these maxima are $5.632 \cdot 10^{-5}$,$7.693 \cdot 10^{-5}$,$8.729 \cdot 10^{-5}$ and $9.367 \cdot 10^{-5}$ V/K respectively. As before, such slight difference occurs due to the validity of the effective mass approximation in the super strong quantizing magnetic fields. These maxima are smaller than the corresponding maxima in the model of constant relaxation time.

The results of thermoelectromotive coefficient calculations in the "constant free path" model with the relaxation time dependent on the magnetic field at $A_m = 15$ in magnetic field range $0.04 \le \mu^* B/\Delta \le 0.1$ for ratios $\zeta_0/\Delta = 0.5;1;1.5;2$ are shown in Figure 13. As in the model

$\tau(x) = l/v_f(x)$, the effects of the layered structure are manifested in phase retardation and in the increase of the thermoelectromotive coefficient amplitude and (or) its derivative oscillations.

The figure also demonstrates that at $A_m = 15$ and $\zeta_0/\Delta = 0.5;1$ only the derivative of thermoelectromotive coefficient represents an oscillating function of the magnetic field. At $A_m = 15$ and $\zeta_0/\Delta = 1.5;2$, however, the actual thermoelectromotive coefficient for the real layered crystal becomes an oscillating function of the magnetic field, while in the effective mass approximation, the thermoelectromotive coefficient remains a monotonously increasing function of the magnetic field. At $\mu^* B/\Delta = 0.1$, $A_m = 15$ and $\zeta_0/\Delta = 0.5;1;1.5;2$ the thermoelectromotive coefficients for the real layered crystal are $4.916 \cdot 10^{-6}$, $2.933 \cdot 10^{-6}$, $1.725 \cdot 10^{-6}$ and $9.487 \cdot 10^{-7}$ V/K respectively. In effective mass approximation these thermoelectromotive coefficients are $4.744 \cdot 10^{-6}$, $2.933 \cdot 10^{-6}$, $1.897 \cdot 10^{-6}$ and $1.380 \cdot 10^{-6}$ V/K respectively.

The results of thermoelectromotive coefficient calculations in the "constant free path" model with the relaxation time dependent on the magnetic field at $A_m = 15$ in magnetic field range $0 \le \mu^* B/\Delta \le 5$ for ratios $\zeta_0/\Delta = 0.5;1;1.5;2$ are shown in Figure 14.

Similarly to the model with $\tau(x) = l/v_f(x)$, the plots in Figure 14 demonstrate that at $A_m = 15$ and $\zeta_0/\Delta = 0.5;1$ the effects of the layered structure after the principal maximum of the thermoelectromotive coefficient are observed better than before this maximum. But at $A_m = 15$ and $\zeta_0/\Delta = 1.5;2$ the effects of the layered structure become well observed before the principal maximum of the thermoelectromotive coefficient.

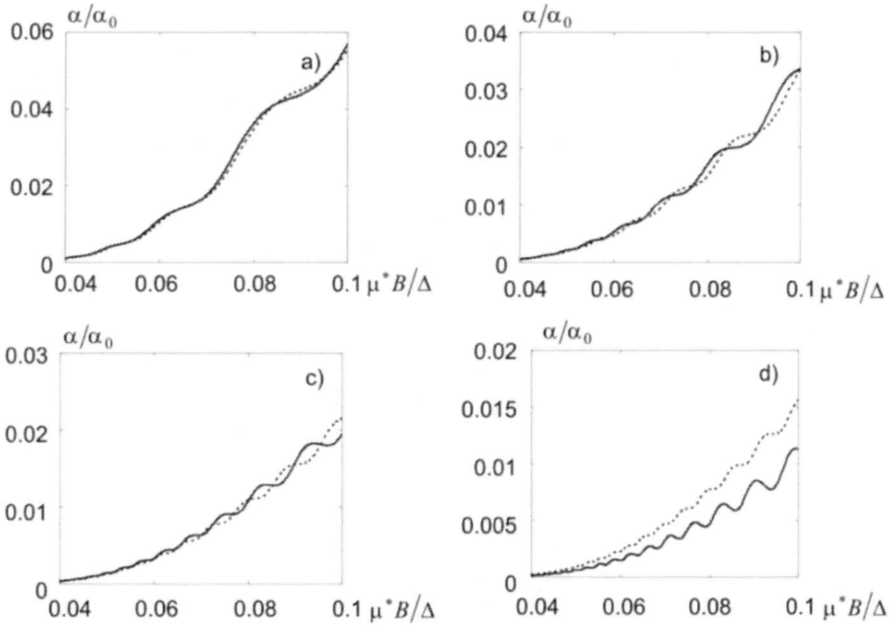

Figure 13. Field dependence of thermoelectromotive coefficient in the model of "constant free path" with the relaxation time, dependent on the magnetic field at $0.04 \le \mu^* B/\Delta \le 0.1$, $A_m = 15$.

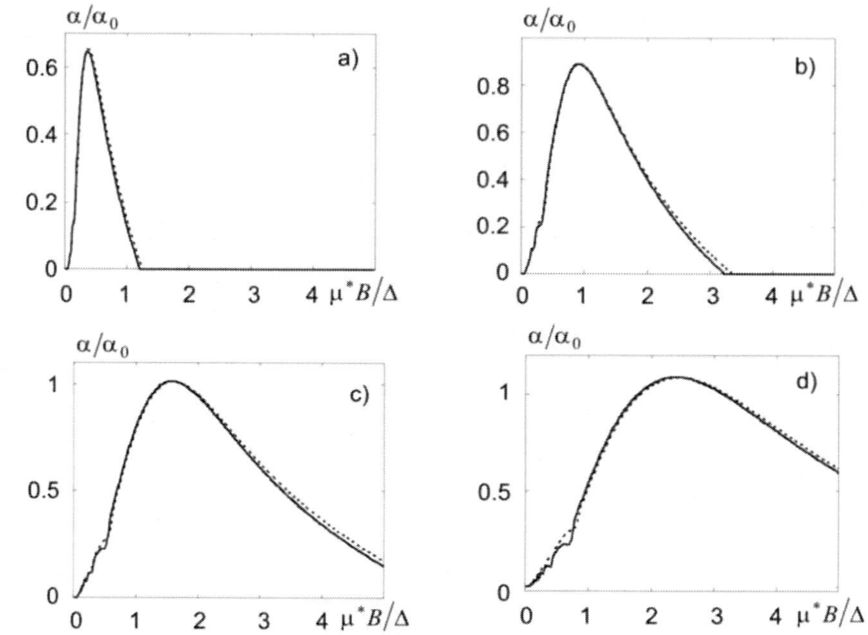

Figure 14. Field dependence of thermoelectromotive coefficient in the model of "constant free path" with the relaxation time, dependent on the magnetic field at $0 \le \mu^* B/\Delta \le 5$, $A_m = 15$.

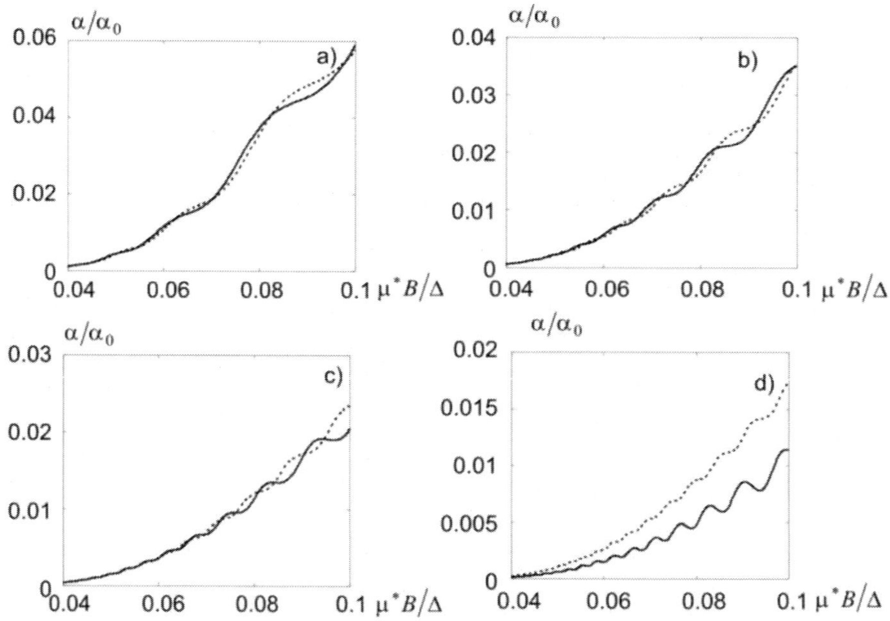

Figure 15. Field dependence of thermoelectromotive coefficient in the model of "constant free path" with the relaxation time, dependent on the magnetic field at $0.04 \le \mu^* B/\Delta \le 0.1$, $A_m = 25$.

As in the model with $\tau(x) = l/v_f(x)$, such behavior of the thermoelectromotive coefficient may be explained by the above mentioned peculiarities of the thermoelectromotive coefficient oscillations and its derivative and the difference between the density of charge carriers' states for the real layered crystal and in the effective mass approximation. The maxima of the thermoelectromotive coefficient for the real layered crystal at $A_m = 15$ and $\zeta_0/\Delta = 0.5; 1; 1.5; 2$ are $5.554 \cdot 10^{-5}$, $7.676 \cdot 10^{-5}$, $8.720 \cdot 10^{-5}$ and $9.375 \cdot 10^{-5}$ V/K respectively. In the effective mass approximation these maxima of the thermoelectromotive coefficient are $5.623 \cdot 10^{-5}$, $7.676 \cdot 10^{-5}$, $8.711 \cdot 10^{-5}$ and $9.358 \cdot 10^{-5}$ V/K respectively.

The results of thermoelectromotive coefficient calculations in the "constant free path" model with the relaxation time dependent on the magnetic field at $A_m = 25$ in magnetic field range $0.04 \le \mu^* B/\Delta \le 0.1$ for ratios $\zeta_0/\Delta = 0.5; 1; 1.5; 2$ are shown in Figure 15.

Similar to the model with $\tau(x) = l/v_f(x)$, the effects of the layered structure are manifested in phase retardation and in the increase of the thermoelectromotive coefficient amplitude and (or) its derivative oscillations. The figure also demonstrates that at $A_m = 25$ and $\zeta_0/\Delta = 0.5;1$ as for the layered crystal as in the effective mass approximation only the derivative of the thermoelectromotive coefficient represents an oscillating function of the magnetic field.

But at $A_m = 25$ and $\zeta_0/\Delta = 1.5;2$ for the real layered crystal the thermoelectromotive coefficient also becomes an oscillating function of the magnetic field. And vice versa, at $A_m = 25$ and $\zeta_0/\Delta = 1.5;2$ in the effective mass approximation only the derivative of the thermoelectromotive coefficient is an oscillating function of the magnetic field, while the actual thermoelectromotive coefficient remains a monotonously increasing function of the magnetic field.

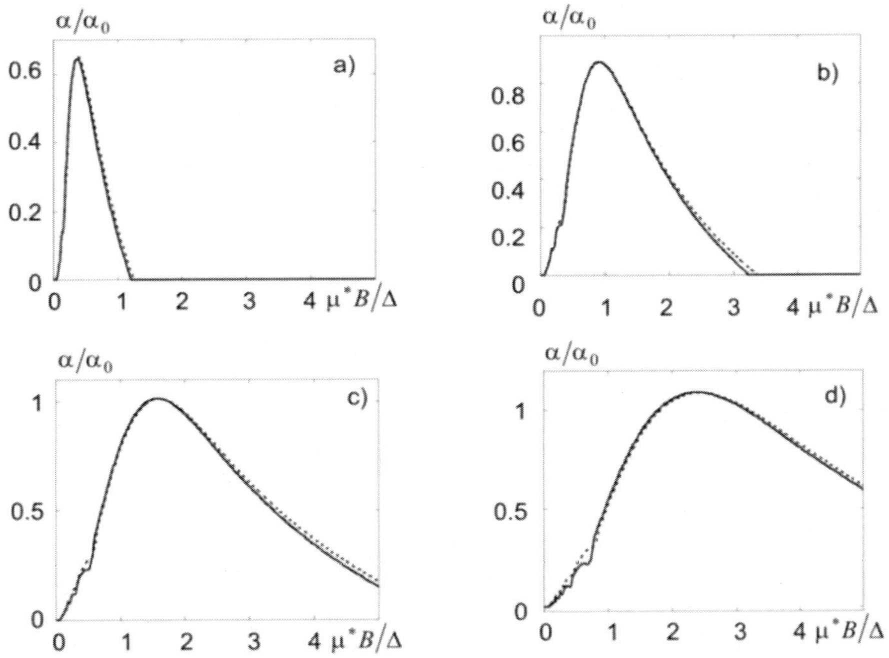

Figure 16. Field dependence of thermoelectromotive coefficient in the model of "constant free path" with the relaxation time, dependent on the magnetic field at $0 \le \mu^* B/\Delta \le 5$, $A_m = 25$.

At $\mu^* B/\Delta = 0.1$, $A_m = 25$ and $\zeta_0/\Delta = 0.5; 1; 1.5; 2$ the thermoelectromotive coefficients for the real layered crystal are $5.089 \cdot 10^{-6}$, $3.019 \cdot 10^{-6}$, $1.725 \cdot 10^{-6}$ and $9.487 \cdot 10^{-7}$ V/K respectively. In the effective mass approximation these thermoelectromotive coefficients are $5.003 \cdot 10^{-6}$, $3.105 \cdot 10^{-6}$, $2.070 \cdot 10^{-6}$, $1.552 \cdot 10^{-6}$ V/K respectively.

The results of thermoelectromotive coefficient calculations in the "constant free path" model with the relaxation time dependent on the magnetic field at $A_m = 25$ in magnetic field range $0 \le \mu^* B/\Delta \le 5$ for ratios $\zeta_0/\Delta = 0.5;\ 1;\ 1.5;\ 2$ are shown in Figure 16.

As in the model with $\tau(x) = l/v_f(x)$, this figure also demonstrates that at $A_m = 25$ and $\zeta_0/\Delta = 0.5; 1$ the effects of the layered structure after the principal maximum of the thermoelectromotive coefficient are observed better than before this maximum. But at $A_m = 25$ and $\zeta_0/\Delta = 1.5; 2$ the effects of the layered structure become well observed before the principal maximum of the thermoelectromotive coefficient. As in the model with $\tau(x) = l/v_f(x)$, such behavior of the thermoelectromotive coefficient may be explained by the above mentioned peculiarities of thermoelectromotive coefficient oscillations and its derivative and the difference between the density of charge carriers' states for the real layered crystal and in the effective mass approximation. The maxima of the thermoelectromotive coefficient for the real layered crystal at $A_m = 25$ and $\zeta_0/\Delta = 0.5; 1; 1.5; 2$ are $5.554 \cdot 10^{-5}$, $7.676 \cdot 10^{-5}$, $8.720 \cdot 10^{-5}$ and $9.375 \cdot 10^{-5}$ V/K respectively. In the effective mass approximation these maxima of the thermoelectromotive coefficient are $5.615 \cdot 10^{-5}$, $7.668 \cdot 10^{-5}$, $8.711 \cdot 10^{-5}$ and $9.375 \cdot 10^{-5}$ V/K respectively.

Now, as in the model with $\tau(x) = l/v_f(x)$, we consider the special case $A_m = 10^5, \zeta_0/\Delta = 2$. The field dependence of the thermoelectromotive coefficient for this case is shown in Figure 17. This figure demonstrates that even at such A_m the thermoelectromotive coefficient and its derivative for the real layered crystal remain the oscillating functions of the magnetic field. In the effective mass approximation for such A_m only the derivative of the thermoelectromotive coefficient is an oscillating function of the magnetic field.

As before, such difference between the dashed and solid curves may be explained by the singularity of the density of states for the real layered crystal at $\zeta_0/\Delta = 2$ and by the absence of planes, tangent to the FS and perpendicular to the magnetic field direction in this case. For the real layered crystal at $A_m = 10^5, \zeta_0/\Delta = 2, \mu^* B/\Delta = 0.1$, as in the model with $\tau(x) = l/v_f(x)$, the thermoelectromotive coefficient is $1.035 \cdot 10^{-6}$ eV/K. In the effective mass approximation the thermoelectromotive coefficient is $2.070 \cdot 10^{-6}$ eV/K.

At $A_m = 10^5$ and $\zeta_0/\Delta = 2$, the effects of the layered structure before the principal maximum of the thermoelectromotive coefficient are well observed. As before, such behavior of the thermoelectromotive coefficient may be explained by the above mentioned peculiarities of the thermoelectromotive coefficient oscillations and its derivative. As before, these peculiarities may be explained by the singularity of the density of states for the real layered crystal at $\zeta_0/\Delta = 2$ and by the absence of planes, tangent to the FS and perpendicular to the magnetic field direction in this case.

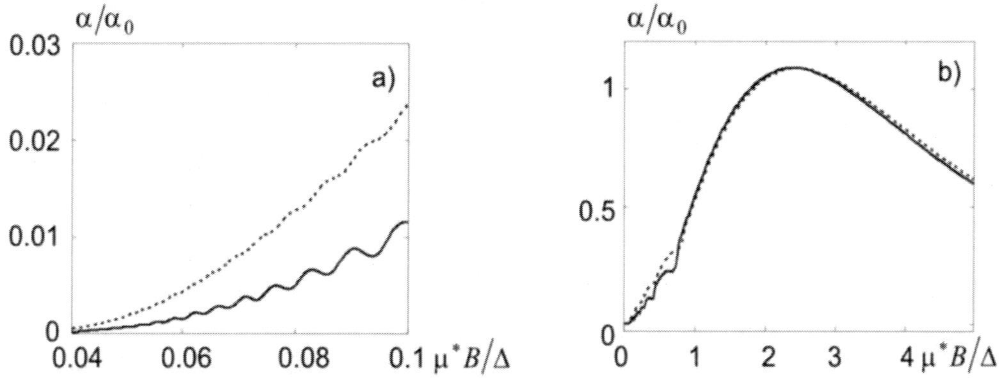

Figure 17. Field dependence of thermoelectromotive coefficient in the model of "constant free path" with the relaxation time, dependent on the magnetic field at $\zeta_0/\Delta = 2$, $A_m = 10^5$ and a) $0.04 \le \mu^* B/\Delta \le 0/1$, b) $0 \le \mu^* B/\Delta \le 5$.

The maximum of thermoelectromotive coefficient for the real layered crystal at $A_m = 10^5$ and $\zeta_0/\Delta = 2$ is $9.375 \cdot 10^{-5}$ V/K . In the effective mass approximation this maximum is $9.349 \cdot 10^{-5}$ V/K .

It means that as $A_m \to \infty$ the field dependence of the thermoelectromotive coefficient, as in the model with $\tau(x) = l/v_f(x)$, tends to some limiting law. But at high magnetic field, i.e. when $\mu^* B/\Delta >> 1$ the thermoelectromotive coefficient, as in the model with $\tau(x) = l/v_f(x)$, is completely determined by formula (30).

4.4. Model of Relaxation Time, Proportional to Fermi Velocity and Independence of Magnetic Field

This section of the chapter considers charge carriers scattering by acoustic phonons. At first, the models of relaxation time $\tau \propto v_f(x)$ and $\tau \propto v_f(x, B)$ [13] is considered. For these models, the following expressions for quantities A, B, C are valid:

$$A = \sum_{l=1}^{\infty} (-1)^{l-1} f_l^{th} \int_0^{(\zeta - \mu^* B)/\Delta} \sqrt{(2y - y^2)[A_m(\zeta\Delta^{-1} - y) + 2y]} \sin\left[\frac{\pi l \Delta}{\mu^* B}(\zeta\Delta^{-1} - y)\right] dy, \tag{37}$$

$$B = 0.5 \int_0^{(\zeta - \mu^* B)/\Delta} \sqrt{(2y - y^2)[A_m(\zeta\Delta^{-1} - y) + 2y]} dy, \tag{38}$$

$$C = \sum_{l=1}^{\infty} (-1)^{l} f_l^{\sigma} \int_0^{(\zeta - \mu^* B)/\Delta} \sqrt{(2y - y^2)[A_m(\zeta\Delta^{-1} - y) + 2y]} \cos\left[\frac{\pi l \Delta}{\mu^* B}(\zeta\Delta^{-1} - y)\right] dy. \tag{39}$$

As in the case of scattering by the charged impurities, in the model with $\tau \propto v_f(x)$ we must take into account the field dependence of chemical potential only in arguments of trigonometric functions and limits of integration. In the model with $\tau \propto v_f(x,B)$ we must consider the field dependence of chemical potential everywhere in formulae (37) – (39). The results of thermoelectromotive coefficient calculations in the model with the relaxation time $\tau \propto v_f(x)$ at $A_m = 5$ in the magnetic field range $0.04 \le \mu^* B/\Delta \le 0.1$ for ratios $\zeta_0/\Delta = 0.5;1;1.5;2$ are shown in Figure 18. As depicted in this figure, the effects of the layered structure are manifested in phase retardation and in the increase of the thermoelectromotive coefficient amplitude and its derivative oscillations. It can be seen that for all considered ratios ζ_0/Δ, the thermoelectromotive coefficient and its derivative are oscillating functions of the magnetic field. If ratio ζ_0/Δ increases, then the frequency of oscillations increases and the difference between the solid and dashed curves becomes more apparent. It is best observed at $\zeta_0/\Delta = 2$. As before, it may be explained by the singularity of the density of states for the real layered crystal at $\zeta_0/\Delta = 2$ and by the absence of planes, tangent to the FS and perpendicular to the magnetic field direction in this case.

At $\mu^* B/\Delta = 0.1$ and $\zeta_0/\Delta = 0.5;1;1.5;2$, for the real layered crystal, the thermoelectromotive coefficients are $2.329 \cdot 10^{-6}$, $1.466 \cdot 10^{-6}$, $2.937 \cdot 10^{-6}$ and $2.657 \cdot 10^{-7}$ V/K respectively. In the effective mass approximation these thermoelectromotive coefficients are $2.415 \cdot 10^{-6}$, $1.897 \cdot 10^{-6}$, $1.121 \cdot 10^{-6}$ and $7.729 \cdot 10^{-7}$ V/K respectively.

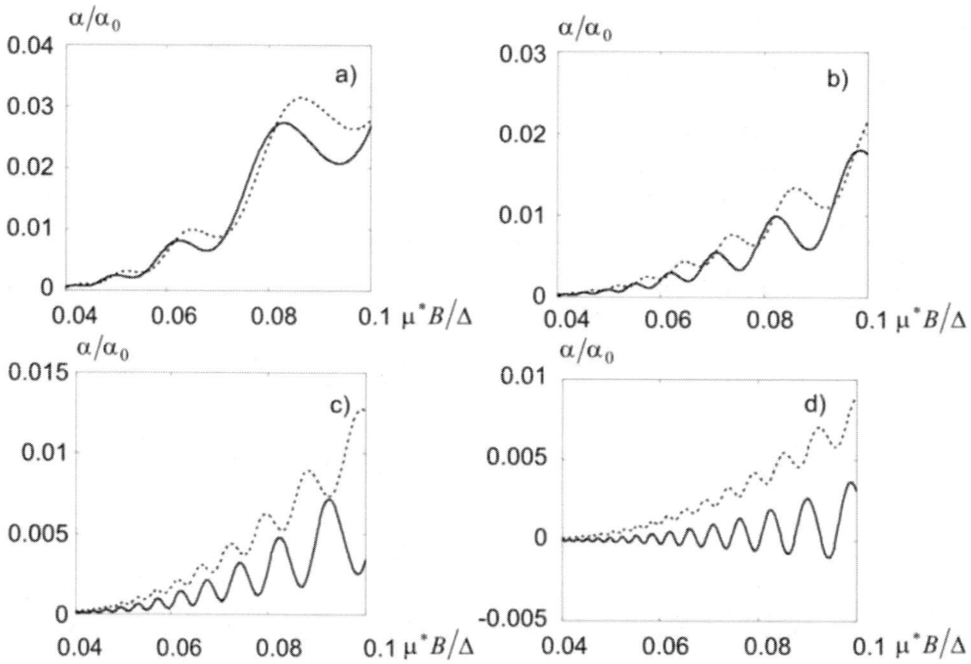

Figure 18. Field dependence of thermoelectromotive coefficient in the model with relaxation time $\tau \propto v_f(x)$, independent of the magnetic field at $0.04 \le \mu^* B/\Delta \le 0.1$

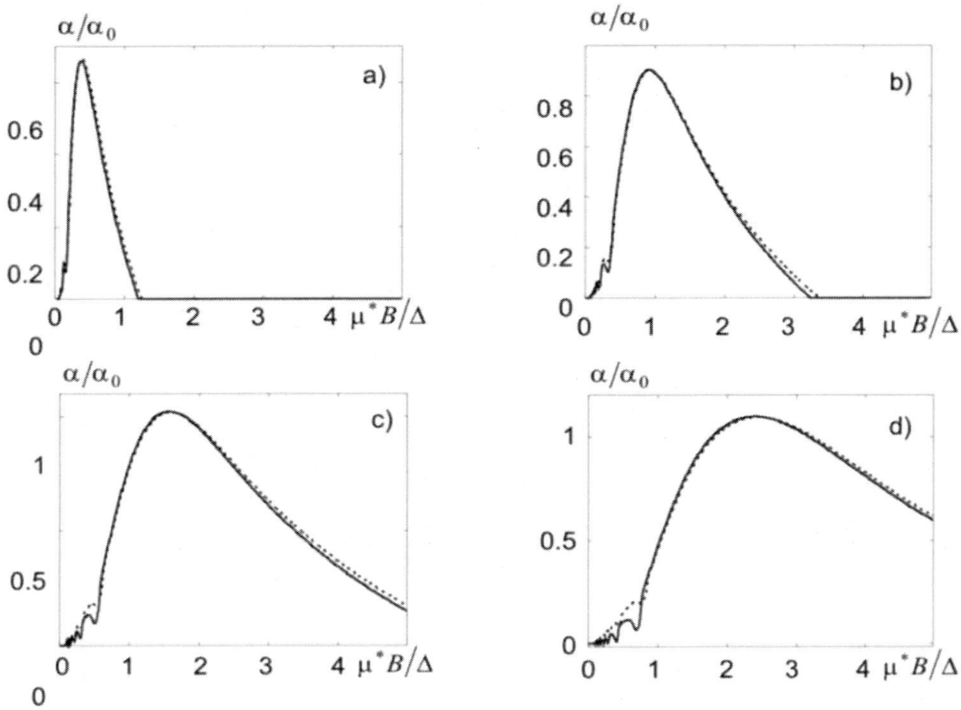

Figure 19. Field dependence of thermoelectromotive coefficient in the model with relaxation time $\tau \propto v_f(x)$, independent of the magnetic field at $0 \le \mu^* B/\Delta \le 5$, $A_m = 5$.

The results of thermoelectromotive coefficient calculations in the model with relaxation time $\tau \propto v_f(x)$ at $A_m = 5$ in magnetic field range $0 \le \mu^* B/\Delta \le 5$ for ratios $\zeta_0/\Delta = 0.5;1;1.5;2$ are shown in Figure 19. As demonstrated by the plots of this figure, the layered structure effects at $\zeta_0/\Delta = 0.5$ are well observed only after the principal maximum, but at $\zeta_0/\Delta \ge 1$ these effects become well observed in both cases: after and before the principal maximum. The maxima of the thermoelectromotive coefficient in the model $\tau(x) \propto v_f(x)$ with the relaxation time independent of the magnetic field for the real layered crystal at $\zeta_0/\Delta = 0.5;1;1.5;2$ are $5.684 \cdot 10^{-5}, 7.780 \cdot 10^{-5}, 8.806 \cdot 10^{-5}$ and $9.444 \cdot 10^{-5}$ V/K respectively. In effective mass approximation, these maxima are $5.744 \cdot 10^{-5}$, $7.762 \cdot 10^{-5}$, $8.780 \cdot 10^{-5}$, $9.418 \cdot 10^{-5}$ V/K respectively. As before, such small differences occur due to the validity of effective mass approximation in super strong quantizing magnetic fields. These maxima are somewhat greater than the corresponding maxima in the model of constant relaxation time.

The results of thermoelectromotive coefficient calculations in the model with the relaxation time $\tau(x) \propto v_f(x)$ at $A_m = 15$ in magnetic field range $0.04 \le \mu^* B/\Delta \le 0.1$ for ratios $\zeta_0/\Delta = 0.5;1;1.5;2$ are shown in Figure 20. The figure demonstrates that, as before, the effects of the layered structure are manifested in phase retardation and the increase of the thermoelectromotive coefficient amplitude and its derivative oscillations. It can be seen that for all considered ratios of ζ_0/Δ the thermoelectromotive coefficient and its derivative, as well as at $A_m = 5$, remain oscillating functions of the magnetic field.

Moreover, the thermoelectromotive coefficient oscillations and its derivative become more apparent, if A_m increases. If ratio ζ_0/Δ increases, then the frequency of oscillations increases and the difference between the solid and dashed curves becomes more apparent. This difference is best observed at $\zeta_0/\Delta = 2$.

As before, it may be explained by the singularity of the density of states for the real layered crystal at $\zeta_0/\Delta = 2$ and by the absence of planes, tangent to the FS and perpendicular to the magnetic field direction in this case. At $\mu^* B/\Delta = 0.1$ and $\zeta_0/\Delta = 0.5;1;1.5;2$ for the real layered crystal the thermoelectromotive coefficients are $1.725\cdot10^{-6}$,$1.294\cdot10^{-6}$,$1.852\cdot10^{-7}$ and $2.678\cdot10^{-7}$ V/K respectively. In effective mass approximation these thermoelectromotive coefficients are $1.725\cdot10^{-6}$, $1.552\cdot10^{-6}$, $8.625\cdot10^{-7}$ and $6.142\cdot10^{-7}$ V/K respectively.

The results of thermoelectromotive coefficient calculations in the model with the relaxation time $\tau \propto v_f(x)$ at $A_m = 15$ in magnetic field range $0 \le \mu^* B/\Delta \le 5$ for ratios $\zeta_0/\Delta = 0.5;1;1.5;2$ are shown in Figure 21. The figure demonstrates that the layered structure effects at $\zeta_0/\Delta = 0.5$ are well observed only after the principal maximum, but at $\zeta_0/\Delta \ge 1$ these effects become well observed both after principal maximum and before it.

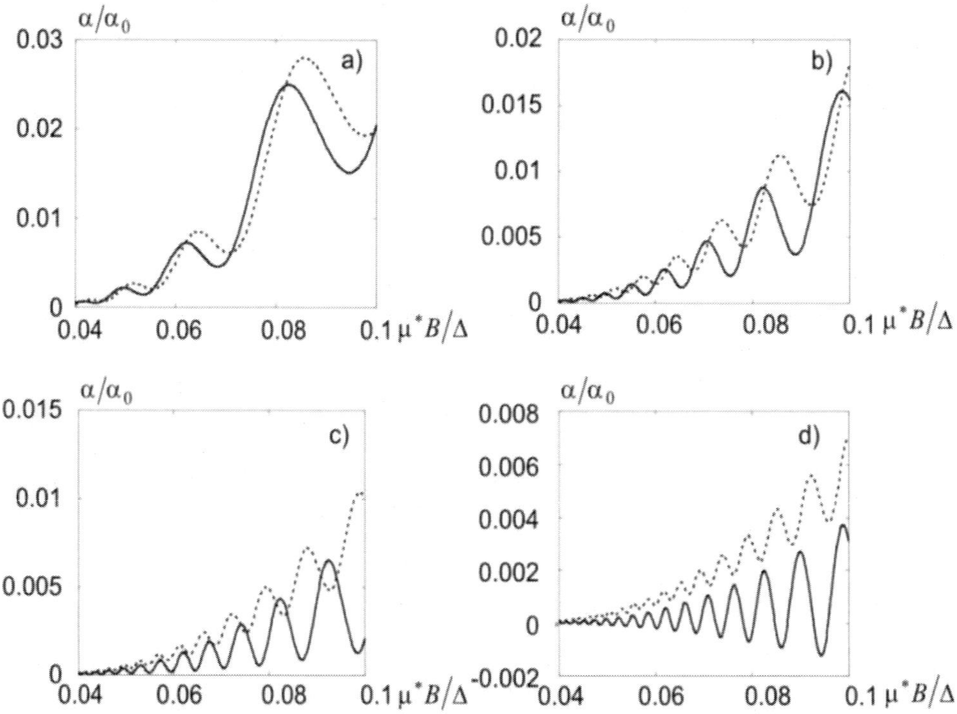

Figure 20. Field dependence of thermoelectromotive coefficient in the model with relaxation time $\tau \propto v_f(x)$, independent of magnetic field at $0.04 \le \mu^* B/\Delta \le 0.1$, $A_m = 15$.

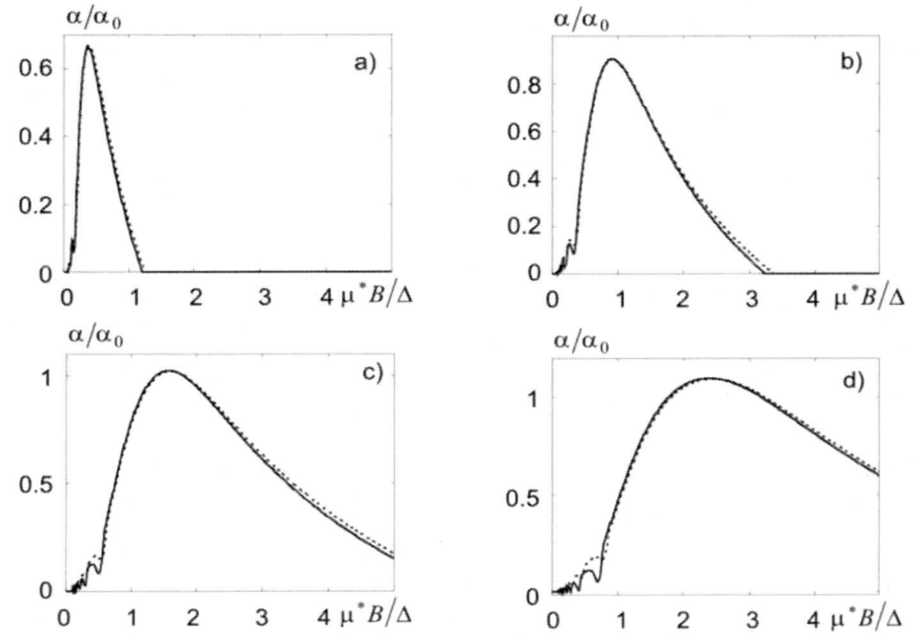

Figure 21. Field dependence of thermoelectromotive coefficient in the model with relaxation time $\tau \propto v_f(x)$, independent of the magnetic field at $0 \le \mu^* B/\Delta \le 5$, $A_m = 15$.

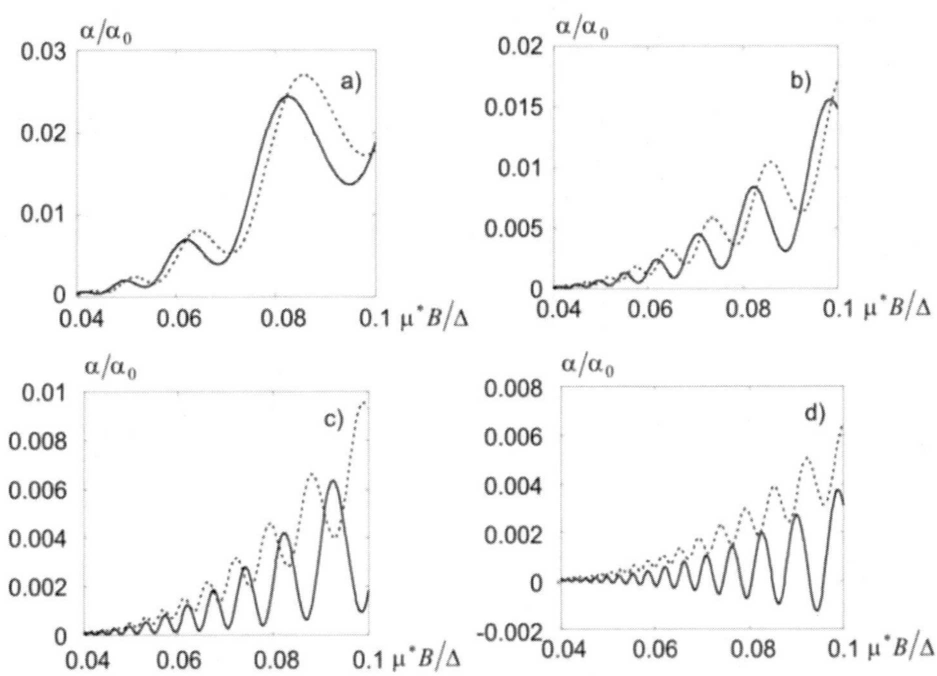

Figure 22. Field dependence of thermoelectromotive coefficient in the model with relaxation time $\tau \propto v_f(x)$, independent of the magnetic field at $0.04 \le \mu^* B/\Delta \le 0.1$, $A_m = 25$.

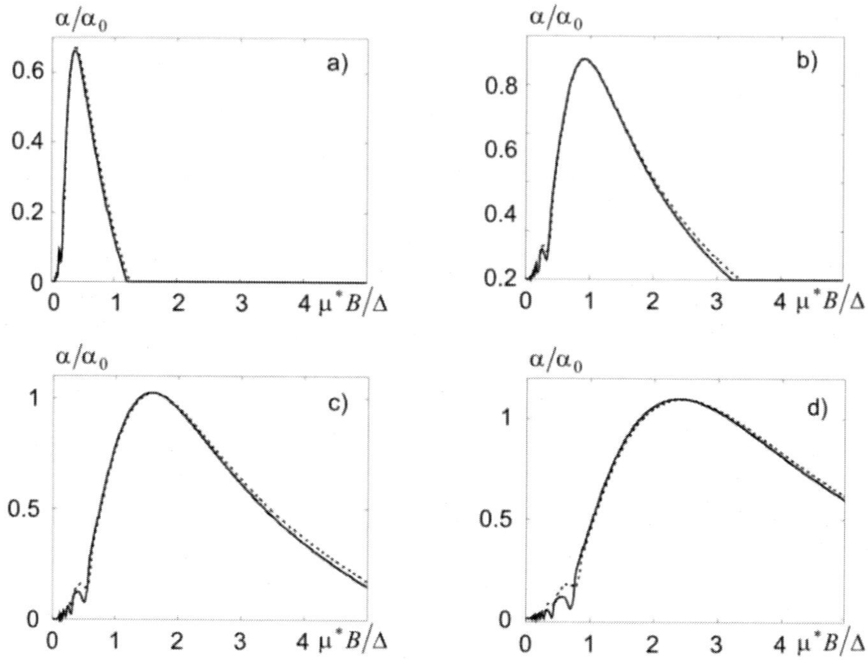

Figure 23. Field dependence of thermoelectromotive coefficient in the model with relaxation time $\tau \propto v_f(x)$, independent of the magnetic field at $0 \le \mu^* B/\Delta \le 5$, $A_m = 25$.

The maxima of the thermoelectromotive coefficient in the model with the relaxation time $\tau \propto v_f(x)$ for the real layered crystal at $\zeta_0/\Delta = 0.5;1;1.5;2$ are $5.701 \cdot 10^{-5}$, $7.788 \cdot 10^{-5}$, $8.815 \cdot 10^{-5}$ and $9.453 \cdot 10^{-5}$ V/K respectively. In effective mass approximation these maxima are $5.770 \cdot 10^{-5}, 7.788 \cdot 10^{-5}, 8.797 \cdot 10^{-5}, 9.427 \cdot 10^{-5}$ V/K respectively. As before, such small difference takes place due to the validity of the effective mass approximation in super strong quantizing magnetic fields. These maxima are also somewhat greater than the corresponding maxima in the model of constant relaxation time. The results of thermoelectromotive coefficient calculations in the model with the relaxation time $\tau \propto v_f(x)$ at $A_m = 25$ in magnetic field range $0.04 \le \mu^* B/\Delta \le 0.1$ for ratios $\zeta_0/\Delta = 0.5;1;1.5;2$ are shown in Figure 22.

As before, the effects of the layered structure are manifested in phase retardation and in the increase of the thermoelectromotive coefficient amplitude and its derivative oscillations. The figure also demonstrates that at $A_m = 25$ and all considered ratios ζ_0/Δ in both cases: for the layered crystal and in the effective mass approximation, the thermoelectromotive coefficient and its derivative remain oscillating functions of the magnetic field.

Moreover, at $A_m = 25$ the thermoelectromotive coefficient oscillations and its derivative become more apparent, than at $A_m = 15$. At $\mu^* B/\Delta = 0.1$, $A_m = 25$ and $\zeta_0/\Delta = 0.5;1;1.5;2$ the thermoelectromotive coefficients for the real layered crystal are $1.639 \cdot 10^{-6}$, $1.294 \cdot 10^{-6}$, $1.565 \cdot 10^{-7}$, $2.682 \cdot 10^{-7}$ V/K respectively.

In the effective mass approximation these thermoelectromotive coefficients are $1.552 \cdot 10^{-6}$, $1.466 \cdot 10^{-6}$, $8.002 \cdot 10^{-7}$ and $5.563 \cdot 10^{-7}$ V/K respectively. The results of thermoelectromotive coefficient calculations in the model with $\tau(x) \propto v_f(x)$ at $A_m = 25$ in the magnetic field range $0 \le \mu^* B/\Delta \le 5$ for ratios $\zeta_0/\Delta = 0.5;1;1.5;2$ are shown in Figure 23. The figure demonstrates that at $A_m = 25$ and all considered ratios ζ_0/Δ the effects of the layered structure are well observed in both cases: after the principal maximum and before this maximum.

As in the case of charge carriers scattering by the charged impurities, such behavior of the thermoelectromotive coefficient may be explained by the above mentioned peculiarities of the thermoelectromotive coefficient oscillations and its derivative and the difference between the density of states of charge carriers for the real layered crystal and in the effective mass approximation. The maxima of thermoelectromotive coefficient for the real layered crystal at $A_m = 25$ and $\zeta_0/\Delta = 0.5; 1; 1.5; 2$ are $5.710 \cdot 10^{-5}$, $7.797 \cdot 10^{-5}, 8.823 \cdot 10^{-5}, 9.453 \cdot 10^{-5}$ V/K respectively. In the effective mass approximation these maxima of the thermoelectromotive coefficient are $5.770 \cdot 10^{-5}$, $7.788 \cdot 10^{-5}, 8.806 \cdot 10^{-5}, 9.427 \cdot 10^{-5}$ V/K respectively.

Similar to the charge carriers scattering by the charged impurities, let's consider the special case $A_m = 10^5, \zeta_0/\Delta = 2$. The field dependence of the thermoelectromotive coefficient for this case at $0.04 \le \mu^* B/\Delta \le 0.1$ is shown in Figure 24.

This figure demonstrates that at such A_m the oscillations of the thermoelectromotive coefficient and its derivative for the real layered crystal and in the effective mass approximation become best observed.

As in the case of charge carriers scattering by the charged impurities, the difference between the dashed and solid curves may be explained by the singularity of the density of states for the real layered crystal at $\zeta_0/\Delta = 2$ and by the absence of planes, tangent to the FS and perpendicular to the magnetic field direction in this case. For the real layered crystal at $A_m = 10^5, \zeta_0/\Delta = 2, \mu^* B/\Delta = 0.1$, the thermoelectromotive coefficient is $2.688 \cdot 10^{-7}$ eV/K. In the effective mass approximation, the thermoelectromotive coefficient is $4.189 \cdot 10^{-7}$ eV/K. At $A_m = 10^5$ and $\zeta_0/\Delta = 2$, the effects of the layered structure in both cases: before the principal maximum of the thermoelectromotive coefficient and after it are well observed.

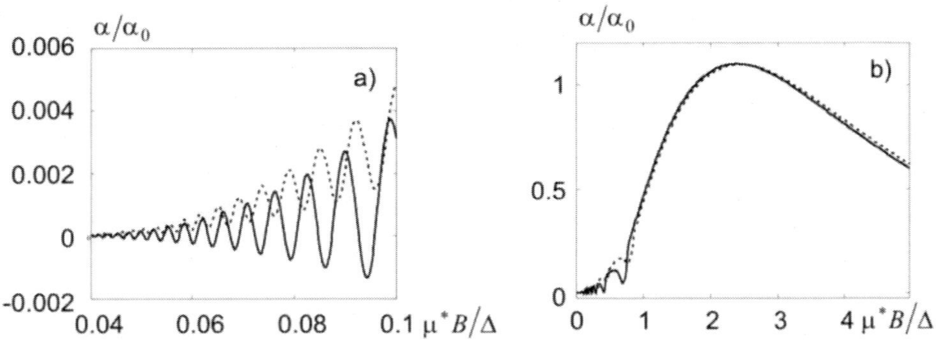

Figure 24. Field dependence of thermoelectromotive coefficient in the model with relaxation time $\tau \propto v_f(x)$ at $\zeta_0/\Delta = 2$, $A_m = 10^5$ and a) $0.04 \le \mu^* B/\Delta \le 0/1$, b) $0 \le \mu^* B/\Delta \le 5$.

As for the charge carriers scattering by the charged impurities, such behavior of the thermoelectromotive coefficient may be explained by the above mentioned peculiarities of thermoelectromotive coefficient oscillations and its derivative. As before, these peculiarities may be explained by the singularity of the density of states for the real layered crystal at $\zeta_0/\Delta = 2$ and by the absence of planes tangent to the FS and perpendicular to the magnetic field direction in this case. The maximum of the thermoelectromotive coefficient for the real layered crystal at $A_m = 10^5$ and $\zeta_0/\Delta = 2$ is $9.462 \cdot 10^{-5}$ V/K .

In the effective mass approximation, this maximum is $9.436 \cdot 10^{-5}$ V/K . It means that as $A_m \to \infty$ the field dependence of thermoelectromotive coefficient, as in the case of the charge carriers scattering by the charged impurities, tends to some limiting law. But at high magnetic field, i.e. when $\mu^* B/\Delta >> 1$ the thermoelectromotive coefficient, as in the case of charge carriers scattering by the charged impurities, is completely determined by formula (30).

4.5. Model of Relaxation Time, Proportional to Fermi Velocity and Dependence on Magnetic Field

The results of thermoelectromotive coefficient calculations in the model with the relaxation time $\tau(x,B) \propto v_f(x,B)$ at $A_m = 5$ in magnetic field range $0.04 \le \mu^* B/\Delta \le 0.1$ for ratios $\zeta_0/\Delta = 0.5;1;1.5;2$ are shown in Figure 25. This figure demonstrates that the effects of the layered structure, as in the model with $\tau(x) \propto v_f(x)$, manifest themselves in phase retardation and the increase of the thermoelectromotive coefficient amplitude and its derivative oscillations. It can be seen that for all considered ratios ζ_0/Δ the thermoelectromotive coefficient and its derivative are oscillating functions of the magnetic field. If ratio ζ_0/Δ increases, then the frequency of oscillations increases and the difference between the solid and dashed curves becomes more apparent. It is best observed at $\zeta_0/\Delta = 2$.

As before, it may be explained by the singularity of the density of states for the real layered crystal at $\zeta_0/\Delta = 2$ and by the absence of planes tangent to the FS and perpendicular to the magnetic field direction in this case.

At $\mu^* B/\Delta = 0.1$ and $\zeta_0/\Delta = 0.5;\, 1;\, 1.5;\, 2$ for the real layered crystal the thermoelectromotive coefficients are $2.329 \cdot 10^{-6}, 1.466 \cdot 10^{-6}, 2.937 \cdot 10^{-6}$ and $2.657 \cdot 10^{-7}$ V/K respectively. In the effective mass approximation, these thermoelectromotive coefficients are $2.415 \cdot 10^{-6}$, $1.897 \cdot 10^{-6}, 1.121 \cdot 10^{-6}$ and $7.729 \cdot 10^{-7}$ V/K respectively. These values are similar to those in the model with $\tau(x) \propto v_f(x)$, because chemical potential in the magnetic field range $0.04 \le \mu^* B/\Delta \le 0.1$ is weakly dependent on the magnetic field.

The results of thermoelectromotive coefficient calculations in the model with relaxation time $\tau(x,B) \propto v_f(x,B)$ at $A_m = 5$ in magnetic field range $0 \le \mu^* B/\Delta \le 5$ for ratios $\zeta_0/\Delta = 0.5;\, 1;\, 1.5;\, 2$ are shown in Figure 26.

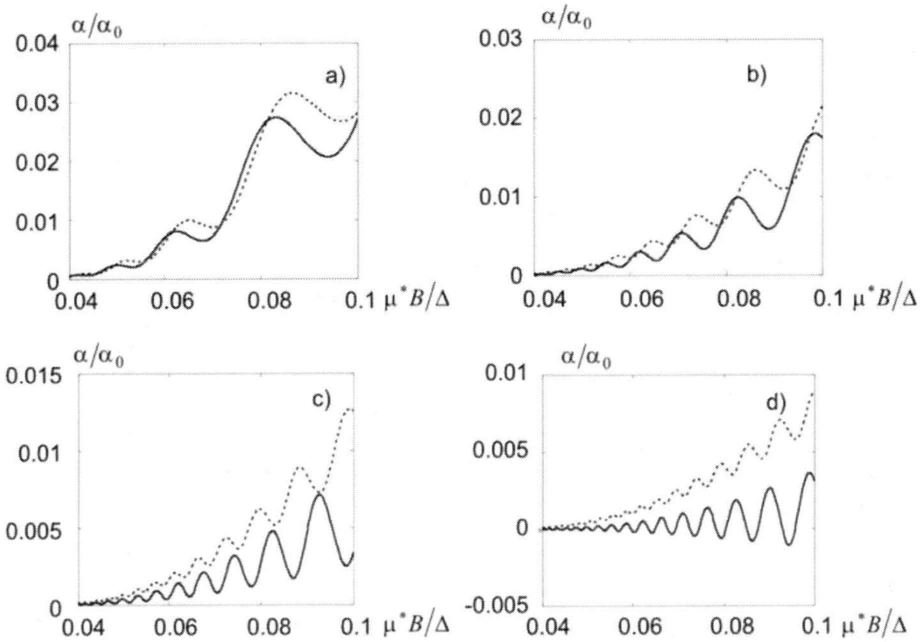

Figure 25. Field dependence of thermoelectromotive coefficient in the model with relaxation time $\tau \propto v_f(x,B)$, dependent on the magnetic field at $0.04 \le \mu^* B/\Delta \le 0.1$, $A_m = 5$.

The figure demonstrates that, as in the model with $\tau \propto v_f(x)$, the layered structure effects at $\zeta_0/\Delta = 0.5$ are well observed only after the principal maximum, but at $\zeta_0/\Delta \ge 1$ these effects become well observed both after the principal maximum and before it. The maxima of the thermoelectromotive coefficient in the model with $\tau(x,B) \propto v_f(x,B)$ with the relaxation time dependent on the magnetic field for the real layered crystal at $\zeta_0/\Delta = 0.5;\ 1;\ 1.5;\ 2$ are $5.684\cdot 10^{-5}$,$7.771\cdot 10^{-5}$,$8.797\cdot 10^{-5}$ and $9.436\cdot 10^{-5}$ V/K respectively.

In the effective mass approximation these maxima are $5.744\cdot 10^{-5}$, $7.762\cdot 10^{-5}$, $8.780\cdot 10^{-5}$ and $9.410\cdot 10^{-5}$ V/K respectively.

As in the model with $\tau \propto v_f(x)$, such small difference takes place due to the validity of the effective mass approximation in super strong quantizing magnetic fields. These maxima are somewhat greater than the corresponding maxima in the model of constant relaxation time.

They are similar to those in the model with $\tau \propto v_f(x)$, because the thermo-diffusive flux and the electric current both are proportional to relaxation time.

The results of thermoelectromotive coefficient calculations in the model with relaxation time $\tau(x,B) \propto v_f(x,B)$ at $A_m = 15$ in magnetic field range $0.04 \le \mu^* B/\Delta \le 0.1$ for ratios $\zeta_0/\Delta = 0.5;\ 1;\ 1.5;\ 2$ are shown in Figure 27. The figure demonstrates that, as in the model with $\tau \propto v_f(x)$, the effects of the layered structure are manifested in phase retardation and in the increase of the thermoelectromotive coefficient amplitude and its derivative oscillations.

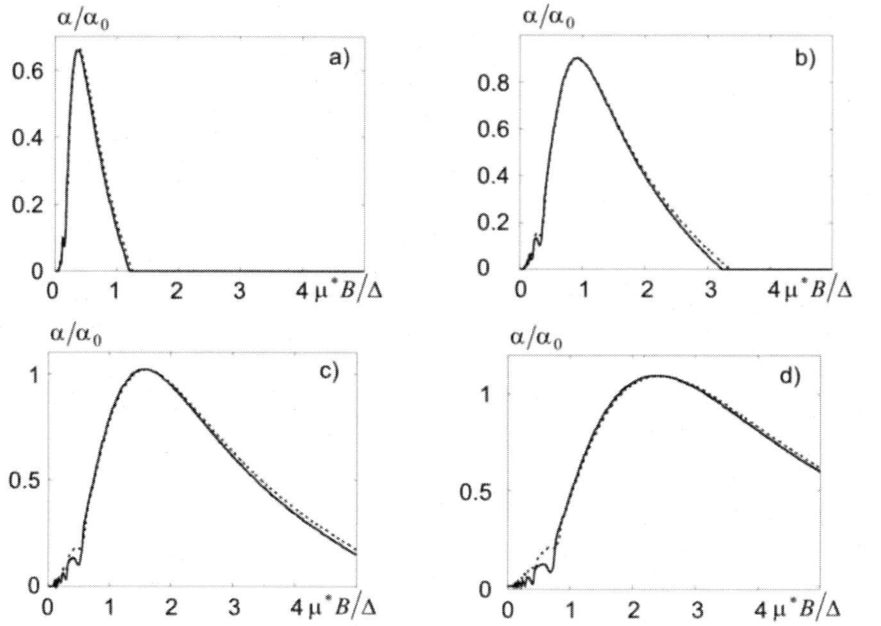

Figure 26. Field dependence of thermoelectromotive coefficient in the model with relaxation time $\tau \propto v_f(x, B)$, dependent on the magnetic field at $0 \le \mu^* B/\Delta \le 5$, $A_m = 5$

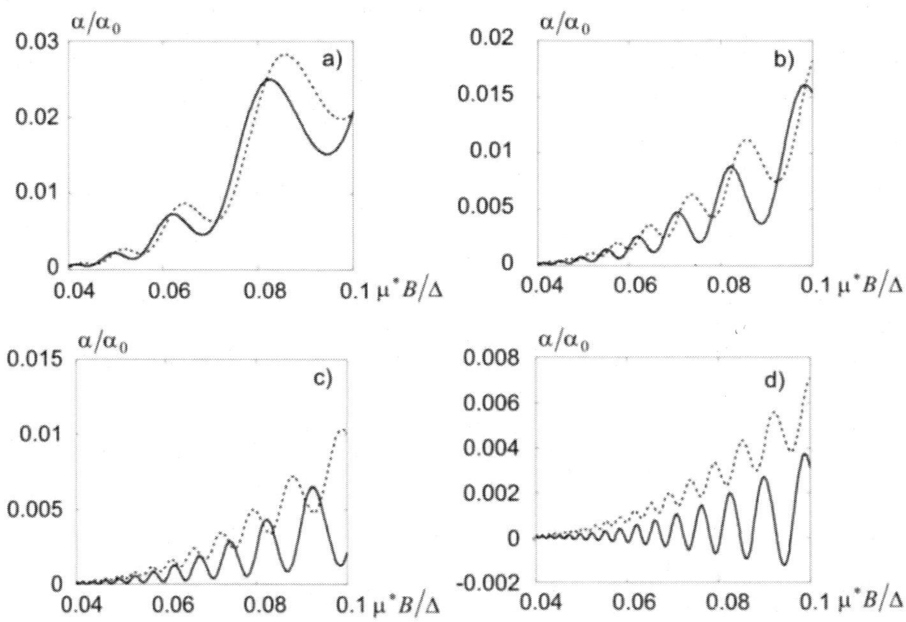

Figure 27. Field dependence of thermoelectromotive coefficient in the model with relaxation time $\tau \propto v_f(x, B)$, dependent on the magnetic field at $0.04 \le \mu^* B/\Delta \le 0.1$, $A_m = 15$.

It can be observed that for all considered ratios ζ_0/Δ the thermoelectromotive coefficient and its derivative, as at $A_m = 5$, remain oscillating functions of the magnetic field.

Moreover, the oscillations of thermoelectromotive coefficient and its derivative become more apparent, if A_m increases. If ratio ζ_0/Δ increases, then the frequency of oscillations increases and the difference between the solid and dashed curves becomes more apparent. It is best observed at $\zeta_0/\Delta = 2$.

As before, it may be explained by the singularity of the density of states for the real layered crystal at $\zeta_0/\Delta = 2$ and by the absence of planes, tangent to the FS and perpendicular to the magnetic field direction in this case.

At $\mu^* B/\Delta = 0.1$ and $\zeta_0/\Delta = 0.5;1;1.5;2$ for the real layered crystal the thermoelectromotive coefficients are $1.811 \cdot 10^{-6}$, $1.294 \cdot 10^{-6}$, $1.871 \cdot 10^{-7}$ and $2.678 \cdot 10^{-7}$ V/K respectively. In the effective mass approximation these thermoelectromotive coefficients are $1.811 \cdot 10^{-6}$, $1.552 \cdot 10^{-6}$, $8.625 \cdot 10^{-7}$ and $6.143 \cdot 10^{-7}$ V/K respectively. Such difference between these values and the values in the model with $\tau \propto v_f(x, B)$ at $A_m = 15$ may be explained by the increase of the thermoeletromotive coefficient oscillations amplitude with the increase of A_m.

The results of thermoelectromotive coefficient calculations in the model with relaxation time $\tau \propto v_f(x, B)$ at $A_m = 15$ in magnetic field range $0 \le \mu^* B/\Delta \le 5$ for ratios $\zeta_0/\Delta = 0.5;1;1.5;2$ are shown in Figure 28. This figure demonstrates that, as in the model with $\tau(x) \propto v_f(x)$, the layered structure effects at $\zeta_0/\Delta = 0.5$ are well observed only after the principal maximum, but at $\zeta_0/\Delta \ge 1$ these effects become well observed both after principal maximum and before it. The maxima of the thermoelectromotive coefficient in the model with relaxation time $\tau(x, B) \propto v_f(x, B)$ for the real layered crystal at $\zeta_0/\Delta = 0.5;1;1.5;2$ are $5.71 \cdot 10^{-5}$, $7.78 \cdot 10^{-5}$, $8.806 \cdot 10^{-5}$ and $9.444 \cdot 10^{-5}$ V/K respectively.

In the effective mass approximation these maxima are $5.770 \cdot 10^{-5}$, $7.780 \cdot 10^{-5}$, $8.789 \cdot 10^{-5}$, $9.418 \cdot 10^{-5}$ V/K respectively. As before, such small difference takes place due to the validity of the effective mass approximation in super strong quantizing magnetic fields. These maxima are also somewhat greater, than the corresponding maxima in the model of constant relaxation time.

The results of thermoelectromotive coefficient calculations in the model with relaxation time $\tau(x, B) \propto v_f(x, B)$ at $A_m = 25$ in magnetic field range $0.04 \le \mu^* B/\Delta \le 0.1$ for ratios $\zeta_0/\Delta = 0.5;1;1.5;2$ are shown in Figure 29.

As in the model with $\tau(x) \propto v_f(x)$, the effects of the layered structure are manifested in phase retardation and in the increase of the thermoelectromotive coefficient amplitude and its derivative oscillations. As demonstrated in Figure 29, at $A_m = 25$ and all considered ratios ζ_0/Δ in both cases: for the layered crystal and in the effective mass approximation the thermoelectromotive coefficient and its derivative remain oscillating function of the magnetic field. Moreover, at $A_m = 25$ the oscillations of the thermoelectromotive coefficient and its derivative become more apparent than they are at $A_m = 15$. At $\mu^* B/\Delta = 0.1$, $A_m = 25$ and $\zeta_0/\Delta = 0.5;1;1.5;2$ the thermoelectromotive coefficients for the real layered crystal are $1.639 \cdot 10^{-6}$, $1.294 \cdot 10^{-6}$, $1.586 \cdot 10^{-7}$ and $2.682 \cdot 10^{-7}$ V/K respectively. In the effective mass approximation these thermoelectromotive coefficients are $1.639 \cdot 10^{-6}$, $1.466 \cdot 10^{-6}$, $7.988 \cdot 10^{-7}$ and $5.565 \cdot 10^{-7}$ V/K respectively.

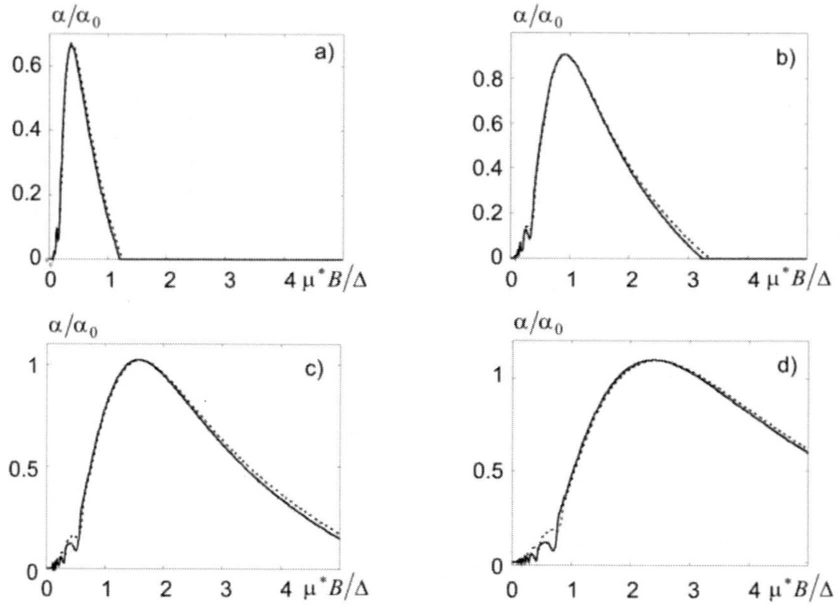

Figure 28. Field dependence of thermoelectromotive coefficient in the model with relaxation time $\tau \propto v_f(x, B)$, dependent on the magnetic field at $0 \le \mu^* B/\Delta \le 5$, $A_m = 15$.

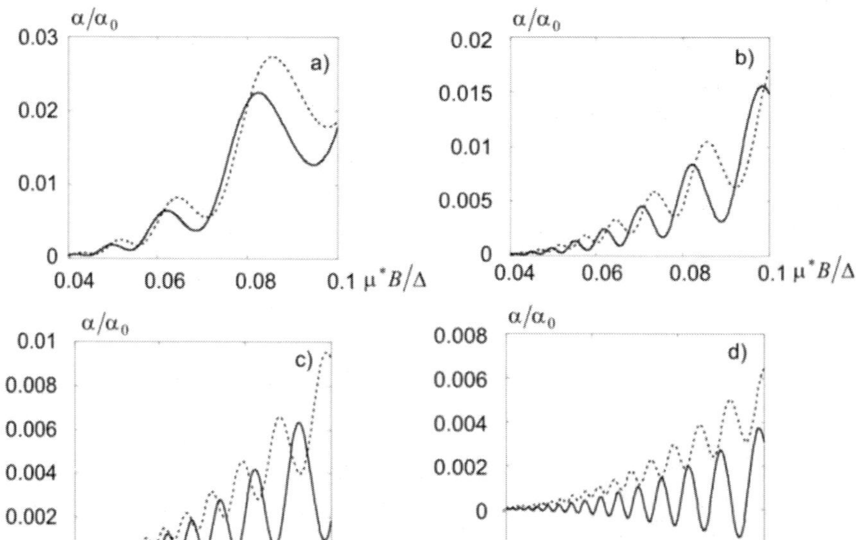

Figure 29. Field dependence of thermoelectromotive coefficient in the model with relaxation time $\tau \propto v_f(x, B)$, dependent on the magnetic field at $0.04 \le \mu^* B/\Delta \le 0.1$, $A_m = 25$.

The results of thermoelectromotive coefficient calculations in the model with $\tau(x,B) \propto v_f(x,B)$ with the relaxation time dependent on the magnetic field at $A_m = 25$ in magnetic field range $0 \le \mu^* B/\Delta \le 5$ for ratios $\zeta_0/\Delta = 0.5; 1; 1.5; 2$ are shown in Figure 30.

As in the model with $\tau \propto v_f(x)$, the plots in Figure 30 demonstrate that at $A_m = 25$ and all considered ratios ζ_0/Δ the effects of the layered structure are well observed both after the principal maximum and before this maximum.

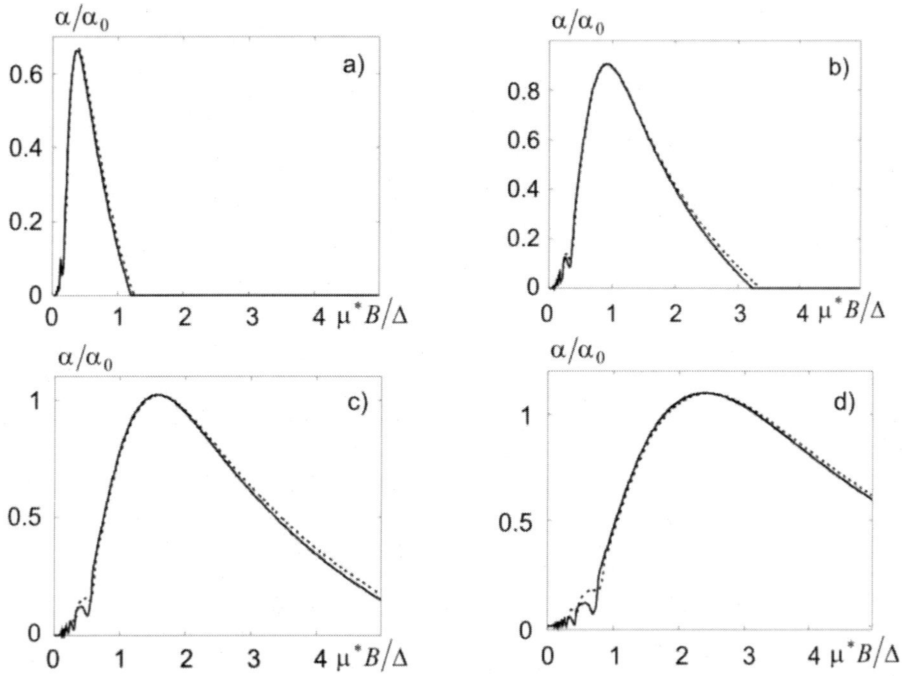

Figure 30. Field dependence of thermoelectromotive coefficient in the model with relaxation time $\tau \propto v_f(x,B)$, dependent on the magnetic field at $0 \le \mu^* B/\Delta \le 5$, $A_m = 15$.

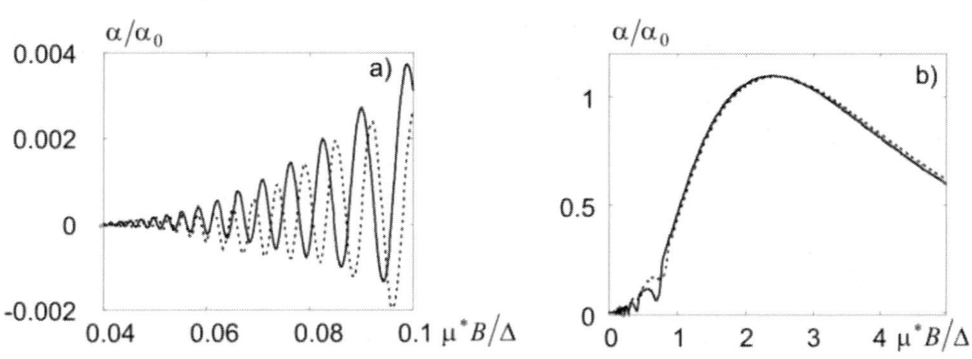

Figure 31. Field dependence of thermoelectromotive coefficient in the model with relaxation time $\tau \propto v_f(x,B)$, dependent on the magnetic field at $\zeta_0/\Delta = 2$, $A_m = 10^5$ and a) $0.04 \le \mu^* B/\Delta \le 0/1$, b) $0 \le \mu^* B/\Delta \le 5$.

As in the case of charge carriers scattering by the charged impurities, such behavior of the thermoelectromotive coefficient may be explained by the above mentioned peculiarities of the thermoelectromotive coefficient oscillations and its derivative and the difference between the density of states of charge carriers for the real layered crystal and in the effective mass approximation. The maxima of the thermoelectromotive coefficient for the real layered crystal at $A_m = 25$ and $\zeta_0/\Delta = 0.5; 1; 1.5; 2$ are $5.710 \cdot 10^{-5}$, $7.797 \cdot 10^{-5}$, $8.815 \cdot 10^{-5}$ and $9.444 \cdot 10^{-5}$ V/K respectively. In the effective mass approximation these thermoelectromotive coefficient maxima are $5.770 \cdot 10^{-5}$, $7.788 \cdot 10^{-5}$, $8.797 \cdot 10^{-5}$, $9.427 \cdot 10^{-5}$ V/K respectively.

Now, as in the model with $\tau \propto v_f(x)$, we shall consider the special case $A_m = 10^5$, $\zeta_0/\Delta = 2$. The field dependence of the thermoelectromotive coefficient for this case at $0.04 \le \mu^* B/\Delta \le 0.1$ is shown in Figure 31. This figure demonstrates that at such A_m the oscillations of thermoelectromotive coefficient and its derivative for the real layered crystal and in the effective mass approximation become best observed. As in the model with $\tau \propto v_f(x)$, the difference between the dashed and solid curves may be explained by the singularity of the density of states for the real layered crystal at $\zeta_0/\Delta = 2$ and by the absence of planes, tangent to the FS and perpendicular to the magnetic field direction in this case. For the real layered crystal at $A_m = 10^5, \zeta_0/\Delta = 2, \mu^* B/\Delta = 0.1$, the thermoelectromotive coefficient is $2.688 \cdot 10^{-7}$ eV/K. In the effective mass approximation the thermoelectromotive coefficient is $4.193 \cdot 10^{-7}$ eV/K.

Similarly to the model with $\tau \propto v_f(x)$, at $A_m = 10^5$ and $\zeta_0/\Delta = 2$ the effects of the layered structure in both cases: before the principal maximum of the thermoelectromotive coefficient and after it are well observed. As for charge carriers scattering by the charged impurities, such behavior of the thermoelectromotive coefficient may be explained by the above mentioned peculiarities of the oscillations of the thermoelectromotive coefficient and its derivative. As before, these peculiarities may be explained by the singularity of the density of states for the real layered crystal at $\zeta_0/\Delta = 2$ and by the absence of planes tangent to the FS and perpendicular to the magnetic field direction in this case. The maximum of the thermoelectromotive coefficient for the real layered crystal at $A_m = 10^5$ and $\zeta_0/\Delta = 2$ is $9.444 \cdot 10^{-5}$ V/K . In the effective mass approximation this maximum is $9.427 \cdot 10^{-5}$ V/K . It means that, as in the model with $\tau \propto v_f(x)$, as $A_m \to \infty$ the field dependence of the thermoelectromotive coefficient, as in the case of charge carriers scattering by the charged impurities tends to some limiting law.

But at high magnetic field, i.e. when $\mu^* B/\Delta >> 1$ the thermoelectromotive coefficient, as in the model with $\tau \propto v_f(x)$, is completely determined by formula (30).

4.6. Model of Relaxation Time, Proportional to Longitudinal Velocity

Finally we consider the model of charge carriers scattering with the relaxation time $\tau \propto v_z(x)$ [13, 14]. This model is a satisfactory approximation for the description of sharply anisotropic scattering of charge carriers by acoustic phonons in the weak magnetic field, induced and spontaneous scattering of charge carriers by acoustic phonons in strong quantizing magnetic field and scattering of charge carriers by the charged impurities and lattice defects in strong quantizing magnetic field. In this model we obtain such explicit expressions for quantities A, B, C :

$$A=\sum_{l=1}^{\infty} f_l^{th}\left[\frac{(\mu^* B)^3}{\pi l \Delta^3}+\frac{\mu^* B \zeta^2}{\pi l \Delta^3}-2\left(\frac{\mu^* B}{\pi l \Delta}\right)^3-\frac{2(\zeta-\Delta)(\mu^* B)^2}{\pi l \Delta^3}-\frac{2\mu^* B \zeta}{\pi l \Delta^2}\right]+ \\ +2\sum_{l=1}^{\infty}(-1)^l f_l^{th}\left[\left(\frac{\mu^* B}{\pi l \Delta}\right)^3 \cos\left(\frac{\pi l \zeta}{\mu^* B}\right)+\left(\frac{\mu^* B}{\pi l \Delta}\right)^2 \sin\left(\frac{\pi l \zeta}{\mu^* B}\right)\right] \tag{40}$$

$$B=\frac{(\zeta-\mu^* B)^2(3\Delta-\zeta+\mu^* B)}{6\Delta^3} \tag{41}$$

$$C=2\sum_{l=1}^{\infty} f_l^{\sigma}\left[\frac{1}{\pi^2 l^2}\left(\frac{\mu^* B}{\Delta}\right)^3+\frac{(\mu^* B)^2(\Delta-\zeta)}{\pi^2 l^2 \Delta^3}\right] \\ +2\sum_{l=1}^{\infty}(-1)^l\left[\left(\frac{\mu^* B}{\pi l \Delta}\right)^3 \sin\left(\frac{\pi l \zeta}{\mu^* B}\right)-\left(\frac{\mu^* B}{\pi l \Delta}\right)^2 \cos\left(\frac{\pi l \zeta}{\mu^* B}\right)\right] \tag{42}$$

Therefore, the oscillating part of the thermoelectromotive coefficient in quasi-classical approximation for the model with $\tau \propto v_z(x)$ may be represented as:

$$\alpha_{os}=4\alpha_0\sum_{l=1}^{\infty}(-1)^l f_l^{th}\left(\frac{\mu^* B}{\pi l \zeta}\right)^2 \sin\left(\frac{\pi l \zeta}{\mu^* B}\right) \tag{43}$$

In the effective mass approximation formulae (40)-(42) may be rewritten as:

$$A=2\sum_{l=1}^{\infty}(-1)^l f_l^{th}\left(\frac{\mu^* B}{\pi l \Delta}\right)^2 \sin\left(\frac{\pi l \zeta}{\mu^* B}\right)-2\sum_{l=1}^{\infty}\frac{\mu^* B(\zeta-\mu^* B) f_l^{th}}{\pi l \Delta^2} \tag{44}$$

$$B=\frac{(\zeta-\mu^* B)^2}{2\Delta^2} \tag{45}$$

$$C=2\sum_{l=1}^{\infty}\left(\frac{\mu^* B}{\pi l \Delta}\right)^2 f_l^{\sigma}-2\sum_{l=1}^{\infty}(-1)^l f_l^{\sigma}\left(\frac{\mu^* B}{\pi l \Delta}\right)^2 \cos\left(\frac{\pi l \zeta}{\mu^* B}\right) \tag{46}$$

As in the case of constant relaxation time, from formula (43) we can see that, even in quasi-classical approximation, the effects of the layered structure will be manifested in the decrease of oscillations frequencies, phase retardation and the increase of oscillations amplitudes.

The results of thermoelectromotive coefficient calculations in the model with relaxation time $\tau \propto v_z(x)$ in magnetic field range $0.04 \le \mu^* B/\Delta \le 0.1$ for ratios $\zeta_0/\Delta = 0.5;1;1.5;\ 2$ are shown in Figure 32. The figure demonstrates that $\zeta_0/\Delta = 0.5;1;1.5$ in the model with $\tau \propto v_z(x)$ the effects of the layered structure are manifested only in phase retardation and increase of the amplitude of the thermoelectromotive coefficient derivative oscillations.

The oscillations of the actual thermoelectromotive coefficient are manifested only for the real layered crystal at $\zeta_0/\Delta = 2$. If ratio ζ_0/Δ increases, then the frequency of oscillations increases too and the difference between the solid and dashed curves becomes more apparent. This is best observed at $\zeta_0/\Delta = 2$.

As before, it may be explained by the singularity of the density of states for the real layered crystal at $\zeta_0/\Delta = 2$ and by the absence of planes tangent to the FS and perpendicular to the magnetic field direction in this case.

At $\mu^* B/\Delta = 0.1$ and $\zeta_0/\Delta = 0.5;1;1.5;2$ for the real layered crystal the thermoelectromotive coefficients are $5.348 \cdot 10^{-6}, 2.933 \cdot 10^{-6}, 9.487 \cdot 10^{-7}$ and $1.720 \cdot 10^{-7}$ V/K respectively. In the effective mass approximation these thermoelectromotive coefficients are $5.348 \cdot 10^{-6}$, $2.933 \cdot 10^{-6}$, $1.552 \cdot 10^{-6}$ and $1.121 \cdot 10^{-6}$ V/K respectively.

The results of thermoelectromotive coefficient calculations in the model with relaxation time $\tau \propto v_z(x)$ in magnetic field range $0 \le \mu^* B/\Delta \le 5$ for ratios $\zeta_0/\Delta = 0.5;\ 1;\ 1.5;\ 2$ are shown in Figure 33.

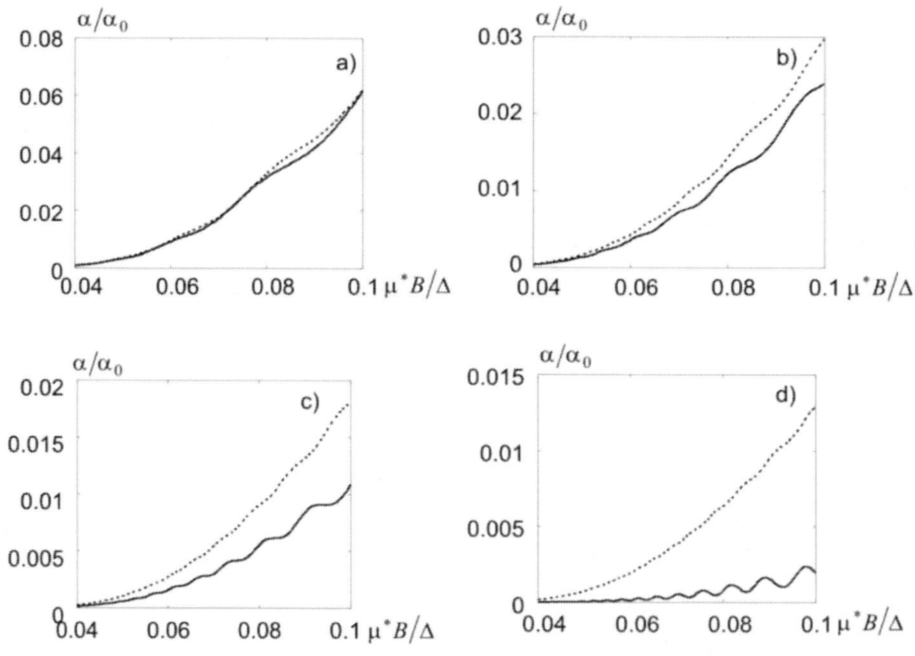

Figure 32. Field dependence of thermoelectromotive coefficient in the model with relaxation time $\tau \propto v_z(x)$ at $0.04 \le \mu^* B/\Delta \le 0.1$, $A_m = 25$.

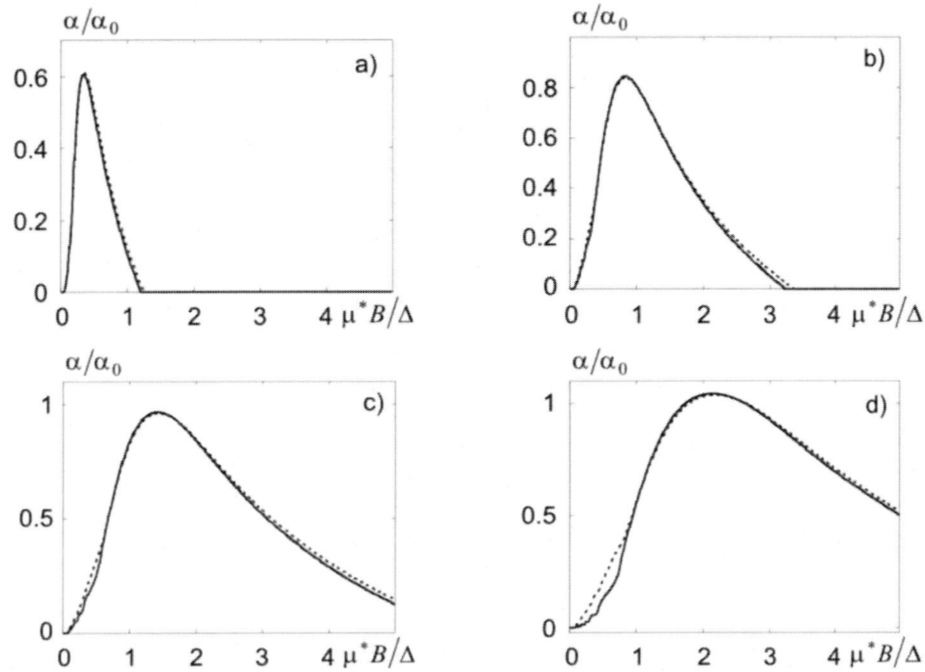

Figure 33. Field dependence of thermoelectromotive coefficient in the model with relaxation time $\tau \propto v_z(x)$ at $0 \le \mu^* B/\Delta \le 5$, $A_m = 25$.

The figure demonstrates that the layered structure effects at $\zeta_0/\Delta = 0.5;1$ are well observed only after the principal maximum, but at $\zeta_0/\Delta = 1.5;2$ these effects become well observed both after the principal maximum and before it. The maxima of the thermoelectromotive coefficient in the model with relaxation time $\tau \propto v_z(x)$ for the real layered crystal at $\zeta_0/\Delta = 0.5;1;1.5;2$ are $5.209 \cdot 10^{-5}, 7.271 \cdot 10^{-5}, 8.323 \cdot 10^{-5}$ and $8.987 \cdot 10^{-5}$ V/K respectively.

In the effective mass approximation, these maxima are $5.261 \cdot 10^{-5}$, $7.245 \cdot 10^{-5}$, $8.271 \cdot 10^{-5}, 8.935 \cdot 10^{-5}$ V/K respectively. As before, the reason for such small difference consists in the validity of the effective mass approximation in super strong quantizing magnetic fields. These maxima are somewhat smaller than the corresponding maxima in the model of constant relaxation time.

As in the case of constant relaxation time, at $\mu^* B/\Delta >> 1$ we can obtain such asymptotic expression for the thermoelectromotive coefficient:

$$\alpha = 0.033\alpha_0 \frac{f^2(\gamma_0)\Delta^3}{(\mu^* B)^2 kT} \tag{47}$$

This formula differs from formula (30) only in the numerical factor. Therefore, if, as before, we assume that $\Delta = 0.01\text{eV}$, $m^* = 0.01 m_0$, $B = 60\text{T}$, $T = 3\text{K}$, then at ratios $\zeta_0/\Delta = 0.5;\ 1;\ 1.5;\ 2$ the values of the thermoelectromotive coefficient α are 10.66,90.90,

332.7 and 897.2 nV/K respectively. Again we can see that such asymptotic law does not lead to incorrect physical consequences, because these values are much less than the corresponding maxima of the thermoelectromotive coefficient in the model with $\tau \propto v_z(x)$.

CONCLUSION

The research work reported in this chapter shows that in the region of quasi-classical approximation the validity for closed FS and all the considered mechanisms of charge carriers scattering the layered structure effects are observed in phase retardation, the decrease of principal frequency and the increase of thermoelectromotive coefficient oscillations amplitude. These effects may be explained by the decrease of Fermi energy (chemical potential of charge carriers) for the real layered crystal in comparison with the effective mass approximation at constant concentration of charge carriers. In quasi-classical approximation that is valid at $\Delta/\mu^* B >> 1$, the amplitude of the thermoelectromotive coefficient oscillations at helium temperatures is proportional to B^2 for the model with $\tau \propto v_z(x)$ and to $B^{3/2}$ for all other considered models of charge carriers scattering. Hence, at low temperatures and $B = 0$ the thermoelectromotive coefficient tends to zero according to general principles of thermodynamics. These results may be used for experimental verification of the validity of different models of charge carriers scattering.

If the concentration of charge carriers, i.e. ratio ζ_0/Δ increases, then in quasi-classical approximation the principal frequency of the thermoelectromotive coefficient and (or) its derivative increase, while the amplitude of thermoelectromotive coefficient oscillations decreases. Such behavior of the thermoelectromotive coefficient is observed both for the real layered crystal and in the effective mass approximation. But if ratio ζ_0/Δ increases, then the difference between the field dependences of the thermoelectromotive coefficient for the real layered crystal and in the effective mass approximation increases too. This difference is the greatest at $\zeta_0/\Delta = 2$ and this may be explained by singularity of the density of charge carriers states in the real layered crystal and by the absence of plane tangent to the FS and perpendicular to the magnetic field direction at $\zeta_0/\Delta = 2$.

Yet the study shows that under the condition of charge carriers scattering by the charged impurities, the increase of anisotropy parameter A_m in the models with $\tau = l/v_f(x)$ and $\tau = l/v_f(x, B)$ smoothes the oscillations of the thermoelectromotive coefficient and (or) its derivative for the real layered crystal, as well as in the effective mass approximation. Hence, for closed FS at high A_m only the oscillations of the thermoelectromotive coefficient derivative are observed. But if $\zeta_0/\Delta = 2$, then even at $A_m = 10^5$ for the real layered crystal both kinds of oscillations of the thermoelectromotive coefficient: the oscillations of derivative and actual thermoelectromotive coefficient oscillations occur. As before, this may be explained by the singularity of the density of charge carriers' states in the real layered crystal and by the absence of the plane tangent to the FS and perpendicular to the magnetic field direction at $\zeta_0/\Delta = 2$.

On the contrary, under the conditions of charge carriers scattering by acoustic phonons the increase of *anisotropy parameter* A_m in the models with $\tau \propto v_f(x)$ and $\tau \propto v_f(x,B)$ intensifies not only the oscillations of the thermoelectromotive coefficient derivative, but also the oscillations of the actual thermoelectromotive coefficient. In quasi-classical approximation the dependence of relaxation time on the magnetic field almost has no influence on the field dependence of the thermoelectromotive coefficient. But in all scattering models at high A_m the field dependences of the thermoelectromotive coefficient tend to some limiting laws.

In the region of strong quantizing magnetic fields for every ratio ζ_0/Δ there is the "optimal" interval of the magnetic fields, where the oscillations of the thermoelectromotive coefficient and (or) its derivative are most intensive. If ratio ζ_0/Δ increases, then the frequency and amplitude of the thermoelectromotive coefficient and (or) its derivative oscillations increases in the "optimal" interval of the magnetic fields. At this condition, the oscillations of the thermoelectromotive coefficient for the real layered crystal are more intensive than in the effective mass approximation.

Similar to quasi-classical approximation, the increase of anisotropy parameter A_m smoothes the thermoelectromotive coefficient oscillations in the "optimal" interval of the magnetic fields under the conditions of charge carriers scattering by the charged impurities and intensifies these oscillations under the conditions of charge carriers scattering by acoustic phonons. It should be pointed out that in the model with $\tau \propto v_z(x)$ the field dependence of the thermoelectromotive coefficient has a linear region, which is determined by the linear "smooth" component of the thermoelectromotive coefficient.

The ratio ζ_0/Δ, which determines the maximal value of the thermoelectromotive coefficient, is the principal parameter. The mechanisms of charge carriers scattering and the effects of the layered structure have almost no influence on the maximum of thermoelectromotive coefficient.

In particular, for the real layered crystal at $\zeta_0/\Delta = 0.5$ the maximal value of the thermoelectromotive coefficient for all considered models of charge carriers scattering varies within the limits $(5.209 \div 5.710) \cdot 10^{-5}$ V/K and the average maximum of the thermoelectromotive coefficient is $5.602 \cdot 10^{-5}$ V/K In the effective mass approximation at $\zeta_0/\Delta = 0.5$ the maximal value of the thermoelectromotive coefficient for all considered models of charge carriers scattering varies within the limits $(5.209 \div 5.710) \cdot 10^{-5}$ V/K and the average maximum of the thermoelectromotive coefficient is $5.663 \cdot 10^{-5}$ V/K . For the real layered crystal at $\zeta_0/\Delta = 1$ the maximal value of the thermoelectromotive coefficient for all considered models of charge carriers scattering varies within the limits $(7.271 \div 7.797) \cdot 10^{-5}$ V/K and the average maximum of the thermoelectromotive coefficient is $7.700 \cdot 10^{-5}$ V/K . In the effective mass approximation at $\zeta_0/\Delta = 1$ the maximal value of the thermoelectromotive coefficient for all considered models of charge carriers scattering varies within the limits $(7.245 \div 7.788) \cdot 10^{-5}$ V/K and the average maximum of the thermoelectromotive coefficient is $7.694 \cdot 10^{-5}$ V/K . For the real layered crystal at $\zeta_0/\Delta = 1.5$ the maximal value of the thermoelectromotive coefficient for all considered models of charge carriers scattering varies within the limits $(8.323 \div 8.823) \cdot 10^{-5}$ V/K and the average maximum of the thermoelectromotive coefficient is $8.734 \cdot 10^{-5}$ V/K . In the effective mass approximation at $\zeta_0/\Delta = 1.5$ the maximal value of the thermoelectromotive coefficient for all considered models of charge carriers scattering varies within the limits $(8.271 \div 8.806) \cdot 10^{-5}$ V/K and the average maximum of thermoelectromotive coefficient is $8.720 \cdot 10^{-5}$ V/K . For the real layered crystal at $\zeta_0/\Delta = 2$ the maximal value of the thermoelectromotive coefficient for all considered models of charge carriers scattering varies within the limits $(8.987 \div 9.462) \cdot 10^{-5}$ V/K and the average maximum of the thermoelectromotive coefficient is $9.386 \cdot 10^{-5}$ V/K . In the effective mass approximation at $\zeta_0/\Delta = 2$ the maximal value of the thermoelectromotive coefficient for all considered models of charge carriers scattering varies within the limits $(8.935 \div 9.436) \cdot 10^{-5}$ V/K and the average maximum of the thermoelectromotive coefficient is $9.365 \cdot 10^{-5}$ V/K . In general, the effects of the layered structure are manifested most vividly under the conditions of charge carriers scattering by acoustic phonons. And vice versa, the effects of the layered structure are smoothed under the conditions of charge carriers scattering by the charged impurities. Therefore, the rate of the layered structure effects manifestation may be considered a criterion of purity and perfection of a sample.

For all models, in which the relaxation time depends on the total velocity of current carriers, the maximal value of the thermoelectromotive coefficient slightly, but regularly varies as the function of A_m. Under the conditions of charge carriers scattering by the charged impurities the maximum of the thermoelectromotive coefficient decreases, if A_m increases. And vice versa, under the conditions of charge carriers scattering by acoustic phonons the maximum of the thermoelectromotive coefficient increases, if A_m increases. While A_m increases, in the models with $\tau = l/v_f(x, B)$ and $\tau \propto v_f(x, B)$ the maximum of the thermoelectromotive coefficient tends to saturation faster, than in the models with $\tau = l/v_f(x)$ and $\tau \propto v_f(x)$. After the maximum, for all considered ratios ζ_0/Δ and all considered models of relaxation time, the thermoelectromotive coefficient at $\mu^* B/\Delta >> 1$ decreases up to zero according to the law: $\alpha \propto T^{-1} B^{-2}$. This result cannot be obtained in the framework of the conventional quasi-classical approach, for which the specific form of the FS and its extension along the magnetic field direction are inessential.

It is evident that the theoretical results, presented in this chapter, should be followed with due experimental verification. But the author of this chapter is familiar only with experimental research, which concerns physical characteristics of the layered crystals with open FS. However, the author hopes that this theoretical investigation will stimulate further experimental research of the layered crystals with closed FS.

REFERENCES

[1] R. F. Fivaz, *J. Phys. Chem. Solids.* 28, 839 (1967).
[2] V. G. Peschanskii, *Zh. Exp. Teor. Phys.* 121, 1204 (2002).

[3] P. D. Grigoriev, *Phys. Rev. B*. 67, 144401 (2003).
[4] B. Laikhtman and D. Menashe, *Phys. Rev. B.* 52, 8974 (1994).
[5] D. Andres, M. V. Kartsovnik, W. Biberacher, H. Weiss, E. Balthes, H. Müller and N. Kushch, *Phys. Rev. B.* 64, 161104 (2001).
[6] P. V. Gorskyi and V. M. Nitsovich, *Ukr. Fiz. Zh.* 23, 1647 (1978).
[7] P. V. Gorskyi and V. M. Nitsovich, *Ukr. Fiz. Zh.* 26, 1528 (1981).
[8] P. V. Gorskyi and V. M. Nitsovich, *Fiz. Met. Metalloved.* 54, 1088 (1982).
[9] P. V. Gorskyi and V. M. Nitsovich, *Fiz. Tekh. Poluprovodn.* 17, 936 (1983).
[10] P. V. Gorskyi, *Ukr. J. Phys.* 50, 1252 (2005).
[11] P. V. Gorskyi, *Ukr. J. Phys.* 55, 1296 (2010).
[12] P. V. Gorskyi, *Metallofiz. Noveishie. Tekhnol.* 32, 1545 (2010).
[13] V. F. Gantmakher, I. B. Levinson. *Scattering of Charge carriers in Metals and Semiconductors*, Nauka, Moscow (1984), North Holland, Amsterdam (1987).
[14] P. V. Gorskyi. *Semiconductors,* 40, 296 (2006).

In: Anisotropy Research
Editor: Hirpa Lemu

ISBN 978-1-62081-977-7

Chapter 10

ANISOTROPY IN CUPRATE SUPERCONDUCTORS: PENETRATION DEPTH ANALYSIS

Angsula Ghosh*

Departamento de Física, Universidade Federal de Amazonas,
Manaus, AM, Brazil

ABSTRACT

The pairing symmetry in the cuprate superconductors continues to be a major topic of debate even after twenty-five years of the discovery. The highly anisotropic nature along with the presence of inhomogeneities have made the problem highly controversial and ambiguous. Various experimental measurements have pointed to a d-wave symmetry, a $d_{x^2-y^2}$ pairing state for a number of hole and electron doped cuprates. Moreover, apart from this unconventional state for high temperature superconductors, the order parameter can have various anisotropies as seen in several theoretical models and experimental measurements. In this chapter we present a short review on anisotropic nature of the pairing state. The possible nature of the pairing and its effect on penetration depth are discussed. The anisotropy in the penetration depth is also discussed from the structural view point of cuprates. We conclude with a brief discussion on the present day view-point on the anisotropies present in the cuprates.

1. INTRODUCTION

The discovery of the high temperature superconductors is celebrating its silver jubilee this year (2011) when that of the conventional superconductors completes its centenary. Although the conventional $s-$wave superconductors are well explained by the *electron-phonon coupling*, the high-T_c (critical temperature) cuprates are still a topic of intense debate due to its highly anisotropic nature. The above anisotropic behavior leads the scientific

* E-mail address: angsula@ufam.edu.br.

community to name it as the *unconventional superconductors*. The anisotropic symmetry of the pairing state along with the highly anisotropic in-plane and out-of-plane properties of the cuprates makes the goal of the condensed-matter physicists highly challenging. The anisotropy in various properties has been demonstrated by numerous experiments and also a wide range of theoretical models have been proposed. Various pairing experiments that utilize the phase sensitive techniques have also pointed to a $d-$wave pairing symmetry in these materials. Moreover, the inhomogeneities in the pairing amplitude with increasing disorder make the theoretical problems and experimental works far more complex. In this case the system breaks up into several superconducting islands (large pairing amplitude) separated by an insulating sea. The scanning tunneling spectroscopy has been an inevitable method to provide an insight into this inhomogeneous nature of the cuprates. The quantum tunneling of electrons helps to reveal this spatial inhomogeneity and also the inhomogeneity of the low-energy density of states deep inside the gap.

The chapter is regarding this enormous task that remains for us to be solved regarding the anisotropic nature of the cuprates. We make a survey of the current status of the *anisotropic superconductors* and the implications of various types of pairing symmetries starting from the $d-$wave to the mixed-wave type for our understanding of the anisotropic nature of the cuprates and its applications. The *penetration depth* anisotropy in the momentum **k** space is also considered and the present experimental and theoretical position of the above is detailed in this chapter. Throughout this article we dedicate our effort to show to the readers how the structure of the cuprates along with various pairing symmetries could explain the anisotropic nature of the penetration depth from a theoretical point of view.

In the next section we briefly introduce the conventional superconductors which demonstrate highly isotropic properties. In Section 10.2, the high temperature superconductors are considered and a brief discussion is made on the properties and nature of these new materials. In Section 10.3, the possible symmetries of the order parameter of these new superconductors are presented. One of the most controversial topics of today's research is the *pseudogap phase* which is dealt in Section 10.4. We discuss some of the possible candidates of the pseudogap. In Section 10.5, we make a short survey on the penetration depth both from the experimental and theoretical point of view. Calculations based on simple models are exhibited and compared with the experimental findings. We observe that several experimental findings exhibit a strong agreement with the existing theories. Finally, we conclude our discussion on this topic demonstrating that the huge anisotropy present in the penetration depth of the cuprates is due to the anisotropic nature of the *order parameter* as well as due to the structure of the cuprates.

2. Conventional Superconductors

The discovery of superconductivity in normal metals in 1911 by Kamerling Onnes [1] put forward a new challenge to the scientific community. The problem of these classical superconductors remained at the forefront of theoretical physics till 1957 when the microscopic theory was proposed by Bardeen Cooper and Schrieffer (BCS) [2, 3]. The BCS theory paved a path for new search in finding superconductivity in other elements and compounds. The culmination of this line of discovery was reached in the early 70s with the discovery of A-15 class of compounds. The highest critical temperature T_c was found in Nb_3Ge. This phenomenon was described due to the formation of the Cooper pairs via an attractive phonon-mediated electron-electron interaction. The spin-singlet ground state is composed of large overlapping Cooper pairs that form a BCS condensate. One of the key ideas to the success of this theory was identifying the order parameter of this second-order phase transition as the energy gap $\Delta(k)$ with respect to the ground state. The energy gap in the conventional superconductors has an $s-$wave symmetry and has same magnitude and phase in all the directions in the **k** space. The isotropic nature of the gap for $T < T_c$ is also manifested in the thermodynamic, electromagnetic and transport properties of these materials. The exponential nature of the specific heat of the superconducting state and a sharp increase in the conductance at the voltage $V = \Delta/e$ are direct consequences of the gap parameter. This concept of the order-parameter $|\psi(r)|$ was first introduced by Ginzburg and Landau (GL) based on the classical theory of Landau on the second order phase transitions [4]. In superconductivity, the quantity $|\psi(r)|^2$ gives a measure of the local superfluid density and is related to the BCS gap parameter $\Delta(k)$.

3. High Temperature Superconductors

In 1986, after a long time of silence the scientific community greeted the announcement of superconductivity at 35K in the La-Ba-Cu-O system [7]. It was indeed a historical moment in the history of science that led to a big jump in T_c and the materials became known as *High Temperature Superconductors* (HTS). This new period that began with this event can be called the "*cuprate*" era because the compounds essentially contain copper and oxygen planes.

The compounds are for e.g $YBa_2Cu_3O_{7-\delta}$ (T_c = 91K) [8] where the yttrium can be replaced by various rare earth elements e.g La, Nd, Sm etc. with similar high T_c. The cuprates provoked a great interest in the community due to its enormous technological applicability. The application of pressure elevates the T_c to much higher values. Some 100 different cuprate materials discovered so far share a common structural point, the so-called *perovskite structure*, which appears to be responsible for the occurrence of high temperature superconductivity.

The high temperature cuprates are principally doped ceramics and are completely different by nature from the conventional superconductors. Moreover, these materials also demonstrate an anomalous characteristic in the normal state properties. It is widely believed that the source of the pairing interaction which is responsible for the superconducting transition lies in the CuO_2 planes and hence could be explained by two-dimensional models. They all demonstrate a very rich phase diagram which we discuss in short before going into further details. The parent compound of the La-Sr-Cu-O compound is essentially an insulator (La_2CuO_4).

The phase diagram is extremely varied depending on the compound. Thus to understand these compounds broadly, we need to focus on the simplest aspects of these cuprates, rather than to invoke more exotic and less generic features.

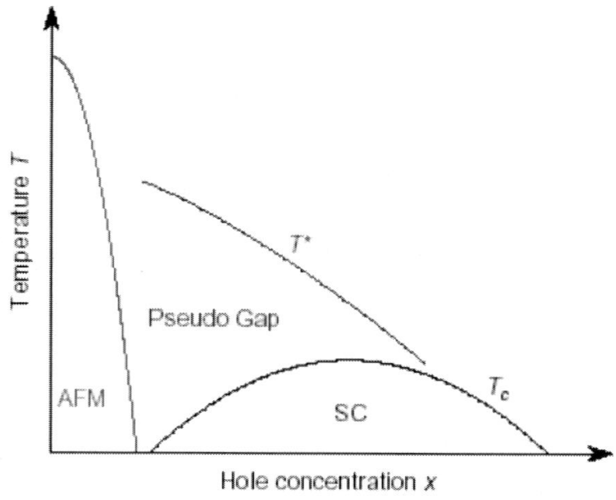

Figure 1. Typical phase diagram of the hole-doped high-T_c superconductors.

In Figure 1, we exhibit a sketch of the phase diagram for the cuprates. Here, x represents the concentration of holes which can be controlled by adding Sr, for e.g., to $La_{1-x}Sr_xCuO_4$. At zero and small values of doping we obtain an *antiferromagnetic phase* (AFM) with the ordering temperature $T_N \sim 300\,K$. Once the AFM order vanishes for a certain value of doping, we observe an insulating phase until a critical concentration above which we observe the insulator-superconductor (SC) transition.

The normal state of the cuprates is highly anomalous that makes the dream of having a unique microscopic theory for these doped ceramics more distant. The resistivity in the cuprates is not only highly anisotropic leading to different characteristics of the in-plane $\rho_{ab}(T)$ and out-of-plane $\rho_c(T)$ resistivities but also highly anomalous due to its variation with doping and temperature. $\rho_{ab}(T)$ in the under-doped (UD) cuprates varies approximately linearly with temperature at high T, whereas a deviation from linearity is suggested at low T [9]. This behavior not only demonstrates a reduced resistivity but also a break in slope at a temperature correlated with the opening of the pseudogap at T^* [10-12]. Optimally-doped (OP) cuprates, on the other hand, are characterized by a linear-T dependence for all $T > T_c$, which evolves to the supralinear contribution on the over-doped (OD) side. The doping induces a more pronounced change in the $\rho_c(T)$ that modifies the metallic behavior on the OD side to the non-metallic behavior in the UD state.

The anisotropy in the resistivity is evident from the ρ_c/ρ_{ab} values that can vary between 30 in OP $YBa_2Cu_3O_7$ to over 10^6 in $Bi_2Sr_2CuO_6$ [13]. On the other hand, c-axis transport in cuprates at optimal doping is very material specific.

The anomalous nature in the normal state properties is also explicit in the *T*-dependence of the *Hall coefficient*. In a normal metal, Hall coefficient ($R_H = 1/nec$) is a constant and depends on the carrier concentration. However, in the cuprates R_H is temperature dependent below the characteristic temperature T^* and above which it ceases to be temperature independent and thus tends to a constant value [14]. Moreover, the *Hall angle* θ_H is also found to depend on T, i.e., $\cot(\theta_H) = AT^2 + B$ below T^*[15].

4. Order Parameter

Order parameter marks the onset of order at the phase transition of a system. The superconducting order parameter is identified as the pairing state that represents the extent of the macroscopic phase coherence and is the well-known energy gap measured from the ground state. The superconducting order parameter was first used in 1950 by Ginzburg and Landau [4] in their phenomenological description of the superconducting free energy based on the classical theory of Landau for second order phase transitions. The position-dependent order parameter is a measure of the superfluid density. Later a link between the phenomenological theory of GL and the microscopic theory of BCS was established by Gorkov [5] in their articles based on the Green's function calculations.

Thus the link between the order parameter and the energy gap was established in the vicinity of T_c. Moreover, the BCS equation could be derived from the GL equation. The gap parameter can be defined as the $\Delta(k) = < c_{k\uparrow} c_{-k\downarrow} >$ in the k space representation.

The conventional superconductor has isotropic gap parameter which is a spin-singlet state with even parity $[\Delta(k) = \Delta(-k)]$. It has the same magnitude in the momentum space and has no nodes in any direction. The parity consideration also brings into the spin-triplet state (total spin 1) which has an odd-parity and has been of extreme importance in the description of liquid Helium [16]. A mixture of the spin-singlet and the spin-triplet states was also found to be useful for the heavy fermion compounds.

After the discovery of the high temperature superconductors, the symmetry of the order parameter posed a new challenge to the condensed-matter physicists.

There is a preponderance of evidence from a number of experimental measurements that pairing in the cuprates is highly anisotropic with a line of nodes of superconducting gap but the exact symmetry of the order parameter is far from reaching a consensus. The gap function can be expressed as a linear combination of the basis functions of the irreducible representation that corresponds to

$$\Delta(k) = \sum \Delta_0 \eta^j(k) \qquad (1)$$

The energy gap in the $s-$wave isotropic symmetry is given by $\eta(k) = 1$ where we have a unique value of the gap parameter for all values of k_x and k_y. The $d_{x^2-y^2}$ state of the energy gap has the form

$$\Delta(k) = \Delta_0[\cos k_x a - \cos k_y b] \qquad (2)$$

where a and b are the lattice constants along the x-axis and the y-axis respectively. The gap is real and is highly anisotropic with nodes along the (110) direction in $k-$space and a sign change in the order parameter occurs between the lobes in the k_x and k_y directions. The other possible candidates for the superconducting gap is the extended $s-$wave with

$$\Delta(k) = \Delta_0[\cos k_x a + \cos k_y b] \qquad (3)$$

In addition to the pure states, the order parameter can also have mixed-paired states such as the $s+id_{x^2-y^2}$, $d_{x^2-y^2}+id_{xy}$ states which break the time-reversal symmetry and are complex. The mixed states which do not break the time reversal symmetry but are possible candidates for the cuprates are $d_{x^2-y^2}+d_{xy}$, $d_{x^2-y^2}+s$ and also $d_{x^2-y^2}+s_{xy}+s$ states. In the following sections we will discuss in detail the possible candidature of all the above states.

Before entering into further details it is important to consider an overview on the major points regarding the structure of the cuprate superconductors that plays a vital role in deciding the type of symmetry of the gap. The crystal structure of cuprates has been intensively investigated. All these high-T_c materials share a common structural point which appears to be responsible for the occurrence of high-T_c superconductivity and that is the general perovskite structure with the planes containing Cu and O atoms separated by charge reservoir blocks. Depending on the choice of stoichiometry, the crystallographic unit cell can contain varying number of CuO_2 planes.

Moreover, the structure can be broadly divided into two categories: a tetragonal as in $Tl_2Ba_2CaCu_2O_8$ (Tl2212) and an orthorhombic lattice as in $YBa_2Cu_3O_{7-\delta}$. The cuprate superconductors such as $La_{2-x}Sr_xCuO_4$ (LSCO), $Tl_2Ba_2CaCu_2O_8$ (Tl2212), $HgBa_2CaCu_2O_6$ (Hg1212) and some $YBa_2Cu_3O_7$ (YBCO) type compounds exhibit a tetragonal crystal structure with D_{4h} point group symmetry whereas the compounds $YBa_2Cu_3O_{7-\delta}$ and $Bi_2Sr_2CaCu_2O_8$ (Bi1212) posses an orthorhombic crystal structure with point group symmetry D_{2h}. In the case of YBCO, the CuO chains result in inequivalent directions a and b. The orthorhombicity in this case is of the order of 2%. In Bi2212 an incommensurate superlattice modulation in BiO layers along the $b-$direction gives rise to the anisotropy in the a and b directions. Group theoretic considerations indicate the allowed pair states to be pure spin-singlet even parity in the case of tetragonal lattice whereas the admixture of the singlet states is allowed in the orthorhombic crystal lattice under the constraint that only a single superconducting transition should be observed. A detail discussion on the allowed symmetries of the gap parameter can be found in [17]. The basic functions related to each point group can lead to order parameter which may or may not exhibit nodes in the k_x-k_y plane. The gap can also be real or imaginary depending on the symmetry. Moreover, the mixed states, nodeless or not can break time-reversal symmetry.

The high $c-$axis to the $a-$axis ratio characterizes the energy bands to be predominantly derived from the CuO_2 planes. This makes the physics of the cuprates to be predominantly a two-dimensional character. Anisotropy in the $c-$axis direction is observed in the normal-state resistivity results. For e.g ρ_c/ρ_{ab} can be as large as 10^5 as observed in Bi2212 [18] that demonstrates the $c-$axis to be insulating. Moreover, large anisotropy in the penetration depth λ_{ab}, λ_c and superconducting coherence length ξ_{ab}, ξ_c have also been observed.

Apart from the large anisotropy several studies in the $c-$axis charge dynamics, dc and ac Josephson effects have convincingly reiterated the bi-dimensional nature of the cuprates.

Hence as superconducting is basically bound to the CuO_2 plane, we therefore mainly concentrate on the two-dimensional structure.

The phase-sensitive experiments have firmly established that high-T_c materials have singlet d-wave Cooper pairs [19]. Measurements [20] of the penetration depth $\lambda(T)$ and superconducting specific heat at different temperatures T and related theoretical analysis [21, 22] also support this point of view.

This idea has also been explored to explain the NMR data in the superconductor YBCO and the Josephson critical current observed in YBCO-SNS and YBCO-Pb junctions [23]. Moreover, a new class of c-axis Josephson tunneling experiments are reported by Kouznetsov *et. al.* [24] which provide a direct evidence of an order parameter of mixed $[d_{x^2-y^2} + \exp(i\theta)s]$ symmetry in YBCO. The microwave complex conductivity measurement in the superconducting state of high quality $YBa_2Cu_3O_{7-\delta}$ single crystals also strongly suggests a multicomponent superconducting order parameter in YBCO [25]. Moreover, Krishana *et. al.* [26] reported a phase transition in the high-T_c superconductor $Bi_2Sr_2CaCu_2O_8$ induced by a magnetic field from a study of the thermal conductivity as a function of temperature and applied field. Possible interpretation of this measurement could be the induction of a minor s or d_{xy} component with a $d_{x^2-y^2}$ symmetry with the application of a weak field [27, 28]. There have also been several recent theoretical studies using mixed s- and d-wave symmetries [29-31] considering different couplings and lattice symmetries that support the above scenario.

Recently the angle-resolved phase-sensitive experiments by Kirtley *et. al.* [32] show that $|\Delta(0,\pi)|/|\Delta(\pi,0)| \sim 1.2$ and the gap nodes shift by about 3° away from the diagonal direction towards the k_x axis whereas the electron tunneling experiment [33] demonstrates a shift of 5° towards the k_x axis. Moreover, the inelastic neutron scattering and the spin dynamics experiments also support the anisotropy in the *superconducting (SC) gap*. The inelastic neutron scattering experiment in detwinned YBCO samples demonstrated that the incommensurated peak of spin fluctuation is stronger at $(\pi+\delta,\pi)$ direction than at $(\pi,\pi+\delta)$ direction. This anisotropy in the gap magnitude at $(0,\pi)$ and $(\pi,0)$ and also the shift in the gap node can be taken into account considering the isotropic s wave along with dominant $d_{x^2-y^2}$ and a subdominant extended s wave [34, 35].

More recently, low temperature scanning tunneling microscopy (STM) experiments have shown that the superconducting energy gap in some cuprates is spatially inhomogeneous [36-42]. However, at present there is no consensus regarding the origin of the inhomogeneities in the cuprates -whether they are in charge density, spin density, local density of states or other properties.

It could be due to the presence of the competing orders or may be attributed to the crystal defects or impurities. An attractive Hubbard Hamiltonian for various inhomogeneous patterns of interacting sites [43] has been considered very recently. A Green's function technique studied the inhomogeneity using a model lattice Hamiltonian in which one small region of lattice has a different (suppressed or enhanced) pairing strength than the rest [44].

5. PSEUDOGAP

Besides, the anomalous nature of various normal state properties, the *underdoped phase* incorporates one of the most controversial topics of contemporary research – the *pseudogap phase*. The depletion of the spectral weight around the Fermi level, popularly known as the pseudogap is one of the most prominent differences that exist between the conventional superconductors and the cuprates. It is observed in the underdoped region of the phase diagram of the cuprates and below a characteristic temperature T^* as shown in Figure 2. Various strange phenomena like the presence of the energy gap, anomalous isotope effect, Hall effect etc. are observed in this region above T_c. However, there is a general consensus regarding the temperature and doping region where it exists. Moreover, the $d_{x^2-y^2}$-wave nature of the *pseudogap* is identical to that of the superconducting gap. In this section, we discuss some of the key experiments that not only provide evidences for the existence of the pseudogap but also reveal important informations about its origin. A detailed report on the experimental evidences can be obtained in the works of Timusk and Statt [45] and Tallon and Loram [46]. The NMR experiments in YBCO were one of the first ones to indicate the existence of a pseudogap. A decrease in the susceptibility in the underdoped phase supports the hypothesis of the gap [47]. The pseudogap can be observed in the Cu nuclei as well as in the O nuclei in the CuO_2 planes of these layered compounds. Many compounds exhibit pseudogap in their magnetic resonance studies. Some of them are $YBa_2Cu_4O_8$ [48], $Y_2Ba_4Cu_7O_{15}$ [49], $Bi_2Sr_2CaCu_2O_8$ [50] and $HgBa_2Ca_2Cu_3O_{8+\delta}$ [51]. The studies, however, indicate no connection between the presence of the pseudogap and the number of adjacent layers in the compound. The effects on superconductivity and on the pseudogap due to the Zn substitution for Cu have revealed some important informations regarding their behavior. Zn has been found to be very effective in suppressing the T_c. However, the NMR Knight shift [52, 53] studies show that the pseudogap remains unaltered by the Zn substitution. This observation and various others have played a key role to show that the superconductivity and pseudogap are independent and competing.

The ARPES (Angle Resolved Photo Emission Spectroscopy) data based on the classical *photoelectric effect* have been vital not only in detecting the pseudogap but also important in revealing the functional form and its connection with the superconducting gap. The studies on Bi-2201 and Bi-2212 [54, 55] are some that deserve to be mentioned. The gap energy decreases with increasing doping unlike the superconducting gap which increases [56]. Moreover, it is maximal near $(0, \pi)$ and vanishes along the line connecting (0,0) and (π, π). Remarkably, the pseudogap has the same magnitude as that of the superconducting gap at each k point and evolve continuously into each other as the doping is reduced. But the energetic competition continues to prevail between these two energy scales.

The tunneling experiments using the scanning tunneling microscope (STM) technique have been performed very successfully on the Bi2212 crystals [55]. They reveal the pseudogap in the underdoped phase of the cuprates. Moreover, they also demonstrate a close relationship between the superconducting gap as observed by the conductance peak and the pseudogap as exhibited by the depression of the conductivity in the normal state. A smooth evolution through T_c from the superconducting state to the pseudogap state is observed.

However, the energy scales are very similar and are temperature independent but depend on the doping level unlike the conventional superconductors.

The electronic specific heat provides an evidence of the gap in the normal state of the cuprates by varying the oxygen doping of YBCO [57] and the Sr content of LBCO [58]. The specific heat coefficient is observed to vary from the underdoped to the overdoped region. We observe a depression in the γ values in the underdoped region below T_c indicating the pseudogap in this region. Similar results have also been found in $La_{2-x}Sr_xCuO_4$ [58] and Bi2212 [54] in the underdoped phase. Moreover, in all the cases we observe that the jump in the specific heat falls sharply at the doping where the pseudogap goes to zero. So we observe that the pseudogap is a generic behavior of cuprates which is evident from the measurement of their thermodynamic properties.

Various experimental works critically examining the behavior of specific heat, photoemission, tunneling and optical conductivity have predicted the existence of two competing energy gaps, a pseudogap and the superconducting gap, both of which exhibit a $d_{x^2-y^2}$ -wave symmetry. This observation motivated the proposal of a d -density wave order (DDW) [59-63] scenario where the particle hole pair of the same orientation form the condensate. The phase diagram of the cuprates is in agreement with this recent picture where the competition between the DDW state and the $d_{x^2-y^2}$ -wave superconducting gap (DSC) state exhibits the quite familiar "superconducting dome".

Moreover, the experimental situation is now quiet promising. The elastic neutron scattering [64] and various other experiments are consistent with the above picture. The pseudogap exhibits a $d_{x^2-y^2}$ -wave symmetry alike the superconducting gap below T_c. However, the broken symmetry of the DDW state makes it different from the DSC of the Cooper pairs. In this scenario the temperature T^* is not a crossover but a transition temperature. Large Nernst effect is a clear signature of DDW. The measured Nernst effect in the pseudogap phase has been successfully described in underdoped LSCO and Bi-2212 [65].

There have also been various other scenarios for this phenomena based on the crossover picture, like the singlet pairing of the spinons [66], phase fluctuations [67] etc. One of the simplest ideas has been the BCS-BEC crossover picture [68] where the pseudogap state has been described as a natural consequence of local, dynamic pairing correlations in the normal state of low-density small pair-size superconductors. The phase coherence is lost at the critical temperature T_c but the pairs dissociate only at a much higher temperature T^*. In ref [66] the pairing of the spinons occur at T^* which describes the magnetic properties within an extended $t-J$ model in the slave-boson representation. The mechanism suggested by Emery and Kivelson [67] indicates a phase fluctuation of the superconducting order parameter in the pseudogap regime. The temperature scale of phase fluctuations is controlled by the zero-temperature superfluid stiffness. Unlike the conventional superconductors, the transition is controlled by phase ordering and explains the Uemura scaling [69]. It is important to mention that in the regime of interest, it is the vanishing of the superfluid density that controls the transition and determines T_c and not the collapse of the gap. The nearly antiferromagnetic Fermi liquid (NAFL) model for this phenomenon was proposed by Pines and collaborators [70]. The close proximity to the antiferromagnetic phase leads to strong anisotropic magnetic interaction which explains the anomalous normal state of the cuprates. The pseudogap in the quasiparticle spectrum is the formation of a precursor to a spin-density wave and is based on the scenario of the dynamic *spin susceptibility* $\chi(q,\omega)$ which is peaked at the commensurate wave vector $Q \approx (\pi,\pi)$.

6. Penetration Depth

The magnetic penetration depth in the conventional superconductors was first addressed by F and H. London in 1950 [3] to explain the penetration of the magnetic field inside a type I superconductor. The $\lambda(0)/\lambda(T)$ of a *s-wave superconductor* saturates exponentially to 1 in the limit $\Delta/K_B T >> 1$. Moreover, the superfluid fraction n_s/n approaches unity as the temperature goes to zero. The results also lead to the explanation of the *Meissner effect*.

There exists a large variety of data on the magnetic penetration depth obtained by various techniques and essential in clarifying the symmetry of the order parameter. A comprehensive review on the property is too lengthy and beyond the scope of this chapter. Hence we would like to concentrate on the data which focus on the symmetry of the order parameter. The anisotropy due to doping and the structure of cuprates will also be presented. Lets consider few examples from the cuprate family to illustrate the anisotropy on the penetration depth results.

6.1. Experimental Overview

In this subsection we provide the behavior of the penetration depth along the a, b and c axes. The in-plane data shall also be presented. As doping plays an important role due to the existence of the pseudogap, the experimental results on doping shall also be considered. Since, YBCO is the most studied of the cuprates due to its availability of large high-quality single crystals, we begin a short review of the present experimental situation of these new materials with the above compound. Just as in other high-T_c cuprates the two-dimensional CuO_2 layers form the essential structural component of this superconductor.

However, the YBCO differs from the other cuprates due to its one-dimensional CuO chains which plays an important role in understanding the electrodynamics of the superconductor. Hence, apart from the difference between the in-plane (ab) and out-of-plane (c) results, the anisotropy is also evident from the measurements along the a and b axes of the CuO_2 planes where the chains are along the $b-$ axis.

The *resistivity anisotropy* $\rho_a/\rho_b \sim 2$ at 100K, where ρ_a and ρ_b are the resistivities parallel to the a and the b axes respectively, demonstrates the conduction along the chains. The λ_{ab} results on twinned $YBa_2Cu_3O_{6.95}$ deviate substantially from the BCS isotropic superconductors both at low and high temperatures [71] as can be seen in Figure 2. However, at low temperatures $\lambda_{ab}^{-2}(T)$ obeys a linear law, rather than a T^2 dependence previously found in thin samples.

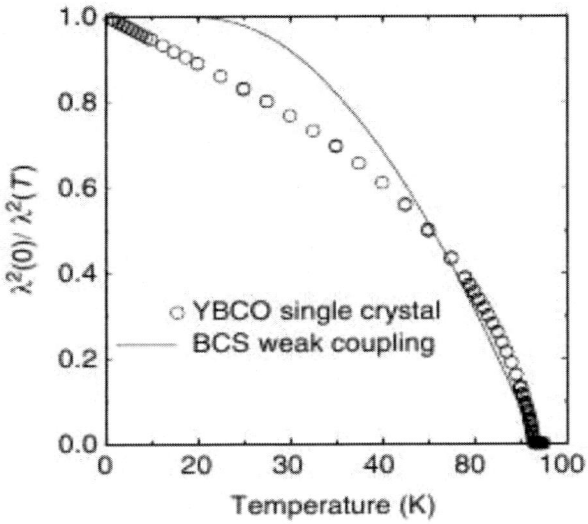

Figure 2. $\lambda_{ab}^2(0)/\lambda_{ab}^2(T)$ vs T for twinned $YBa_2Cu_3O_{6.95}$. The solid line is the weak-coupling s -wave results.

The above linear dependence seen at low temperatures is now accepted for all high temperature superconductors and has been accepted as an evidence of the nodes on the *Fermi surface* in the order parameter. In particular, these data were found to be consistent with the *d- wave theories*. Moreover, measurements on $YBa_2Cu_3O_{6+x}$ also demonstrated features of the presence of two gaps or the mixed-type symmetry of the order parameter [72]. The sample dependence has also been an important factor in the understanding of the penetration depth.

Improved samples has demonstrated a linear behavior of $(1-\lambda^2(0)/\lambda^2(T))$ upto $T_c/3$ and leading to a T^2 dependence at lower temperatures. The crossover temperature depends on the purity and imperfection of the sample [73].

In addition, studies on the YBCO sample with impurities like Zn and Ni, for example, produces an effective change from linear to a T^2 dependence [74, 75].

Various works on YBCO have exhibited substantial anisotropies in the penetration depth measured along the a and b axis. The infra-red measurements yield $\lambda_a(0) = 1{,}600$ Å and $\lambda_b(0) = 1{,}030$ Å [76]. The temperature dependencies of $\lambda_a(T)$ and $\lambda_b(T)$ observed from microwave measurements demonstrate qualitatively the same characteristics. In fig 3 the $\lambda_a(T)$, $\lambda_b(T)$ and $\lambda_c(T)$ are plotted in $YBa_2Cu_3O_{6.95}$ from microwave measurements [71, 74]. $\lambda_a(T)$ and $\lambda_b(T)$ have similar temperature dependencies which clearly demonstrates that the influence of the chains on the temperature dependence should be very little. The behavior of $\lambda_c(T)$ on temperature exhibits a qualitative influence from the a and b axes results. Moreover, the penetration depth is strongly reduced in all samples in the b polarization (along the chains). For the 124 material we have $\lambda_a(T) = 2000$ Å and $\lambda_b(T) = 800$ Å, whereas for the 123 material they are $\lambda_a(T) = 1600$ Å and $\lambda_b(T) = 1000$ Å [77]. The strong anisotropy in the penetration depth results is consistent with the transport anisotropy found in the normal state. In single- chain 123 materials the dc *resistivity anisotropy* in the a-b plane, ρ_a/ρ_b is 2-2.3 [78] whereas in the double chain 124 it increases to 3.3 [79].

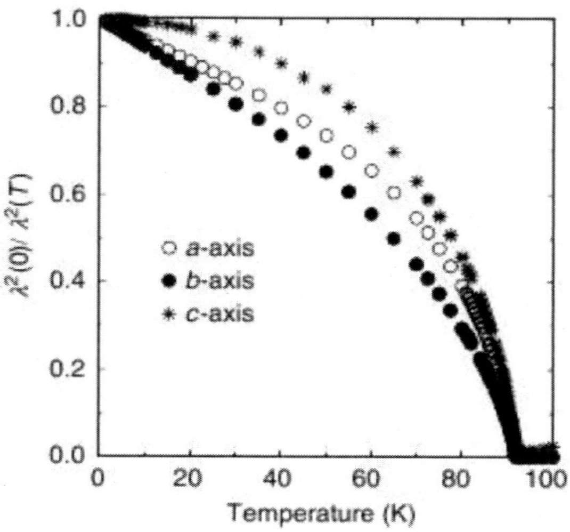

Figure 3. $\lambda_i^2(0)\lambda_i^2(T)$ vs T in $YBa_2Cu_3O_{6.95}$ for $i = a,b,c$ obtained from [71, 76].

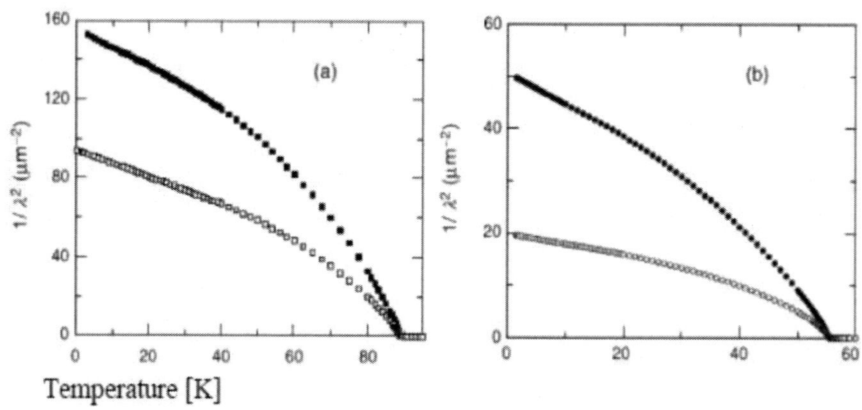

Figure 4. Anisotropic in-plane superfluid density vs temperature for (a) overdoped $YBa_2Cu_3O_{6.99}$ and (b) underdoped for $YBa_2Cu_3O_{6.52}$. Solid symbols are for the b -direction measurements whereas open symbols are for a direction.

Hence it is clear that the anisotropies of both the normal state properties and $\lambda(T)$ are more pronounced in the double-chain 124 system, suggesting that the condensate resides on both the chains and the planes.

Apart from the temperature dependence, the doping effect is extremely important in order to disentangle the effects of chains, planes and defects in the understanding of the anisotropy of the cuprates. The data from the in-plane superfluid density of the ordered phases in Figure 4 demonstrate the dependence on temperature in the overdoped $YBa_2Cu_3O_{6.99}$ (with T_c = 89K) and underdoped $YBa_2Cu_3O_{6.52}$ (with T_c = 59K) [80] strengthening the conviction regarding the large in-plane anisotropy in both samples. Apart from the chains, the presence of the pseudogap in the single excitation spectrum plays an important role in the doping dependent results. Understanding the doping-dependent behavior of the superconducting properties would provide important clues to unravel the origin of pseudogap.

Penetration depth results for other members of the cuprate family also demonstrate similar characteristic. For $Bi_2Sr_2CaCu_2O_8$ a linear T dependence of in-plane data was observed [81, 82]. The behavior is very similar to the $YBa_2Cu_3O_{6+x}$, but exhibiting a strong d wave influence. Similar linear temperature dependence was also observed in $Tl_2Ba_2Cu_3O_{4+\delta}$ [83], a characteristic of the *d-wave symmetry*. For $HgBa_2Ca_2Cu_3O_8$ [84] the anisotropy could be observed in the in-plane and $c-$ axis results. $\lambda_{ab}(0) = 2100$ Å and $\lambda_c(0) = 61000$ Å. The temperature dependence is linear and also very similar to the $YBa_2Cu_3O_{6+x}$ measurements.

The c-axis results in all the members of the cuprate family differs significantly from the in-plane superfluid density. In the $YBa_2Cu_3O_{6+x}$ system a power law close to $T^{2.5}$ [85] has been observed. Moreover, the results at different doping overlap with one another and also coincide with the measurements at higher hole doping. The quadratic c-axis penetration depth has also been observed in other members like $La_{2-x}Sr_xCuO_4$ [86], $Bi_2Sr_2CaCu_2O_8$ [87], $HgBa_2Ca_2Cu_3O_8$ [88].

The difference in the c-direction indicates towards a tunneling dependence on the in-plane wavevector $k_{\|}$. A T^5 dependence of the penetration depth has been seen in the tetragonal compound $HgBa_2CuO_{4+\delta}$.

6.2. Theoretical Models

The interest in understanding the behavior of the *penetration depth* in the new materials goes back to the initial period of their discovery. The *d-wave symmetry* of the cuprates invoked a new interest in the study of various properties of the cuprates. However, till date there is no suitable microscopic theory for the high T_c superconductors. Although it is accepted that *Cooper pairs* are formed in such materials, there is a great deal of controversy about a proper description of the normal state and the pairing mechanism for such materials [17]. Hence, in the absence of a suitable microscopic theory, a phenomenological tight-binding model in two dimensions which incorporates the proper lattice symmetry has been suggested [31]. This model has been very successful in describing many properties of high-T_c materials and is often used in the study of these compounds. The *Hamiltonian* is given by

$$H = \sum_{k\sigma}(\varepsilon_k - \mu)c^{\dagger}_{k\sigma}c_{k\sigma} + \sum_{k}\Delta_k c^{\dagger}_{k\uparrow}c^{\dagger}_{k\downarrow} + h.c \tag{4}$$

where $c^{\dagger}_{k\sigma}$ is the *fermion creation operator* for wavevector k and spin σ. The energy dispersion relation is given by $\varepsilon_k = -2t(\cos k_x a + \beta \cos k_y b - \gamma \cos k_x a \cos k_y b)$ which gives the band structure. t and βt are the nearest-neighbor hopping integrals in the a and b directions respectively. γt is the second nearest hopping integral. This simple model can be used to understand the change in the behavior of the penetration depth of the d superconductors as well as the mixed-wave superconductors. All the parameters are given in terms of t unless otherwise mentioned. The penetration depth [3] can be written as

$$\lambda^2(0)/\lambda^2(T) = 1 - \frac{2}{NT}\sum_k f_k(1-f_k) \tag{5}$$

where f_k is the Fermi function. The $d_{x^2-y^2}$ gap can be written as in eq. 2. The results are exhibited in the Figure 5 within a mean-field scenario. A pure s -wave superconducting (BCS type) penetration depth that exhibits an usual exponential dependence on temperature is also plotted for comparison. Unlike the s wave, the d gap leads to a *power law* dependence $\lambda^2(0)/\lambda^2(T) \sim (T/T_c)^{1.3}$ [89] as observed in various experimental studies.

In the case of mixed ($d+is$)-wave symmetry, we find that it is possible to have a crossover at $T = T_{c1}$ from an exponential to power-law dependence for $T_{c1} < T_c$ and a second second-order phase transition is observed. The pronounced change in the nature of the superconducting state at $T = T_{c1}$ becomes very apparent from a study of the entropy [89]. The ratio of the strengths of the s- and d-wave interactions should lie in a narrow region in order to have a coexisting s and d wave phases in the case of $d+is$ symmetry. As the $s-$ wave ($d-$ wave) interaction becomes stronger,the d-wave (s-wave) component of the order parameter quickly reduces and disappears and a pure $s-$ wave ($d-$ wave) state emerges. The behavior of the $d_{x^2-y^2}+is$ pairing are presented in Figure 5 where we observe the modification due to change in symmetry of the order parameter.

The anisotropy in the zero-temperature penetration depth along the a and b axes indicates the presence of the condensate in the CuO chains along with the CuO_2 planes. The above study of a simple two-dimensional tight-binding model of the superconducting state of the cuprates fails to demonstrate the contribution along the three axes of the cuprates.

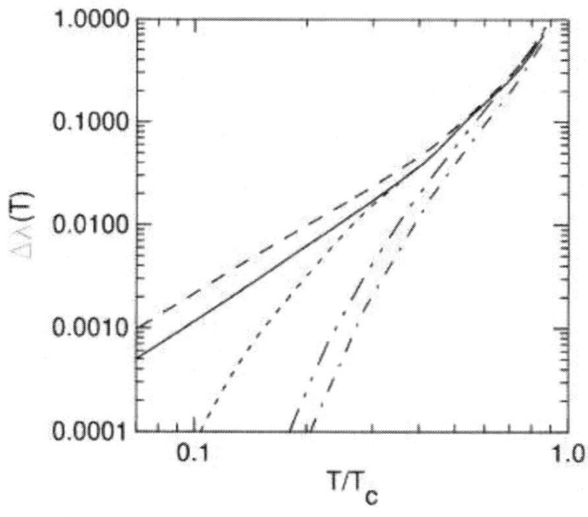

Figure 5. Penetration depth ratio $\Delta\lambda(T) \sim [\lambda(T)-\lambda(0)]/\lambda(0)$ vs T/T_c .: d-wave (solid line, square lattice), d-wave (dashed line, orthorhombic distortion, $\beta = 0.95$), s-wave (dashed-dotted line, square lat- tice), $d_{x^2-y^2}+is$ -wave (dotted line, square lattice) and $d_{x^2-y^2}+is$ -wave (dashed-double-dotted line, orthorhombic distortion, $\beta = 0.95$.

However, the knowledge regarding the d – wave nature of $\lambda(T)$ is of utmost importance before proceeding to the understanding of the detailed anisotropic nature that depends on the structure of the cuprates. Moreover, the doping dependence of the penetration length also deserves attention from the theoretical point of view. The consideration of the pseudogap in the underdoped becomes essential for the understanding of these materials.

In various works [59-63], the DDW wave has been considered as the pseudogap along with $d_{x^2-y^2}$ -wave superconducting phase (DSC) to study the *phase diagram* in order to understand this new phase in the underdoped region of the cuprates. We observe the familiar phase-diagram with the DDW state and superconducting dome which is in perfect accordance with the experimental predictions.

Various other properties were also calculated. The essence of this idea is that the pseudogap observed in underdoped cuprates is a real gap in the one-particle excitation spectrum at the wave-vector ($\frac{1}{4}$, 0).

However, it is pseudo in experiments due to the difficulty in its detection. As it is widely believed that the pairing interaction lies in the CuO_2 planes, the only active pieces of the crystal are the CuO_2 planes and the remaining ions act as charge reservoirs. But in some cuprates it is essential to consider the one-dimensional chain structures apart from the planes. Experiments on the dc resistivity,[90, 91] the polarized reflectance,[92, 93] apart from the penetration depth results have already found large anisotropies between the a and the b directions which suggest that currents also flow along the chains in both the normal and superconducting states. But little is known about the coupling of the planes and the chains and how the planar superconductivity affects the physics in the chains.

Recently, the proximity model [94] has been demonstrated to explain consistently the scanning tunneling microscopy, angle-resolved photoemission, and infrared-spectroscopy experiments.

This model that includes the DDW gap and the CuO chains should reveal the qualitative difference of the penetration depths in the underdoped regime [95]. In ref. [96], Trumin *et. al* have experimentally measured in-plane penetration depth and verified that the results are in accord with the predictions of the DDW phase. They have exhibited that the slope of the $n_s(p,T)/n_0$ curves increases at low T and for underdoped samples. However, the precursor pairing model for the pseudogap predicts a decrease in slope and contradicts the experimental findings. The tunneling in the c -direction is incorporated to account for the the proximity effect.

In the first moment the aim is to emphasize both the doping and the temperature dependence of the a and b -axes penetration depth in the cuprates. We consider the effect of the DDW gap in the underdoped systems and also the effect of the coupling between the different layers on the temperature dependence of the penetration depth.

The *modified Hamiltonian* can then be written as

$$H = \sum_{k\sigma\lambda}(\varepsilon_{\lambda k} - \mu_\lambda)c^\dagger_{\lambda k\sigma}c_{\lambda k\sigma} + \sum_{k\sigma}t(k)c^\dagger_{1k\sigma}c_{2k\sigma} + i\sum_{k\sigma}W_k c^\dagger_{1k\sigma}c_{1k+Q\sigma}$$
$$+ \sum_k \Delta_k c^\dagger_{1k\uparrow}c^\dagger_{1k\downarrow} + h.c \qquad (6)$$

where $c^{\dagger}_{\lambda k\sigma}$ is the fermion creation operator for wavevector k and spin σ in the planes. W_k is the DDW gap and Q is the nesting vector (π,π). μ_λ is the chemical potential for the layers.

The energy dispersion relations are given by $\varepsilon_{1k} = -2t_1(\cos(k_x a) + \beta\cos(k_y b))$ and $\varepsilon_{2k} = -2t_2\cos(k_y b)$ which gives the band structure. t_1 and βt_1 are the nearest neighbor hopping integrals along the in-plane a and b axes respectively. $t(k)$ is the coupling between the chains and the planes. λ corresponds to the chain and/or plane bands. All the parameters are given in terms of t_1 unless otherwise mentioned. The penetration depths can be calculated using the Green's function formalism and is given by

$$\lambda_{ij}^{-2} = \frac{8\pi e^2}{\hbar^2 c^2}\frac{1}{\Omega}\sum_{k,\alpha\beta}\hat{\gamma}^{(i)}_{k\alpha\beta}\hat{\gamma}^{(j)}_{k\beta\alpha}\left(\hat{G}_{k\alpha\beta}\,|_{\Delta=0} - \hat{G}_{k\alpha\beta}\right) \tag{7}$$

where

$$\hat{G}_{k\alpha\beta} = \frac{df(E_{k,\alpha})}{dE_{k,\alpha}}\delta_{\alpha\beta} + \frac{f(E_{k,\alpha}) - f(E_{k,\beta})}{E_{k,\alpha} - E_{k,\beta}}(1-\delta_{\alpha\beta}) \tag{8}$$

The matrices $\gamma_\mu(k,k)$ are determined by the same approach as used in the ref [97] where the matrices $\hat{\gamma}_\mu(k,k)$ are given by

$$\hat{\gamma}_\mu(k,k) = U^{\dagger}(k)\gamma_\mu(k,k)U(k) \tag{9}$$

Ensuring the Hamiltonian and the formalism for the calculation of the penetration depth we would like to consider the dependence of the penetration depth on the presence of the chains, *plain-chain coupling* and also on doping. The presence of the pseudogap is fundamental in this aspect.

We neglect the k dependence of the plain-chain coupling, $t(k) = t_0$ since this term is important in a very small k space region. The penetration depth on the plane along the a and b directions are presented in Figure 6 and Figure 7 to study the anisotropy due to the presence of the chain bands.

Figure 6 exhibits the anisotropy in the overdoped regime and also in the underdoped phase in the absence of the interaction between the two layers. The underdoped phase is marked by the presence of the pseudogap and also the superconducting gap ($T_c = 0.021$ and $T^* = 0.16$) whereas in the overdoped phase we have only the superconducting gap $(T_c = 0.12)$. The figure exhibits the penetration depth anisotropy due to the presence of the one-dimensional chains.

The effect of the interaction between the layers can be observed in Figure 7. In Figure 7a we plot λ_a and λ_b in the pure superconducting phase for $t_0 = 0.10$ with $T_c = 0.12$ and in Figure 7b we obtain the penetration depth for the coexisting phase in the underdoped regime. In the coexisting phase we have $T_c = 0.017$ and $T^* = 0.15$. The ratio of $\lambda_a^2/\lambda_b^2 = 1.8$ at T = 0.0 which is close to the value observed in experimental measurements for the overdoped state [93].

Hence it can be seen that the anisotropy in the penetration in the a and b directions occurs with doping as well as with temperature. Hence, the consideration of the chains and the proximity effect between the planes and chains is fundamental to exhibit the anisotropy in the a and b directions [95].

In the next moment, we consider the above model with a momentum dependent chain-plane to account for the proximity effect. $t(k) = -t_0 \cos(k_z d/2)$ is the coupling between the chains and the planes, where $d/2$ is the distance between the chains and the planes [98]. $t(k)$ not only determines the gap in the chains but also determines the ratio of $\lambda_c(0)/\lambda_{ab}(0)$. The dependence of the λ_a, λ_b and λ_c on t_0 is studied.

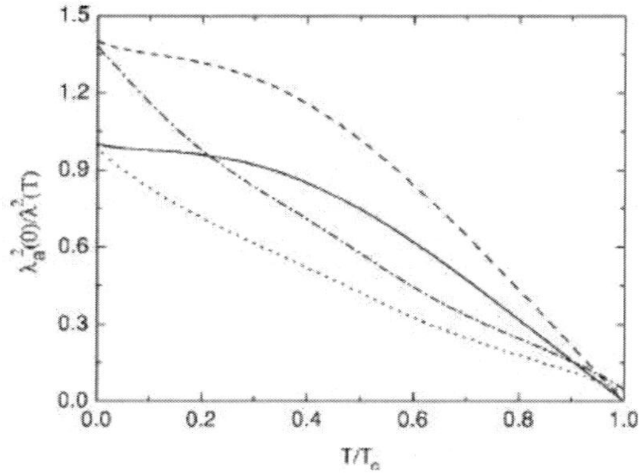

Figure 6. $\lambda_a^2(0)/\lambda^2(T)$ vs T/T_c for $t = 0.0$. The solid line (dashed line) indicates the λ_a (λ_b) for the underdoped phase, whereas the dotted (dashed-dotted) line indicates the λ_a (λ_b) for the overdoped phase.

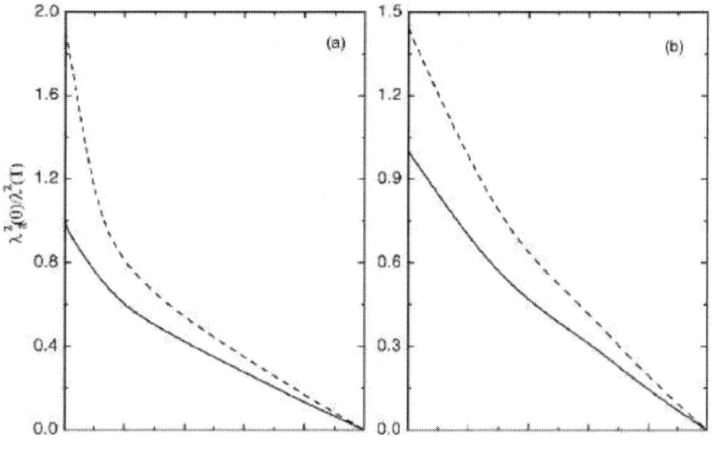

Figure 7. $\lambda_a^2(0)/\lambda^2(T)$ vs T/T_c for $t = 0.1$. (a) Pure DSC state (b) DDW+DSC state. The solid line(dashed line) indicates the λ_a (λ_b).

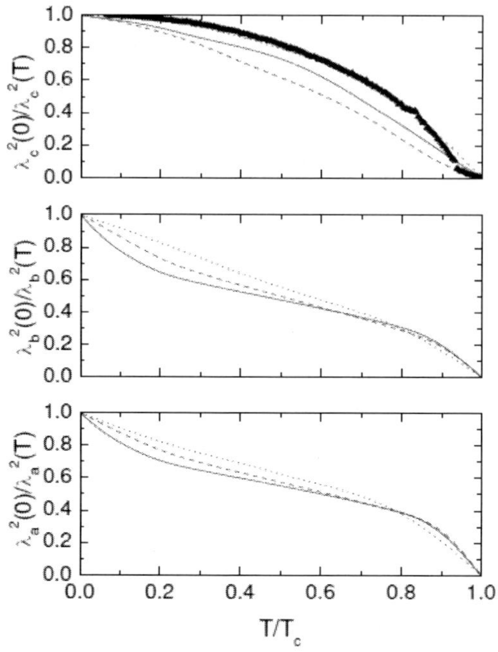

Figure 8. $\lambda^2(0)/\lambda^2(T)$ vs T/T_c along the a, b and c directions for $T_c = 61.2\,K$ (solid line) $T_c = 36.0\,K$ (dashed line) and $T_c = 18.0\,K$ (dotted line) for the tunneling parameter $t_0 = 0.05$. Experimental results for $T_c = 18.2$ K for $c-$axis results are plotted [99].

The temperature dependence of λ_a, λ_b and λ_c for $t_0 = 0.05$ for various values of T_c are exhibited in Figure 8. The values of T_c are 18K, 36K and 61.2K which depends on the doping and can be obtained by varying the parameter μ_1 [98]. The experimental results [99] for $T_c = 18.2$ K are also plotted to compare with the theoretical predictions and we observe a very good agreement. The plots for $\lambda_c^{-2}(T)$ when plotted with T satisfies the same qualitative behavior as observed in [99] for the three values of T_c. Comparing the T_c of the three sets and the $\lambda_c^{-2}(0)$ we observe a similar relation between the experimental measurements and this model. Close to T_c we observe a deviation from the experimental results where critical phase fluctuations dominate the low frequency electromagnetic response which has not been considered in the model. Besides, it can be observed that when $T_c^{(1)}/T_c^{(2)} \sim 2,\ 1.7$ and 3.4 the ratio $\lambda_{c1}^{-2}(0)/\lambda_{c2}^{-2}(0) \sim 3.4,\ 2.5$ and 8.28 respectively. Comparing it with the experimental values for nearly the same ratios of T_c-'s we have similar values. From ref. [99] we obtain the ratio to be $\lambda_{c1}^{-2}(0)/\lambda_{c2}^{-2}(0) \sim 3.0,\ 1.7$ and 8.0 when $T_c^{(1)}/T_c^{(2)} \sim 2.2,\ 1.23$ and 2.7 respectively. We observe a similar modification in the zero-temperature penetration depth with the increase or decrease of the critical temperature. So not only the temperature dependencies but also the absolute values follows the experimental predictions. So this simple tight-binding model calculation shows a good agreement with the experimental result as can be seen from Figure 6 and Figure 8. The theoretical calculation also demonstrates that the experimental results could be used to distinguish between the underlying phases- a DDW state, DSC state or a coexisting DDW+DSC region.

The in-plane penetration depth in the underdoped phase of the superconductors can also be obtained from the above model in the presence of the momentum dependent coupling between the sublattices. In this model, two conducting layers per unit cell stacked alternately one above the other in the z-direction, in which the quasiparticles propagate freely, is considered [100]. Superconducting pairing only takes place within the CuO_2 (S) plane and the other layer (N) is nominally driven superconducting by a proximity effect. Between the S and the N layers we consider that the quasiparticles hop with the matrix element $t(k)$ in accord with the experimental prediction [101]. Several works have been carried out where this type of model, more popularly known as the S/N model, has been involved, for e.g., the Raman spectra [102], density of states [97] etc. Calculations of various properties (optical conductivity, magnetic relaxation rate, knight shift) have been performed even in more complex systems where five layers per unit cell has been considered [103]. It is important to note that to do so the energy dispersion relations is given by $\varepsilon_{\lambda k} = -2t_\lambda(\cos(k_x a) + \beta\cos(k_y b))$ [100] which gives the band structure. t_λ and βt_λ are the nearest neighbor hopping integrals in the a and b directions respectively. In Figure 9 we exhibit the temperature dependencies of the penetration depths along the ab plane for $t_0 = 0.1$. Various values of μ_1 are considered. The lines represent the dependencies for $\mu_1 = -0.37$ (solid line), -0.38 (dashed line) and -0.39 (dotted line) in Figure 9. The choices are made in such a manner so as to give approximately the same values of transition temperature T_c. A low-temperature linear dependence of $\lambda_{ab}(T)$ on T/T_c is observed as found in the experimental works of refs. [88, 96, 104]. The experimental measurements for the underdoped YBCO crystal $T_C = 41K$ and $T_c = 70K$ [96] are plotted in the figure for a comparison of the theoretical results. With proper conversions it is seen that the respective T_c's are 44K and 77K [100]. Hence, a very good agreement between the above results is obtained, even though the calculations are based on a simple model. Moreover, it can be seen that the results vary depending on the level of doping of the sample. It is also worthy to note that $1/\lambda_{ab}^{-2}(0)$ depends linearly on T_c which is further linearly dependent on μ_1 – a parameter related directly with doping. So the study exhibits the in-plane penetration depth anisotropy with doping.

In Figure 10 we demonstrate the variation of λ_c with temperature for various values of μ_1. The T^2 behavior in the optimally doped samples is the most universally accepted power law for the cuprates [74, 101]. T [88], T^3 [85] behavior and T^5 [88, 105] have also been seen in several cuprates. In Figure 10 we observe that the temperature dependence varies from $T^{3/2}$ dependence for extremely underdoped samples to a T^3 dependence in the moderately underdoped samples. We also observe a T^2 dependence in the samples with $\mu_1 = -0.38$. So we see that the temperature dependence is sensitive to the level of doping. DDW order therefore becomes a decisive factor for the temperature dependence of the penetration depth along $c-$axis.

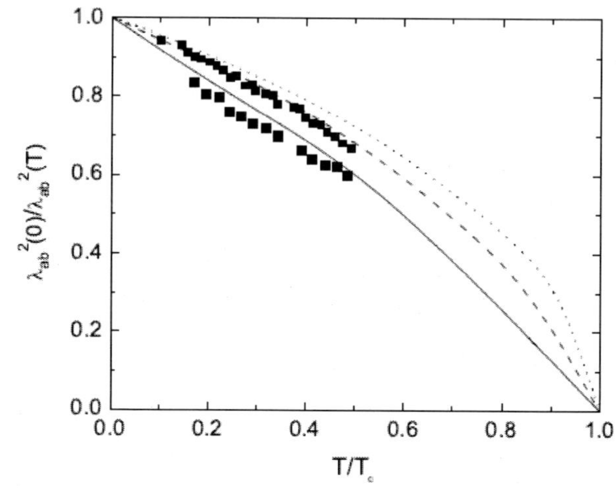

Figure 9. $\lambda_{ab}^2(0)/\lambda_{ab}^2(T)$ vs T/T_c along the ab direction for $\mu_1 = -0.37$ (solid line) $\mu_1 = -0.38$ (dashed line) and $\mu_1 = -0.39$ (dotted line) for the tunneling parameter $t_0 = 0.10$. Experimental results denoted by squares [96].

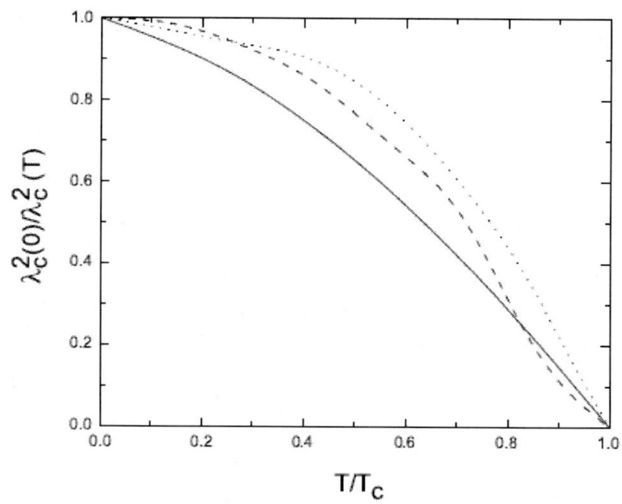

Figure 10. $\lambda_c^2(0)/\lambda_c^2(T)$ vs T/T_c for $\mu_1 = -0.37$ (solid line) $\mu_1 = -0.38$ (dashed line) and $\mu_1 = -0.39$ (dotted line) for the tunneling parameter $t_0 = 0.10$.

So, in general we observe a $\sim T^{\alpha}$ dependence, where α is the exponent dependent on doping. As we approach the optimally doped samples we observe a increase in superconducting gap and a decrease in the DDW gap leading to a much faster decrease of the *superfluid density*.

The anisotropy of the penetration depth has been demonstrated clearly using the proximity model. In all the above cases the superconducting parameter has a $d_{x^2-y^2}$ symmetry. We observe that not only the physics in the CuO chains but also the existence of several layers become an essential feature in the process of understanding the cuprates. Yet, it is far from being complete. Recently the angle-resolved phase-sensitive experiments by Kirtley *et. al.* [32] show that $|\Delta(0,\pi)|/|\Delta(\pi,0)|\sim 1.2$ and the gap nodes shift by about 3° away from the diagonal direction towards the k_x axis whereas the electron tunneling experiment [33] demonstrates a shift of 5° towards the k_x axis. Moreover, the inelastic neutron scattering experiments and the spin dynamics experiments also support the anisotropy in the superconducting gap. The inelastic neutron scattering experiments in detwinned YBCO samples demonstrated that the incommensurated peak of spin fluctuation is stronger at $(\pi+\delta,\pi)$ direction than at $(\pi,\pi+\delta)$ direction.

To consider the above asymmetry, the superconducting gap has been proposed to be a mixture of $d_{x^2-y^2}$, an extended s, and an isotropic s-wave order parameter, essential for the existence of the anisotropy in the gap magnitude between $(0,\pi)$ and $(\pi,0)$ points [34, 35]. Henceforth, the theoretical calculations need to incorporate the anisotropy observed in the experimental results. The gap function with asymmetric Δ along the two axis, for e.g $\Delta(k)=\Delta_x\cos k_x-\Delta_y\cos k_y$, exhibits a shift in the node but does not show the anisotropy in the magnitude at $(\pi,0)$ and $(0,\pi)$. Moreover, depending on the phase between the d- and the s-wave, a second jump of the specific heat has been observed due to the vanishing of the s and d-wave components at different temperatures. The non-observance of the second jump, along with the anisotropy in the gap magnitude at $(0,\pi)$ and $(\pi,0)$ for $t_x/t_y>1$ and also the shift in the gap node can only be taken into account considering the isotropic s wave along with dominant $d_{x^2-y^2}$ and a subdominant extended s wave. In this model, the unit cell is considered to consist of a CuO_2 plane and a CuO chain, coupled through single coherent coupling. The basis function that determines the symmetry of the superconducting order parameter is considered to consists of a mixture of $d_{x^2-y^2}$ wave, an extended s-wave and an isotropic s wave i.e.

$$\eta_k=\alpha(\cos k_x-\cos k_y)+\beta(\cos k_x+\cos k_y)+\eta\,.$$

The angle resolved tunneling experiment demonstrates $\alpha=83\%$, $\beta=2\%$ and $\eta=15\%$ [33], whereas the work [106] obtains the weights of the irreducible representations as $\alpha=75\%$, $\beta=10\%$ and $\eta=15\%$. The energy dispersion relations are given by

$$\varepsilon_{1k}=-2t_1(1+\xi)\cos k_x-2t_1(1-\xi)\cos k_y-4r_2t_1\cos k_x\cos k_y$$

and $\varepsilon_{2k}=-2t_2\cos k_y$ which gives the band structure with the the nearest neighbor hopping integrals in the a and b directions along with the next neighbor hopping term. In the case of tetragonal lattice, we consider $\xi=0$ and also a symmetric $d_{x^2-y^2}$ SC gap. In case of the orthorhombic distortion, we consider $\xi=0.025$ i.e. $b/a=1.05$ which is close to the experimental result of $b/a \sim 1.015$ [107]. Moreover, $r_2 = 0.1$ is considered as the next-neighbor interaction. We consider two sets of anisotropic state in the presence of the orthorhombicity (i) $\alpha = 78\%$, $\beta = 9\%$ and $\eta = 13\%$ which leads to the anisotropy between the $(0, \pi)$ and $(\pi,0)$ points as 1.2, comparable with the experiments of Kirtley *et. al.* [32] and (ii) considers $\alpha = 86.3\%$, $\beta = 4.5\%$ and $\eta = 9.2\%$ where the anisotropy in the $(0, \pi)$ and $(\pi,0)$ points is 1.1.

In Figure 11 the temperature dependencies of the penetration depth in the overdoped region along the a and b axis are exhibited. For set (i) λ_a (λ_b) is depicted by the solid line (dashed line), and for set (ii) by the dotted line (dashed-dotted line). It is important to note that only the anisotropy in the gap of the superconducting state in the 2D state alone is unable to describe the characteristics of this state and thereby requires the inclusion of the second layer. Moreover, due to the orthorhombicity and the anisotropy a much stronger interaction in the chains is required to obtain the value of $\lambda_a(0)/\lambda_b(0)$ close to the experimental results. This can be observed comparing the present results with that of ref. [63]. $\lambda_a(T)$ and $\lambda_b(T)$ have rather similar temperature dependencies just as observed in various experimental results [76].

Moreover, we observe a power law dependence of the penetration depth irrespective of the choice of the parameters. $\lambda_a^2(0)/\lambda^2(T) \sim 1-(T/T_c)^{0.75}$. It is worthy to note that the exponent is different from the value of the exponent of 1.3 obtained in ref. [29] for pure d waves. The accurate determination of the temperature dependence of the penetration depth is still a challenge to the experimental community. Hence controversy prevails till date regarding the dependencies. Moreover, doping plays a significant role and a wide variety of exponents has been obtained.

The above work is an effort to predict the exponents from mean field calculation and observe the changes in the values expected due to various parameters like doping, presence of chains, anisotropic and asymmetric property of the superconducting gap and also the next-nearest neighbor parameter.

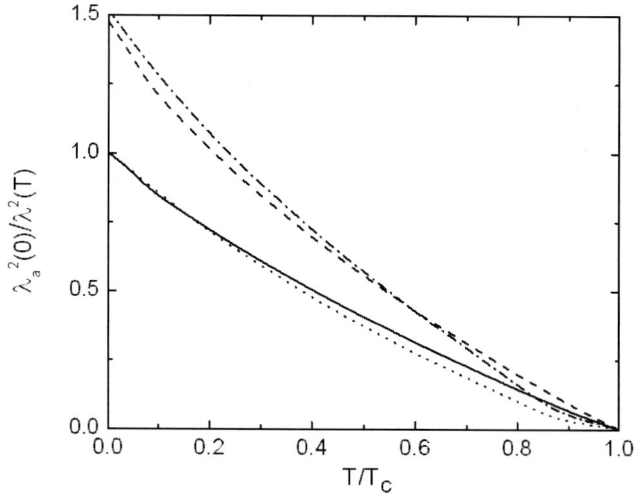

Figure 11. $\lambda_a^2(0)/\lambda^2(T)$ vs T/T_c in the a and b direction.

Apart from the other studies we also observe the suppression of the superfluid density with underdoping by augmenting a $d-$wave BCS model incorporating tunneling between the CuO_2 planes along with a phenomenological charge renormalization factor that is vanishingly small for states away from the nodes of the $d-$wave pair potential but close to unity in the vicinity of the nodes [108]. The unified theory of ab plane and the $c-$axis penetration depth allowed them to fit the experimental data of the $c-$axis penetration depth very well except near T_c where fluctuations starts to dominate.

The Hamiltonian for the penetration depth calculations can be given by

$$H = H_{pair} + H_{int} \tag{10}$$

where H_{pair} is the standard *BCS Hamiltonian* as in Eq. (4) and H_{int} refers to interactions that describes the physics of the proximity to the insulating phase at half filling. The $c-$axis penetration depth has been calculated employing

$$\frac{1}{\lambda_c^2} = 8e^2 d \sum_{k,p} T_{k-p}^2 T \sum_{i\omega} F(k,\omega) F(p,\omega) \tag{11}$$

where T_k is the *tunneling matrix* which is proportional to $t_\perp$, an energy scale characterizing the strength of tunneling and also dependent on the inverse length scale characterizing the degree of *momentum conservation* Λ, and $F(K,\omega)$ is the in-plane *Matsubara Green's function.*

The c-axis penetration depth has been calculated to fit numerically with the experiments of refs. [109] and [99]. In Figure 12 the low-temperature values of $\delta\lambda_c(T) = \lambda_c^{-2}(0) - \lambda_c^{-2}(T)$ are plotted along with the experimental measurements of Refs. [109] and [99].

The diamonds represent data curves for a particular doping value. Each doping value is characterized by a particular T_c which has been taken to be proportional to doping. For clarity the highest doping values are displayed (i.e., T_c = 20.2, 19.5, 18.2, 17.8, 16.4, 15.1 K). The solid line is the best fit with the parameters $\hbar/\Lambda = 120$Å and $t_\perp = 26$ meV. The anisotropy parameter, $\eta = v_F/v_\Delta = 6.8$ that decides in which way $\delta\lambda_c(T)$ changes its behavior as quasipartcles are excited to the condensate where v_F and v_Δ are the Fermi and gap velocities at the node. The deviations at higher temperatures can be ascribed to the fluctuation effects and also due to the effect of the charge renormalization factor on the finite temperature correction.

Finally, to illustrate the overall agreement of the model with the data of Hosseini *et. al.* [99, 109] for several representative doping values are plotted in Figure 13. The agreement is strikingly good at low temperatures for all doping levels. All data set are fit with a single set of parameters as listed in Figure 12, the only parameter that varies is the *cutoff energy* according to $E_c = 0.49\,T_c$ meV/K with T_c being the actual measured critical temperature.

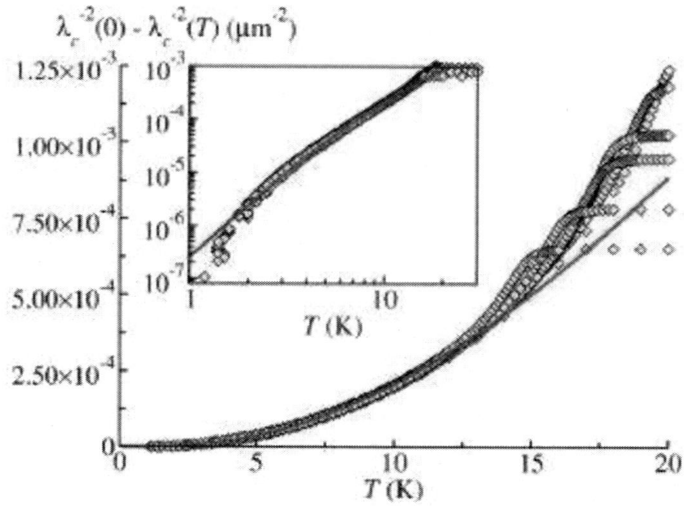

Figure 12. $\lambda_c^{-2}(0)-\lambda_c^{-2}(T)$ vs T (K) for various dopings having experimental T_c values as 20.2, 19.5, 18.2, 17.8, 16.4, 15.1 K. The solid curve is the fit with the parameters given in the text. Experimental data of refs [109] and [99] are also plotted.

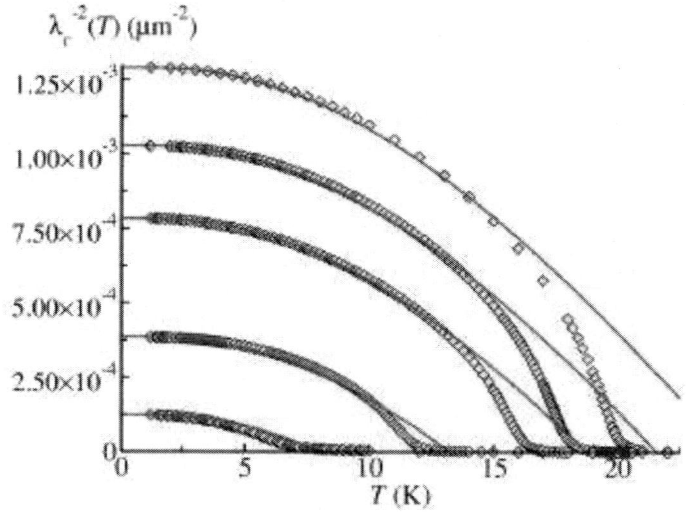

Figure 13. $\lambda_c^{-2}(T)$ vs T for various values of T_c = 20.2, 19.5, 18.2, 17.8, 16.4, 15.1K representing decreasing effective doping. The parameters are provided in the text. Experimental data of refs [109] and [99] are also plotted.

CONCLUSION

In this chapter the anisotropy in the cuprate superconductors has been addressed from the view point of the penetration depth measurements. An overview of the present experimental situation is presented which clearly demonstrates a large anisotropy in the penetration depth along the three axes. The in-plane and the $c-$axis measurements also reveal highly anisotropic nature of cuprate materials. Theoretical model calculations are exhibited based on a tight-binding model within a mean-field scenario. The chains have been found to be an important component in understanding the anisotropy along the a and b directions. The in-plane penetration depth has been calculated using a model that incorporates two-layers per unit cell coupled through proximity effect. The anisotropy between the in-plane and out-of-plane results has also been shown. Moreover, the asymmetry in the superconducting gap is considered to understand its influence on the anisotropy of the penetration depth results. A phenomenological charge renormalization factor that is vanishingly small for states away from the nodes of the $d-$wave pair potential but close to unity in the vicinity of the nodes has also played an important role in explaining the above anisotropy. Yet, the understanding of the anisotropy is far from being complete. The anomalous normal state and the difficulty in experimental measurements due to disorder and impurities have made the task much more challenging.However, with the advent of new techniques, we hope to have a better situation in understanding the anisotropy of these materials.

ACKNOWLEDGMENT

The author would like to thank the Brazilian funding agency CNPq for its financial support. The author also thanks Dr. P. Chaudhuri for helpful discussions.

REFERENCES

[1] K. Onnes, *Leiden Comm.* 120b, 122b,124c (1911) .

[2] J. Bardeen, L.N. Cooper, J.R. Schrieffer, *Phys. Rev.* 108, 1175 (1957); Schrieffer J.R., *Theory of Superconductivity*, New York Benjamin, (1964).

[3] M. Tinkham, *Introduction to Superconductivity*, New York Dover (1996).

[4] V.L. Ginzburg, L.D. Landau, *J. Expt. Theo. Phys.* 20, 1064 (1950).

[5] L.P. Gorkov, *Sov. Phys. JETP* 34 505 (1959); L.P. Gorkov, *Sov. Phys. JETP* 36 1364 (1959).

[6] A.A. Abrikosov, L.P.Gorkov, I.Ye. Dzyaloshinskii, *Quantum Field Theoretical Methods in Statistical Physics*, Paris Pergamom Press Ltd. (1965)

[7] J.G. Bednorz, K.A. Muller, *Z. Phys. B.* 64, 189 (1986).

[8] M.K. Wu *et. al. Phys. Rev. Lett.* 58, 908 (1987).

[9] A.P. Mackenzie, S.R. Julian, D.C. Sinclair, C.T. Clin, *Phys. Rev. B* 53, 5848 (1996).

[10] T. Ito, K.Takenaka, S. Uchida, *Phys. Rev. Lett.* 70, 3995 (1993).

[11] B. Bucher, P. Steiner, J. Karpinski, E. Kaldi, P. Watcher, *Phys. Rev. Lett.,* 70, 2012 (1993).

[12] H. Takagi, B. Batlogg, H.L. Kao, J. Kwo, R.J. Cava, J.J. Krajewski, W.F. Jr Peck, *Phys. Rev. Lett.*, 69, 2975 (1992).

[13] N.E. Hussey, *Handbook of High Temperature Superconductivity Theory and Experiment*, Editor Schrieffer J.R., Springer(2007).

[14] H.Y. Hwang, B. Batlogg, H. Takagi, H.L. Kao, J. Kwo, R.J. Cava, J.J. Krajewski, W.F. Jr. Peck, *Phys. Rev. Lett.*, 72, 2636 (1994).

[15] T. R. Chien, Z. Z. Wang, N. P. Ong, *Phys. Rev. Lett.* 67, 2008 (1991).
[16] A.J. Leggett, *Rev. Mod. Phys.* 47 331(1975).
[17] C.C. Tsuei, J. R. Kirtley, *Rev. Mod. Phys.* 72, 969 (2000).
[18] M. Giura, R. Fastampa, S. Sarti, E. Silva, *Phys. Rev. B* 68, 134505 (2003).
[19] H. Ding, *Nature* 382, 51 (1996); D. J. Scalapino, *Phys. Rep.* 250, 329 (1995); J. P. Carbotte, C. Jiang, *Phys. Rev. B* 48, 4231(1993); B. G. Levi, *Phys. Today* 49 (5), 19 (1996).
[20] W. Hardy *et. al.*, (1993) Phys. Rev. Lett. 70, 3999; K. A. Moler *et. al.*, (1994) *ibid.* 73, 2744; K. Gofron *et. al.*, *ibid.* 73, 3302 (1994) .
[21] M. Prohammer, A. Perez-Gonzalez, J.P. Carbotte, *Phys. Rev. B* 47, 15152 (1993); J. Annett, N. Goldenfeld, S.R. Renn, *ibid.* 43 2778 (1991); Q.P. Li., B.E.C. Koltenbah, R. Joynt, *Phys. Rev. B* 48, 437 (1993).
[22] S. K. Adhikari, A. Ghosh, *Phys. Rev. B* 55, 1110 (1997); J. Phys.: Cond. Mat. 10, 135 (1998); A. Ghosh, S.K. Adhikari, *Euro. Phys. J. B* 2, 31 (1998).
[23] M. Covington, *et. al.*, *Phys. Rev. Lett.* 79, 277 (1997).
[24] K.A. Kouznetsov *et. al.*, *Phys. Rev. Lett.* 79, 3050 (1997).
[25] S. Sridhar, *Physica C* 282, 256 (1997).
[26] K. Krishana, *Science* 277, 83 (1997).
[27] R.B. Laughlin, *Phys. Rev. Lett.* 80, 5188 (1998); *Physica C* 234, 280 (1994).
[28] M. Franz, Z. Te s′ anovi c′ , *Phys. Rev. Lett.* 80, 4763 (1998).
[29] Ghosh, S.K. Adhikari, *J. Phys. CM* . 10, L319 (1998).
[30] M. Mitra, H.Ghosh, S.N. Behera, *Euro. Phys. J. B* 2, 371 (1998); J.-X. Zhu, W. Kim, C.S. Ting, *Phys. Rev. B* 57, 13410 (1998); M. Liu, D.Y. Xing, Z.D. Wang, *Phys. Rev. B* 55, 3181 (1997); C. O' Donovan, J.P. Carbotte, *Physica C* 252, 87 (1995); E.A. Shapoval, *JETP Lett.* 64, 625 (1996).
[31] Ghosh, S.K. Adhikari, *Physica C* 355, 77 (2001).
[32] J.R. Kirtley, *et. al.*, *Nat. Phys.* 215, 1 (2006)
[33] H.J.H. Smilde, *et. al.*, *Phys. Rev. Lett.* 95, 257001 (2005).
[34] M. Gomes, A. Ghosh, *Eur. Phys. J. B* 66, 461 (2008).
[35] T.Zhou, J-X. Li, *Phys. Rev. B* 72, 134512 (2005).
[36] T.Cren, *et. al.*, *Phys. Rev. Lett.* 84, 147 (2000).
[37] C. Howald, P. Fournier, A. Kapitulnik, *Phys. Rev. B* 64, 100504(R) (2001).
[38] S.H. Pan, *et. al.*, *Nature* 413, 282 (2001).
[39] K.M. Lang, *et. al.*, *Nature* 415, 412 (2002).
[40] C. Howald, *et. al.*, *Phys. Rev. B* 67, 014533 (2003).
[41] T. Kato, S. Okitsu, H. Sakata, *Phys. Rev. B* 72, 144518 (2005).
[42] H. Mashima, *et. al.*, *Phys. Rev. B* 73, 060502(R) (2006).
[43] K. Aryanpour, *et. al.*, *Phys. Rev. B* 73, 104518 (2006).
[44] A.C. Fang, *et. al.*, *Phys. Rev. Lett.* 96, 017007 (2006).
[45] T. Timusk, B. Statt, *Rep. Prog. Phys.* 62, 61 (1999).
[46] J.L. Tallon, J. W. Loram, *Physica C* 349, 53 (2001).
[47] Jr, W.W. Warren, R.E. Walstedt, J.F. Brennert, R. J. Cava, R. Tycko, R.F. Bell, G. Dabbagh, Phys. *Rev. Lett.* 62, 1193 (1989) .
[48] M. Bankay, M. Mali, J. Roos, D. Brinkmann, *Phys. Rev. B* 50, 6416 (1994).
[49] R. Stern, M. Mali, I. Mangelschots, J. Roos, D. Brinkmann, J-Y Genoud, T. Graf, J. Muller, *Phys. Rev. B* 50, (1994) 426 (1994).

[50] R.E. Walstedt, R.F. Bell, D.B. Mitzi, *Phys. Rev. B* 44, 7760 (1991).
[51] M.H. Julien, P. Caretta, M. Horvatic, M. Berthier, C. Berthier, Y. Segransan, P. Carington, D. Colson, *Phys. Rev. Lett.* 76 4238 (1996).
[52] H. Alloul, *et. al. Phys. Rev. Lett.* 67, 3140 (1990).
[53] G.V.M. Williams, *et. al. Phys. Rev. B* 51, 16503 (1995).
[54] J.W. Loram, *et al. Physica C* 341-348, 831 (2000).
[55] Ch. Rennner, B. Revaz, J-Y. Genoud, K. Kadowaki, O. Fischer, *Phys. Rev. Lett.* 80 149 (1998).
[56] M. Norman, *et. al. Nature* 392, 157 (1998).
[57] J.W. Loram, K.A. Mirza, J.R. Cooper, J.L. Tallon, *Physica C* 282-287, 1405 (1997).
[58] J.W. Loram, K.A. Mirza, J.R. Cooper, N. Athanassopoulou, W.Y. Liang, *Proc 10th HTS Anniversary workshop on Physics, Materials and Applications* ed. B. Batlogg *et al*, World Scientific.
[59] S. Chakravarty, R.B. Laughlin, D.K. Morr, C. Nayak, *Phys. Rev. B* 63 9450 (2001).
[60] Ghosh, *Int. J. of Mod. Phys B* 15 4161 (2001).
[61] Ghosh, *Physica B* 373 240 (2006).
[62] Ghosh, *Phys. Rev. B* 73 012504 (2006).
[63] Ghosh, *DDW order and its role in Cuprates in Progress in Superconductivity research*, Editor: O Chang, Nova publishers (2007).
[64] H. A. Mook, P. Dai, F. Dogan, *Phys. Rev. B.* 64, 012502 (2001).
[65] K. Maki, B. Dóra, A. Viroszrek, A. Ványolos, *Curr. Appl. Phys.* 4, 693 (2004); H. Won, S. Haas, D. Parker, K. Maki, *Phys. Status Solidi B* 242, 363 (2005).
[66] H. Fukuyama, *Prog. Theo. Phys. Suppl.* 108, 287 (1992).
[67] V.J Emery, S.A. Kivelson, *Phys. Rev. Lett.* 74, 3253 (1995).
[68] M. Randeria, Cond-mat/9710223 (unpublished) and references therein; S C. A. R. a de Melo,; M. Randeria, J. R Engelbrecht, *Phys. Rev. Lett.,* 71, 3202 (1993).
[69] Y.J. Uemura, *et. al. Phys. Rev. Lett.* 62, 2317 (1989).
[70] D. Pines, *Z. Phys. B* 103, 129 (1997); D. Pines, *Physica C* 282, 28 (1997).
[71] W.N. Hardy, D.A. Bonn, D. Morgan, R. Liang, K. Zhang, *Phys. Rev. Lett.* 70, 3999 (1993).
[72] N. Klein, Trans Tech Publications; in *Materials Science Forum*, 130-132 (1993).
[73] P.J. Hirschfeld, N. Goldenfeld, *Phys. Rev. B* 48, 4219 (1993).
[74] D.A. Bonn *et al. Czech. J. Phys* 46 3195 (1996).
[75] D. Achkir *et al. Phys. Rev. B* 48, 13184 (1993).
[76] K. Zhang *et al. Phys. Rev. Lett.* 73, 2484 (1994).
[77] D.N. Basov *et al, Phys. Rev. Lett.* 74, 598 (1995).
[78] T.A. Friedmann *et al. Phys. Rev. B* 42, 6217 (1990).
[79] J. Schoenes *et al. Phys. Rev. B* 48, 16869 (1993).
[80] D.A. Bonn, W. N. Hardy, *Handbook of High Temperature Superconductivity, Theory and Experiment*, Editor J. R. Schrieffer, Springer (2007),
[81] T. Jacobs, S. Sridhar, Q. Li, G.D. Gu, N. Koshizuka, *Phys. Rev. Lett.* 75, 4516(1995).
[82] S. F. Lee *et al. Phys. Rev. Lett* 77, 735 (1996).
[83] D.M. Broun *et al. Phys. Rev. B* 56, 11443 (1997).
[84] Panagopoulos *et al. Phys. Rev. B* 53, 2999 (1996).
[85] Hosseini *et al. Phys. Rev. Lett.* 81, 1298 (1998).
[86] T. Shibauchi *et al. Phys. Rev. Lett*, 72, 2263(1994).

[87] D.M. Broun, Ph. D Thesis, Cambridge University (2000).
[88] Panagopoulos *et al. Phys. Rev. Lett.* 79, 2320 (1997).
[89] Ghosh, S. K. Adhikari, *Physica C* 322, 37 (1999).
[90] Friedmann, *et al.*, *Phys. Rev. B*, 42, 6217 (1990).
[91] R. Gagnon, C. Lupien, L. Taillefer, *Phys. Rev. B,* 50, 3458 (1994).
[92] B. Koch, H. P. Geserich, Th. Wolf, *Solid State Commun.* 71, 495 (1989).
[93] D.N. Basov, *et al*, *Phys. Rev. Lett.* 74, 598(1995).
[94] D.K. Morr, A.V. Balatsky, *Phys. Rev. Lett.* 87, 247002 (2001).
[95] Ghosh, *Phys. Rev. B* 73, 012504 (2006).
[96] M.R. Trumin, Y. A. Nefyodov, A. F. Shevchun, *Phys. Rev. Lett.* 92, 067006 (2004).
[97] W.A. Atkinson, *J. P. Carbotte Phys. Rev. B* 51 1161 (1995).
[98] Ghosh, *Phys. Stat. Sol. B* 245, 149 (2008).
[99] Hosseini *et. al. Phys. Rev. Lett.* 93, 107003(2004).
[100] Ghosh, *Euro. Phys. J B* 62, 271(2008).
[101] T. Shibaguchi *et al.*, *Phys. Rev. Lett.* 72, 2263(1994).
[102] A.A. Abrikosov, *Physica C* 182, 191 (1991); A.A. Abrikosov, R.A. Klemm, *Physica C* 191, 224 (1992).
[103] M. Tachiki, S. Takahashi, F. Steglich, H. Adrian, *Z. Phys. B* 80, 161 (1990); M. Tachiki, T. Koyama, S. Takahashi, *Prog. Theo. Phys. Suppl* 108, 297 (1992)
[104] C. Homes, T. Timusk, D.A. Bonn, R. Liang, W.N. Hardy, *Physica C* 254, 265(1995).
[105] M.B. Gaifullin, *et. al.*, *Phys. Rev. Lett.* 77, 4632 (1999).
[106] X-S. Ye, J-X. Li, *Phys. Rev. B* 76, 174503 (2007).
[107] M. C. Schabel, *et al., Phys. Rev. B* 57, 6090 (1998); *Phys. Rev. B* 57, 6107(1994).
[108] D.E. Sheehy, T.P. Davis, M. Franz, *Phys. Rev. B* 70, 054510 (2004).
[109] Hosseini, *Ph. D Thesis*, University of British Columbia (2002).

In: Anisotropy Research
Editor: Hirpa Lemu

ISBN 978-1-62081-977-7

Chapter 11

MAGNETICALLY ACTIVE ANISOTROPIC FLUIDS BASED ON LIQUID CRYSTALS

Natália Tomašovičová[1,*], Peter Kopčanský[1] and Nándor Éber[2]

[1]Institute of Experimental Physics, Slovak Academy of Sciences, Watsonova, Košice, Slovak Republic

[2]Institute for Solid State Physics and Optics, Wigner Research Centre for Physics, Hungarian Academy of Sciences, Budapest, Hungary

ABSTRACT

Ferronematics are magnetically active anisotropic fluids. They are a manifestation of the idea of Brochard and de Gennes, who suggested that doping liquid crystals by fine magnetic particles may enhance their sensitivity to magnetic fields. Experiments on ferronematics confirmed on the one hand that magnetic particles affect considerably the behaviour in external fields; on the other hand they proved that the interaction between the host liquid crystal and the embedded magnetic particles is more complex than assumed initially. In the present chapter we will introduce the basic field-induced reorientational effect of liquid crystals (the Fréedericksz transition) which serves also as a basis for most experiments on ferronematics. We will sketch the mechanism how magnetic particles may affect the magnetic response. We will describe the preparation of magnetic particles and of the ferronematics made of them. An overview will be given of the experimental results obtained on the behaviour of ferronematics depending on the host liquid crystal, on the type, shape, size and concentration of the magnetic particles. A special emphasis will be given to the magnetic field induced shift of the isotropic-nematic phase transition temperature, both in ferronematics and in polymer dispersed liquid crystals doped with magnetic particles.

Keywords: Magnetic nanoparticles, liquid crystals, ferronematics

PACS: 61.30-v, 61.30.Gd, 61.30.Hn, 75.50.Mm, 64.70.M-

---

[*] Corresponding author: nhudak@saske.sk.

1. INTRODUCTION

The beginning of the study of liquid crystals is related with the year 1888, when Austrian botanic Friedrich Reinitzer [1] found that cholesteryl benzoate does not melt in the same manner as other compounds, but has two melting points. At 145.5 °C it melts into a cloudy liquid, and at 178.5 °C it melts again and the liquid becomes clear.

The intermediate phase between the two melting points was later coined liquid crystalline or mesophase. It is thermodynamically stable in the given temperature range and can be characterized by a long-range orientational ordering of the molecules and, as a consequence, by the anisotropy of its properties.

Liquid crystals may be divided into two classes: *lyotropic* and *thermotropic*. Lyotropic liquid crystals consists of two or more components (e.g. amphiphilic molecules in a solvent) that exhibit liquid-crystalline phase in certain concentration and temperature ranges. Thermotropic liquid crystals, the class upon which this chapter will focus, are compounds whose liquid crystalline properties are induced purely by temperature variation. They possess one or more mesophases at temperatures between their melting point, below which the material exists as a crystalline solid, and the "clearing point", above which the material exists as an isotropic liquid with properties of a conventional liquid phase, characterized by random and isotropic molecular ordering.

Anisotropy of liquid crystals is a property not observed in other fluids. This anisotropy makes liquid crystals behave differently compared to ordinary fluids when subjected to mechanical stresses or to an electromagnetic field. One of the most spectacular effects is that either applied fields or flow may affect the orientation of liquid crystals.

For a long time, however, liquid crystals have not attracted great attention among scientists and have remained a pure scientific curiosity for about 80 years. In 1968 the potential to utilize an electro-optic effect of liquid crystals for display applications was first demonstrated [2]. In 1969, Hans Kelker succeeded in synthesizing the compound 4'-methoxybenzylidene-4-n-butylaniline (MBBA) that had a liquid crystalline phase at room temperature [3].

The next milestones towards the commercialization of liquid crystals and displays made of them were the invention of the twisted nematic display mode [4] and the synthesis of further chemically stable substances with low melting temperatures by George Gray *et al.* [5]. From that on the rapid adoption of liquid crystals for electronic products has started, which by now had substantial influence on our everyday life.

Though in the past decades many liquid crystal phases have been identified [6], here we will focus exclusively on the nematic phase in which the long axes of individual molecules are oriented in about the same direction, however, without any positional ordering of their centres of mass.

Liquid crystals can be oriented under electric or magnetic fields due to the anisotropy of dielectric permittivity or diamagnetic susceptibility [6]. As the former is in the order of unity, in conventional devices the driving voltages are in the order of a few volts.

However, because of the small value of the anisotropy of the diamagnetic susceptibility ($\sim 10^{-7}$), the magnetic field $\boldsymbol{H}$ necessary to align liquid crystals have to reach rather large values ($\boldsymbol{B} \sim 1$ T). We note here that while we keep talking about the magnetic field $\boldsymbol{H}$ at describing various phenomena, all experimental data in this chapter will be given for the magnetic induction $\boldsymbol{B}$, which is related to $\boldsymbol{H}$ via the equation $\boldsymbol{B} = \mu_0(1 + \boldsymbol{\chi})\, \boldsymbol{H}$. As the elements of the susceptibility tensor $\boldsymbol{\chi}$ are of $\sim 10^{-7}$, the second term (being much smaller than typical measurement errors) means a negligible correction only. That justifies the approximation $B = \mu_0 H$ which will be used throughout the chapter when comparing theory with experiments.

In an effort to enhance the magnetic susceptibility of liquid crystals, the idea of doping them with small amount of tiny magnetic particles was theoretically introduced by Brochard and de Gennes [7]. They have developed the continuum theory of magnetic suspensions in nematic liquid crystals (ferronematics). In their theoretical work they had considered rod-like magnetic particles with the length $L \gg a$ (where a is the molecular size of the nematic liquid crystal) and with a diameter of $d \approx L/10$.

The volume concentration of magnetic particles was supposed to be sufficiently small ($\phi \sim 10^{-3}$) in order to be able to ignore the inter-particle magnetic dipole-dipole interaction. They predicted that a rigid anchoring with $\boldsymbol{m} \parallel \boldsymbol{n}$ would result in the ferromagnetic behaviour of the mixture. Here the unit vector $\boldsymbol{n}$ (the director) describes the preferential direction of the nematic molecules and the unit vector $\boldsymbol{m}$ denotes the orientation of the magnetic moment of the magnetic particles.

In the first experimental paper, Rault *et al.* [8] reported about the basic magnetic properties of a suspension of rod-like γ-Fe_2O_3 particles in the liquid crystal MBBA. In order to stabilize the suspension, the particles were first coated with a surfactant and then mixed with the nematic liquid crystal. The obtained results confirmed the existence of a ferromagnetic state in such suspension.

Additionally they have found that the temperature of the nematic-isotropic transition decreases with increasing volume concentration of particles. Later, based on the estimations given in [7], first lyotropic [9, 10] and then further *thermotropic* [11] *ferronematics* have been prepared and studied. These experiments confirmed the existence of considerable orientational and concentrational effects in liquid crystals doped with magnetic particles, but raised a lot of questions as well.

In the following section (Sec. 11.2) we first discuss the basics of the field induced reorientational (Fréedericksz) transitions in liquid crystals (both in single and in combined electric and magnetic fields) as they provide the easiest way to observe the change in the magnetic response due to doping. In Sec. 11.3 we give a short introduction to *ferronematics*, to their theoretical description by the *Burylov-Raikher theory* and will sketch the mechanism how magnetic particles may affect the magnetic response. Sec. 11.4 is devoted to the preparation of nanoparticles and ferronematics as well as to introducing, via examples, the capacitance method for monitoring reorientation. Experimental results on ferronematics are divided into three sections. In Sec. 11.5 the influence of spherical nanoparticles on various host nematics (having different sign combination of their anisotropies) is discussed, followed by Sec. 11.6 where the effect of doping the same host with various nanoparticles (differing in size, shape or type) is described. In Sec. 11.7 we show that ferronematics and polymer dispersed liquid crystals doped with magnetic nanoparticles allow the detection of the magnetic field induced shift of the isotropic to nematic phase transition which has theoretically been predicted long time ago. Finally we close the chapter in Sec. 11.8 with some summarizing remarks.

2. FREEDERICKSZ TRANSITIONS

As it was already mentioned in Sec. 11.1, the ability of liquid crystals to become oriented by electric or magnetic fields is due to the anisotropy of their *dielectric permittivity* (ε_a) or *diamagnetic susceptibility* (χ_a). These anisotropies are defined by the relations

$$\varepsilon_a = \varepsilon_{||} - \varepsilon_{\perp}, \chi_a = \chi_{||} - \chi_{\perp}, \tag{1}$$

where $\varepsilon_{||}$ and $\chi_{||}$ denote the dielectric permittivity and the diamagnetic susceptibility, respectively, measured along the director, while the subscript $\perp$ stands for values measured perpendicular to $\boldsymbol{n}$. What is the preferred orientation of the director under the influence of the field depends on the sign of the anisotropy. For *positive anisotropy* the director favours to be parallel with the field, while for *negative anisotropy* $\boldsymbol{n}$ prefers to orient normal to the field.

The influence of the field is usually tested experimentally in sandwich cells, where a nematic layer fills the gap of thickness D between two parallel flat glass plates. The plates are often coated by transparent conductive layers (e.g. indium-tin-oxide, ITO) which serve as electrodes and allow to create an electric field in the nematic layer by applying an electric voltage. In the absence of external fields the orientation of the director in the bulk is determined by the properties of the cell surfaces. By proper surface treatment the orientation can be made *planar*, i.e. the director is parallel with the surface (Figure 1a and b), or *homeotropic*, i.e. $\boldsymbol{n}$ is perpendicular to the surface (Figure 1c and d). In the following we will assume strong anchoring condition at the cell surfaces which means that the director is fixed at the surface even if there is a strong distortion in the bulk of the layer.

When the cell is subjected to an external field, the field can be either parallel with or perpendicular to the cell substrates. Figure 1 depicts four basic experimental geometries obtained as a combination of the directions of the surface director and the external field. Whenever the initial director alignment coincides with that preferred by the field, the orientation in the bulk of the layer remains stable in the field. That occurs in the case of Figure 1a and d for nematics with positive anisotropy, as well as for the geometries in Figure 1b and c with negative anisotropy.

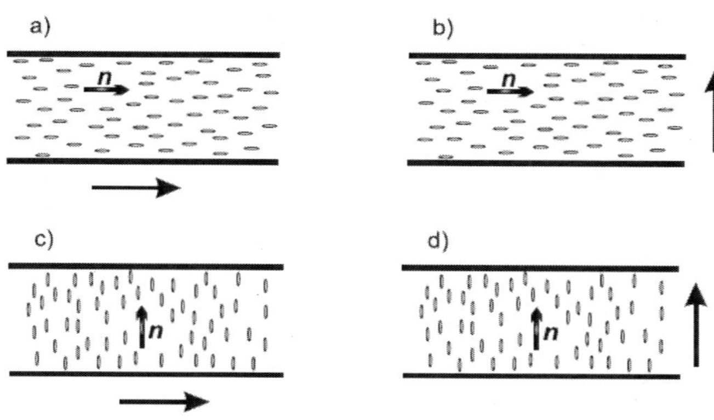

Figure 1. Experimental geometries with planar (a, b) and homeotropic (c, d) surface alignments in an external field (long arrow) parallel with (a, c) or perpendicular to (b, d) the cell substrates.

In the other cases, the field exerts a destabilizing torque on the director which counteracts the stabilizing elastic torques due to the surface alignment. The final state is determined by the torque balance requirement. Interestingly, there is a threshold value of the field, below which no change in the director alignment occurs; however, for fields above the threshold the director rotates away in the bulk of the nematic layer from its initial orientation forming a splay-bend distortion. This phase transition, i.e. the change of the orientation of the director due to the presence of an electric or magnetic field, is called *Fréedericksz transition* [6, 12].

In order to calculate the threshold fields (and/or the final $\boldsymbol{n}(\boldsymbol{r})$ in the cell) one has to consider the *free energy density* of the system, which is composed of two contributions. The elastic one is given by

$$f_K = \frac{1}{2}[K_1(\mathrm{div}\boldsymbol{n})^2 + K_2(\boldsymbol{n}\cdot\mathrm{curl}\boldsymbol{n})^2 + K_3(\boldsymbol{n}\times\mathrm{curl}\boldsymbol{n})^2], \tag{2}$$

where K_1, K_2 and K_3 are the corresponding *Frank elastic moduli* for splay, twist and bend, respectively [6]. The contribution of an electric field $\boldsymbol{E}$ can be expressed as

$$f_E = -\frac{1}{2}\varepsilon_0\varepsilon_a(\boldsymbol{E}\cdot\boldsymbol{n})^2, \tag{3}$$

while that of the magnetic field $\boldsymbol{H}$ reads

$$f_H = -\frac{1}{2}\mu_0\chi_a(\boldsymbol{H}\cdot\boldsymbol{n})^2. \tag{4}$$

Here ε_0 is the *permittivity*, μ_0 is the *magnetic permeability* of vacuum.

Assuming that the confining plates of the nematic layer are located at $z = 0$ and $z = D$, the total free energy of the sample can be expressed as the volume integral:

$$F = \Sigma\int_0^D (f_K + f_E + f_H)dz, \tag{5}$$

where Σ is the substrate area. After minimization of the total free energy the threshold fields are obtained in the limit of infinitesimal deformation. If only a magnetic field is present one gets the *magnetic Fréedericksz threshold field* as

$$H_c = \frac{\pi}{D}\sqrt{\frac{K_i}{\mu_0\,|\,\chi_a\,|}}. \tag{6}$$

The calculation is analogous for the electric case, however, one has to take into account that instead of an electric field one can apply only an electric voltage to the cell; therefore we obtain an *electric Fréedericksz threshold* voltage:

$$U_c = E_c D = \pi \sqrt{\frac{K_i}{\varepsilon_0 \mid \varepsilon_a \mid}}. \tag{7}$$

In the threshold formulas K_i is the elastic modulus relevant to the given geometry; namely for planar cells K_1 (splay Fréedericksz transition), and for homeotropic ones K_3 (bend Fréedericksz transition) should appear.

The situation is becoming slightly more complicated if the liquid crystal is subjected to both electric and magnetic fields, as the two fields may be either perpendicular to (e.g. Figure 2a) or parallel with each other (e.g. Figure 2b), thus doubling the number of possible experimental geometries. For the understanding of the influence of combined fields, it is enough to discuss two basic cases.

The first case is if the preferred director orientation for both fields are the same; then the two fields alone would induce the same type of distortion. Therefore their combined action leads to an enhancement of the deformation; moreover, distortion starts already at a voltage $U < U_c$ and a magnetic field $H < H_c$ for which the relation

$$\left(\frac{U}{U_c}\right)^2 + \left(\frac{H}{H_c}\right)^2 = 1 \tag{8}$$

holds. This scenario occurs, for example, in the geometry shown in Figure 2b using a nematic with $\varepsilon_a > 0$, $\chi_a > 0$.

In the second case the two fields favour different director alignment. It means that only one of the fields tends to destabilize the initial director orientation, the other acts against and stabilizes the initial state. Consequently, in the presence of a stabilizing field the threshold of the destabilizing field increases. Alternatively, if the Fréedericksz transition has already been induced by the destabilizing field, applying the other field results in the reduction or even in the disappearance of the distortion. Under such conditions the relation between the voltage and the magnetic field at the deformation threshold reads

$$\left(\frac{U}{U_c}\right)^2 - \left(\frac{H}{H_c}\right)^2 = 1 \tag{9}$$

if the magnetic field is stabilizing (as it would occur e.g. in a nematic with $\varepsilon_a > 0$, $\chi_a > 0$) in the geometry of Figure 2a), or

$$\left(\frac{H}{H_c}\right)^2 - \left(\frac{U}{U_c}\right)^2 = 1 \tag{10}$$

if the electric field is stabilizing.

In the following sections we will see examples for both scenarios.

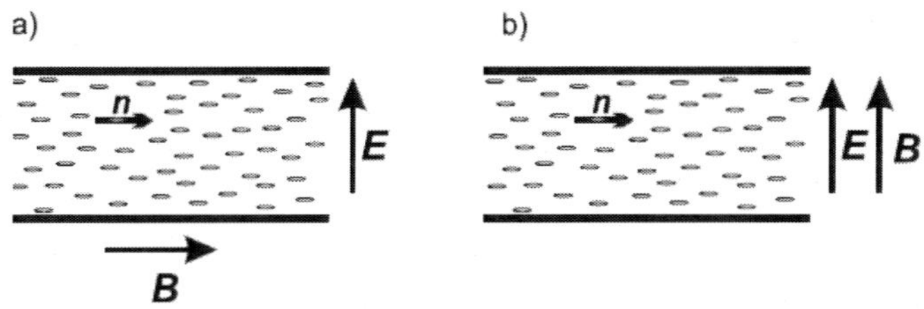

Figure 2. Examples of experimental geometries for studying combined electric and magnetic fields. (a) The two fields are perpendicular, (b) the two fields are parallel.

3. FERRONEMATICS

Ferronematics, i.e. stable colloidal suspensions of monodomain ferro- or ferrimagnetic particles in nematic liquid crystals, have been a subject of extensive theoretical as well as experimental studies [13 - 19]. They attract considerable interest because their response to an external magnetic field exceeds substantially that of pure nematics. The most essential feature of these systems is a strong orientational coupling between the magnetic particles and the liquid crystal matrix. The applied magnetic field changes the orientation of the magnetic particles and due to the coupling between magnetic particles and liquid crystal molecules the director follows it. By that way the presence of magnetic nanoparticles in the nematic liquid crystal enhances its *magnetic susceptibility*. Later we will see, however, that ferronematics cannot completely be described just as a homogeneous continuum with a modified susceptibility.

Though the first theory of ferronematics [7] pretended $\boldsymbol{m} \parallel \boldsymbol{n}$, experiments in some *thermotropic ferronematics* [11] have proved that in the studied case the presence of a parallel orientation of $\boldsymbol{m}$ and $\boldsymbol{n}$ can be excluded. Therefore, the theory has been revised and expanded by Burylov and Raikher [20 - 24] considering a finite value of the surface density of *anchoring energy* W at the nematic - magnetic particle boundary. The anchoring energy, as well as the parameter ω defined as the ratio of the anchoring energy to the elastic energy of the liquid crystal ($\omega = Wd/K$, where d is the size of the magnetic particles and K is the orientational-elastic Frank modulus), characterize the type of anchoring of the nematic director on the surface of the magnetic particle. The condition $\omega >> 1$ corresponds to the rigid anchoring. When $\omega \leq 1$ the anchoring is soft which, unlike the rigid anchoring, permits both $\boldsymbol{m} \parallel \boldsymbol{n}$ and $\boldsymbol{m} \perp \boldsymbol{n}$ boundary conditions.

The expression for the *free-energy density* of ferronematics derived by Burylov and Raikher in [22, 23], which allows all anchoring options mentioned above, reads

$$f = \frac{1}{2}\left\{\left[K_1(\mathrm{div}\boldsymbol{n})^2 + K_2(\boldsymbol{n}\cdot\mathrm{curl}\boldsymbol{n})^2 + K_3(\boldsymbol{n}\times\mathrm{curl}\boldsymbol{n})^2\right] - \chi_a(\boldsymbol{H}\cdot\boldsymbol{n})^2\right\}$$
$$-M_s\phi(\boldsymbol{m}\cdot\boldsymbol{H}) + \frac{AW\phi}{d}(\boldsymbol{n}\cdot\boldsymbol{m})^2 + \frac{\phi k_B T}{V} ln\phi. \qquad (11)$$

Here the first row corresponds to the free energy density of the nematic host (as given in Eqs. (2) and (4)), while the second row lists the additional terms related to the magnetic particles. The first term of the second row corresponds to the free energy of the particles in the magnetic field ($\boldsymbol{M}_s$ is the saturation magnetization, ϕ is the volume fraction of the particles). The second term describes the interaction between the host nematic and the magnetic particles via surface anchoring. Here there are two parameters characterizing the anchoring: the *anchoring energy* W and the parameter $A = 1 - 3\cos^2\alpha$ which depends on the easy-orientation angle α the nematic director $\boldsymbol{n}$ makes with the particle's surface. For *homeotropic anchoring* (i.e. $\alpha = \phi/2$), which will mostly be assumed in the following, one obtains $A = 1$. Finally the last term comes from the mixing entropy (V is the volume of the particle and $k_B T$ is the thermal energy) which is independent of the orientation and therefore is irrelevant for the forthcoming discussion.

By these extensions, the *Burylov-Raikher theory* could be successfully applied for thermotropic ferronematics. In its frame the destabilization of the uniform texture of ferronematics exposed to external magnetic or electric field (the Fréedericksz transitions) [20 - 24] could be described and the expressions for their critical fields in different geometries could be derived. We will also use their results later at the interpretation of our experiments.

In order to get a qualitative picture, how the magnetic particles may influence the response of the host liquid crystal and why is the mutual orientation of $\boldsymbol{n}$ and $\boldsymbol{m}$ important, let us discuss the behaviour of ferronematics in an experimental geometry for combined electric and magnetic fields as shown in Figure 2a.

We consider a cell with planar alignment and a liquid crystal whose diamagnetic susceptibility anisotropy, as well as the dielectric permittivity anisotropy are positive. First we assume that the director $\boldsymbol{n}$ and the magnetic moment of the magnetic particles $\boldsymbol{m}$ are initially oriented parallel (see Figure 3, left). If we apply an electric field (perpendicular to the electrode surfaces) higher than the electric Fréedericksz threshold value, the director will start to reorient toward the electric field (Figure 3, middle). Due to the coupling between $\boldsymbol{n}$ and $\boldsymbol{m}$, the magnetic moment of particles follows the director. Now we apply an external magnetic field perpendicular to the electric one. The increasing magnetic field will turn the magnetic moment of magnetic particles toward its direction. Via the coupling between the director and the magnetic moment of the magnetic particles a torque appears (in addition to the one coming from the diamagnetic anisotropy of the host liquid crystal) which also tends to turn the director toward the magnetic field. We can say that in this case the magnetic particles help the reorientation of the director in the applied magnetic field.

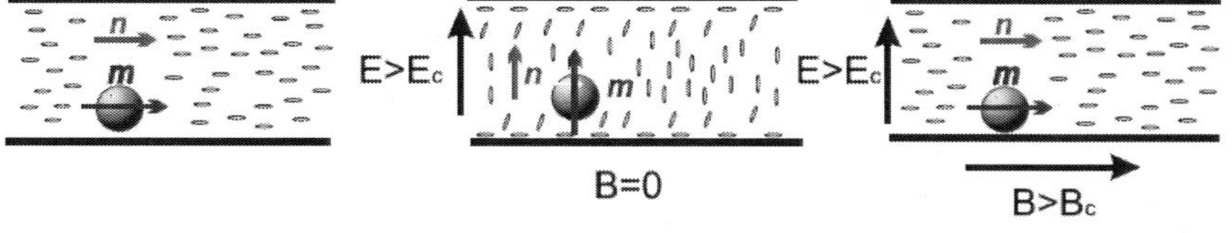

Figure 3. Cross section of a planar cell filled with a ferronematic ($\varepsilon_a > 0$, $\chi_a > 0$) with initially parallel **n** and **m**; in the field free basic state (left), in electric field only (middle), and in electric and magnetic fields (right).

Consequently, the critical magnetic field necessary to restore the initial planar state (see Figure 3, right) becomes smaller in ferronematics than in the host nematic (which latter is given by Eq. (9)). Moreover, the reduction of the critical field will become larger if the liquid crystal is doped with a higher volume concentration of the magnetic nanoparticles.

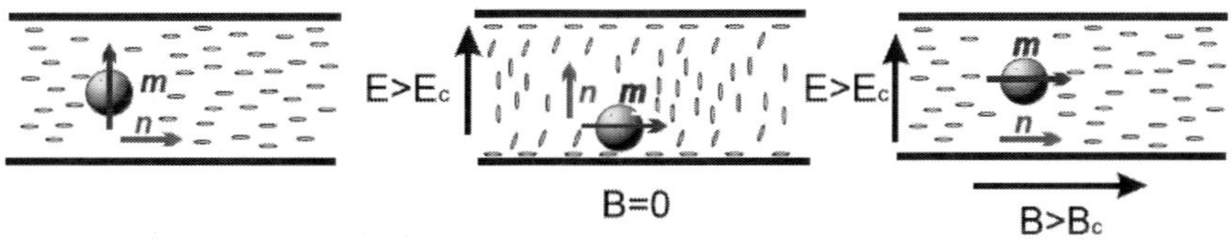

Figure 4. Cross section of a planar cell filled with a ferronematic ($\varepsilon_a > 0$, $\chi_a > 0$) with initially perpendicular **n** and **m**; in the field free basic state (left), in electric field only (middle), and in electric and magnetic fields (right).

Now, let us check the same situation, but with the perpendicular initial orientation between the director ***n*** and the magnetic moment of magnetic particles ***m*** (Figure 4, left). In a high enough electric field, the director rotates and turns the magnetic moment too (Figure 4, middle). As a result, however, ***m*** is now already mostly parallel with the magnetic field, when it is switched on. The magnetic field, on the one hand, still tends to restore the planar state via the diamagnetic coupling with the host nematic, but on the other hand does not allow the rotation of the magnetic moment. Therefore, the torque originating in the coupling between ***m*** and ***n*** acts against the diamagnetic one and hinders the director rotation. As a consequence, the critical magnetic field necessary for the restoration of the basic state becomes higher in the ferronematics than for the pure host liquid crystal; in this case with increasing volume concentration of the magnetic particles the critical magnetic field of the ferronematic increases. In this scenario at large ***H*** we find $\boldsymbol{m} \parallel \boldsymbol{n}$ (just as in the previous case), i.e. the high magnetic field overwrites the initial boundary condition at the particle's surface.

These two examples show that the critical magnetic fields of the Fréedericksz transition may be either decreased or increased by doping with magnetic particles. Which situation occurs, depends on the mutual initial direction of ***n*** and ***m***. In the forthcoming Secs. 11.5 and 11.6 we will show that besides this, the sensitivity of ferronematics to the external magnetic field depends also on the type, size, shape and concentration of the magnetic particles, as well as on the anisotropy of diamagnetic susceptibility of the host liquid crystal material.

4. Experimental Methods

Experimental studies of ferronematics rely on the availability of magnetic nanoparticles. The nanoparticles of required size and shape are typically obtained by an inorganic chemical synthesis. As nanoparticles we have used magnetite (Fe_3O_4) as well as carbon nanotubes. Below we present some recipes as examples for the preparation of nanoparticles.

Synthesis of the spherical magnetic nanoparticles was based on the co-precipitation of Fe^{2+} and Fe^{3+} salts by NH_4OH at 60 °C. To obtain the Fe_3O_4 precipitate, $FeCl_2{\cdot}4H_2O$ and $FeCl_3{\cdot}6H_2O$ were dissolved in deionized water by vigorous stirring (the ratio Fe^{3+}/Fe^{2+} was 2:1). The solution was heated to 80 °C and 25% NH_4OH was added. The precipitate was isolated from the solution by magnetic decantation by washing with water.

Magnetite nano-rods were synthesized through hydrolysis of $FeCl_3$ and $FeSO_4$ solutions (the Fe^{3+}/Fe^{2+} molar ratio was 2:1) containing urea. In a typical experiment $FeCl_3{\cdot}6H_2O$, $FeSO_4{\cdot}7H_2O$ and $(NH_2)_2CO$ were all dissolved in purified deoxygenated water. This mixture was added to a flask with reflux condenser. They were heated in water bath for 12 hours at 90-95 °C. During this procedure a dark precipitate has been formed. The sample was cleaned several times by purified and deoxygenated water, and then it was dried at a lower pressure at 50 °C for 3 hours.

The chain-like particles have been extracted from the magnetotactic bacteria *Magnetotacticum Magnetospirillum* (AMB-1). For the cultivation of AMB-1 a medium consisting of Wolfe's vitamin solution, Wolfe's mineral solution, KH_2PO_4, sodium succinate hexahydrate, sodium tartrate dihydrate, sodium acetate trihydrate, 0.2% (w/v) resazurin (aqueous), $NaNO_3$, ascorbic acid, and 0.01 M ferric quinate was used. Resazurin played the role of a colorimetric indicator of the redox potential. The pH was adjusted to 6.75 with NaOH. This medium was pre-reduced by copper under nitrogen for a period of 1 hour and was subsequently dispensed into culture bottles in an anaerobic hood. The inoculated bottles were incubated at 25 °C for 4 days.

The isolation and purification of the magnetosomes from bacteria cells should be made by a method which leaves the membrane surrounding the magnetosome intact and allows removal of all contaminating material. As a first step of the isolation of the magnetosomes 10 mg of the Magnetotacticum Magnetospirillum cells suspended in 100 ml of the mixture of 20 mM 4-(2-hydroxyethyl)-1-piperazineethanesulfonic acid (HEPES) - 4 mM ethylene-diaminetetraacetic acid (EDTA) (pH 7.4) were disrupted by sonification. A subsequent centrifugation (10 min, 3036 rpm) allowed removal of unbroken cells and cell debris. The cell extract was then placed on NdFeB magnets for 1 hour. The black magnetosomes sedimented at the bottom of the tube while the residual contaminating cellular material was retained in the upper part of the tube and was then removed.

As a next step, to eliminate the electrostatically bound contamination, the magnetic particles were rinsed first with 50 ml of the mixture 10 mM HEPES - 200 mM NaCl (pH 7.4) and subsequently with 100 ml of 10 mM HEPES (pH 7.4). After removal from the magnet the magnetic particles were flushed with 10 mM HEPES buffer. The magnetosome suspension (the black sedimentation) was centrifuged (18000 rpm, 30 min, 4 °C), then the cell extract was placed on the magnet again for 30 minutes in order to separate the magnetic particles sedimenting at the bottom from the residual contaminating cellular material retained in the upper part of the tube. Repeating the last procedure for 10 times, one could obtain well purified magnetosomes, i.e. about 10-15 spherical magnetic particles coated with surfactant and connected in a chain.

The prepared magnetic nanoparticles can easily attach to each other forming larger grains. In order to suppress this tendency for aggregation, the magnetic Fe_3O_4 particles are usually coated with a surfactant (for example oleic acid and dodecyl benzene sulfonic acid). In magnetosomes (chain-like particles) this surfactant cover is naturally there.

Single-wall nanotubes (SWCNTs), both neat and functionalized with carboxyl group (SWCNT-COOH) are commercially available from Cheap Tubes Inc. They were produced by a catalytic chemical vapor deposition technique and were purchased as purified materials. SWCNTs (of mixed chirality) exhibit in about 60% semiconducting property and in about 40% metallic one.

The length of the nanotubes ranged from 0.5 μm to 2 μm, their outer diameter from 1 nm to 2 nm and the inner diameter: 0.8 nm - 1.6 nm. To prepare SWCNTs functionalized with magnetic nanoparticles, about 0.1 g of SWCNT-COOH was dissolved in 60 ml of water under ultrasonic irradiation of 20 min.

The mixture was further stirred vigorously for 30 min at 60 °C under nitrogen atmosphere. Then, 153 mg of $FeCl_3{\cdot}6H_2O$ was added under stirring and let them reacting for 30 min. Thereafter, 80 mg of $FeSO_4{\cdot}7H_2O$ was added and stirred for additional 30 min. Next, 30 ml of 6% NH_4OH aqueous solution was added into the mixture drop by drop at 60 °C during 1 hour and was let to react for another 2 hours. The whole process was performed under nitrogen atmosphere. The reaction mixture was then several times centrifuged; the product was washed with water to neutral pH and dried at 50 °C for 24 hours. The total iron concentration in the SWCNT/Fe_3O_4 product was determined spectrophotometrically: after HCl/H_2O_2 induced oxidation of Fe^{2+} to Fe^{3+} and an addition of 1% ammonium thiocyanate followed by absorption measurement of the thiocyanate complex at λ = 495 nm. The amount of ferromagnetic particles in the prepared samples was 25% w/w.

The morphology and size distribution of the prepared nanoparticles were determined with transmission electron microscopy (TEM Tesla BS 500). Fourier-transform infrared (FTIR) spectra were collected using KBr pellets with an FTLA2000-100 instrument (from ABB) acquiring 32 scans per specimen at a nominal resolution of 4 cm^{-1}. The spectrum of a pure KBr pellet was used as a background to generate the sample spectra. Magnetization measurements were performed with a *SQUID magnetometer* (Quantum Design MPMS 5XL) at room temperature.

Ferronematics finally can be prepared by adding the (surfactant covered) magnetic nanoparticles to the thermotropic nematic liquid crystal in the required concentration in the isotropic phase under continuous stirring. Homogeneity of the resulting suspension is ensured by ultrasonic agitation.

Experimental studies of the ferronematics described in this chapter are based on detecting the field-induced orientational (Fréederiksz) transitions described in Sec. 11.2. The director distortions occurring in the field were monitored by precise capacitance measurement. In order to allow that, the ferronematic sample is filled into a flat capacitor of area Σ and gap thickness D. The method is then based on the fact that the local value of the dielectric permittivity in the bulk of a nematic or ferronematic layer (along the z axis) changes with the director orientation:

$$\varepsilon(z) = \varepsilon_{\perp} + \varepsilon_a n_z^2. \tag{12}$$

Here n_z denotes the z component of the director $\boldsymbol{n}$. The total capacitance of the cell can be expressed by an integral over the cell thickness:

$$C = \varepsilon_0 \Sigma \left(\int_0^D \frac{dz}{\varepsilon(z)} \right)^{-1}. \tag{13}$$

Actually measurements have been carried out both in the ferronematic samples and in their host nematics. Mostly a capacitor made of antiparallelly rubbed ITO-coated glass electrodes (LINCAM Co.) was used which ensured a planar initial orientation. This capacitor, having an electrode area of $\Sigma \approx 1\,\text{cm} \times 1\,\text{cm}$, was positioned into a regulated thermostat system; the temperature was stabilized with the accuracy of 0.05 °C. The capacitance was measured at the frequency of 1 kHz by the high precision capacitance bridge Andeen Hagerling with a maximum accuracy of fF.

In order to illustrate how the method is working, in Figure 5a we show the capacitance of a planar nematic 4-(trans-4'-n-hexylcyclohexyl)-isothiocyanatobenzene (6CHBT) in the function of a magnetic field normal to the cell substrates, i.e. in the experimental geometry shown in Figure 1b.

As 6CHBT has $\varepsilon_a > 0$ and $\chi_a > 0$, the minimal value of capacitance is observed for the initial planar alignment while the maximal value is obtained after reorientation, in the infinitely high magnetic field limit when the director $\boldsymbol{n}$ becomes perpendicular to the capacitor electrodes. The Fréedericksz threshold magnetic field can be obtained from the $C(H)$ curve by linear extrapolation from the initial rise of the capacitance as shown in the figure.

In order to allow an easier comparison of the behaviour of various samples, it is convenient to introduce a dimensionless *reduced capacitance* C_r by the formula

$$C_r = \frac{(C - C_0)}{(C_{max} - C_0)}, \tag{14}$$

where C is the capacitance measured at a given magnetic field H, C_{max} is the capacitance at the maximal value of H and C_0 corresponds to the capacitance at $H = 0$ (see Figure 5b).

By another example we illustrate the behaviour of 6CHBT in combined electric and magnetic fields in the experimental geometry of Figure 2a. Figure 11.6 shows how the capacitance of the cell depends on the external magnetic field at different bias voltages. Here the *Fréedericksz transition* is induced by the electric field normal to the substrates (if the voltage exceeds the threshold value), while a magnetic field parallel to the substrates turns the director back toward the initial planar alignment, as seen by the decrease of the capacitance.

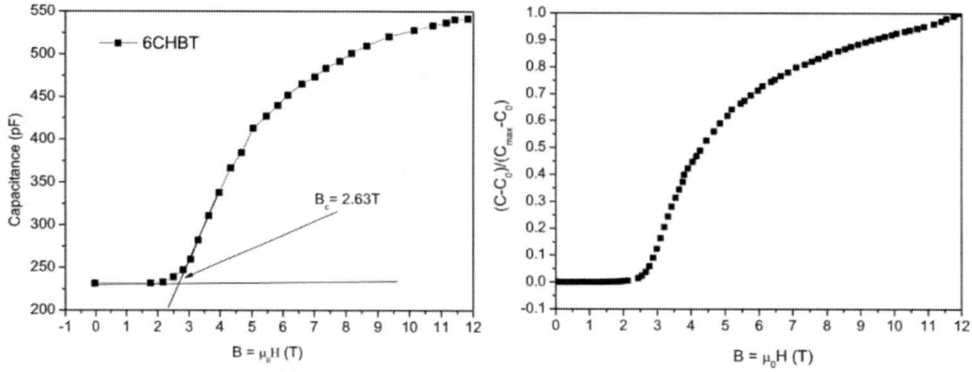

(a) (b)

Figure 5. Dependence of (a) the capacitance, (b) the dimensionless reduced capacitance of the liquid crystal 6CHBT on the external magnetic field - a classical magnetic Fréedericksz transition in the geometry of Figure 1a.

5. Structural Phase Transitions in Liquid Crystals Doped with Spherical Magnetic Particles

In this section the influence of doping liquid crystals with spherical nanoparticles is addressed. The nematics selected for the experimental tests, 4'-methoxybenzylidene-4-n-butylaniline (MBBA), 4-(trans-4'-n-hexylcyclohexyl)-isothiocyanatobenzene (6CHBT) and the 4-alkyl-4'-cyanobicyclohexyl mixture (ZLI1695), represent host materials with different combination of their material parameters (ε_a and χ_a). All liquid crystals were doped with the same spherical magnetite nanoparticles with a mean diameter of 11.4 nm.

During the experiments electric and/or magnetic fields were applied to control the orientation of the director and the critical magnetic fields belonging to the structural (reorientational) transitions in the given experimental geometry were determined, both for the ferronematics and for the host nematic. The *Burylov-Raikher theory* was adopted to interpret the findings and to deduce conclusions on the anchoring at the nematic--magnetic particle interface.

5.1. MBBA ($\varepsilon_a < 0$, $\chi_a > 0$)

Let us begin with MBBA which is a room-temperature nematic liquid crystal with the isotropic-nematic phase transition temperature T_{NI} = 48.0 °C. This compound has a *negative dielectric anisotropy* and a *positive diamagnetic anisotropy*. Three ferronematic suspensions were prepared with different volume concentrations ϕ of the spherical magnetic nanoparticles: $1\cdot10^{-4}$, $2\cdot10^{-4}$ and $5\cdot10^{-4}$ [25]. All measurements were made at the temperature T = 35 °C.

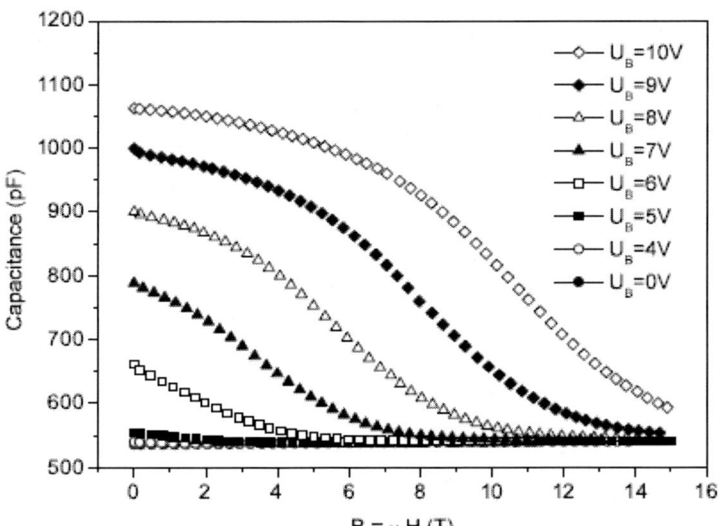

Figure 6. Dependence of the capacitance of the liquid crystal 6CHBT on the external magnetic field at different values of the bias voltage in the geometry of Figure 2a.

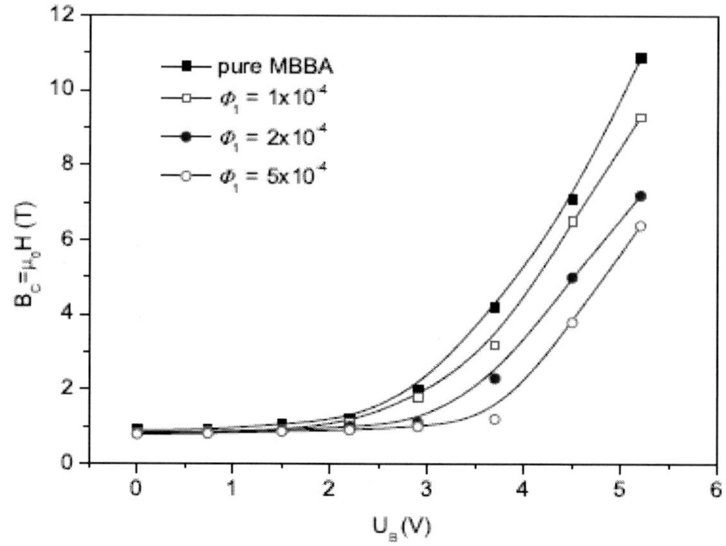

Figure 7. Dependence of the critical magnetic field on the bias voltage for pure MBBA and for MBBA doped with magnetic nanoparticles.

The structural transitions provoked by combined electric and magnetic fields were monitored in the experimental geometry of Figure 2b. The external bias electric field $\boldsymbol{E}_b$ as well as the magnetic field $\boldsymbol{H}$ was applied perpendicular to the initial director orientation. Though the two fields are parallel, due to the opposite signs of the dielectric and diamagnetic anisotropies, they act against each other; while $\boldsymbol{H}$ induces a *Fréedericksz transition*, the applied bias voltage inhibits it and stabilizes the initial planar state. Therefore the bias voltage dependent magnetic threshold field in a nematic follows Eq. (10).

Figure 7 shows how the critical magnetic fields of pure MBBA and of MBBA based ferronematics depend on the bias voltage. It is seen that the predicted bias dependence of the threshold field holds for MBBA. It is also evident from this figure that with increasing volume concentration of the magnetic nanoparticles the critical magnetic field decreases. For the interpretation of these results one has to recall the Burylov-Raikher theory. From the expression for the free energy of ferronematics in Eq. (11) they have derived an approximate relation between the critical field H_C of the nematic host and the critical field H_{CFN} of the ferronematic [22, 23]:

$$H_{CFN}^2 - H_C^2 = -\frac{2W\phi}{\mu_0 \chi_a d}. \tag{15}$$

Here d is the mean diameter of the magnetic particles, ϕ is their volume concentration in the liquid crystal, μ_0 is the permeability of vacuum and χ_a is the anisotropy of the diamagnetic susceptibility of the host liquid crystal.

This formula was obtained using two basic assumptions. First is the homeotropic alignment of the liquid crystalline molecules on the surface of the magnetic particle; the second is a parallel initial orientation between the director and the magnetic moments of the particles.

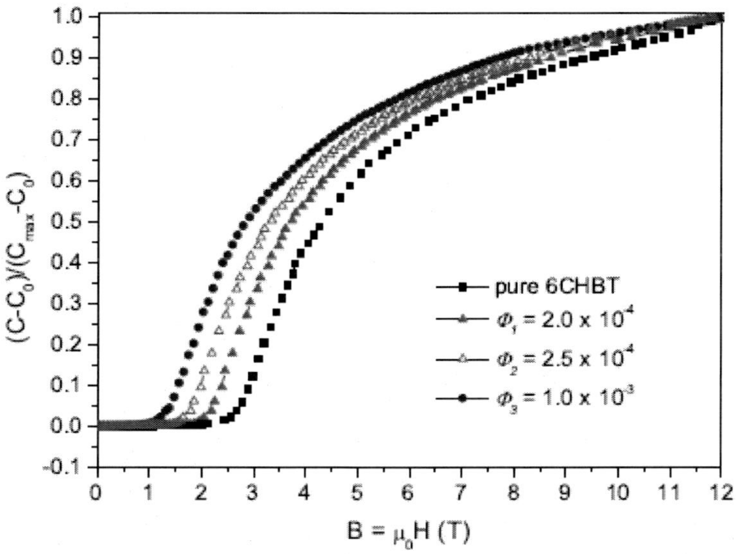

Figure 8. Reduced capacitance versus magnetic field for pure 6CHBT and for 6CHBT doped with different volume concentrations of magnetic nanoparticles.

Eq. (15) implies that for positive χ_a the magnetic particles reduce the critical magnetic field in ferronematics, in perfect agreement with the experiments. This fact actually justifies the assumption on $\boldsymbol{m} \parallel \boldsymbol{n}$, used in the derivation.

As both H_{CFN} and H_C were measured, Eq. (15) can be used to estimate the anchoring energy density W at the nematic - magnetic particle boundary. Using the value of $\chi_a = 1.32 \cdot 10^{-7}$ for MBBA (at $T = 35$ °C) and knowing the concentration of the particles one obtains $W \sim 10^{-7}$ Nm^{-1}. Now the parameter $\omega = Wd/K_1$ can also be calculated. As for MBBA at the measuring temperature $K_1 = 6.7$ pN, $\omega \sim 10^{-3} = 1$ is obtained [25]; that characterizes a soft anchoring of the nematic director on the surfaces of the magnetic particles in MBBA-based ferronematics. Though the soft anchoring might allow both $\boldsymbol{m} \parallel \boldsymbol{n}$ and $\boldsymbol{m} \perp \boldsymbol{n}$ boundary conditions, the decrease of the critical magnetic field on doping, as we have seen already, selects the initially parallel orientation of the director and the magnetic moment in MBBA-based ferronematics.

5.2. 6CHBT ($\varepsilon_a > 0$, $\chi_a > 0$)

6CHBT is a low-melting *enantiotropic*, room temperature nematic liquid crystal with high chemical stability [26]. Its nematic-isotropic phase transition temperature (clearing point) is $T_{NI} = 42.8$ °C. 6CHBT has $\varepsilon_a > 0$ and $\chi_a > 0$.

Three ferronematics were prepared with volume concentrations of the magnetic particles of $\phi_1 = 2 \cdot 10^{-4}$, $\phi_2 = 2.5 \cdot 10^{-4}$ and $\phi_3 = 1 \cdot 10^{-3}$, respectively [27].

In ferronematics based on 6CHBT the simple magnetic Fréedericksz transition was monitored in the experimental geometry of Figure 1b. The reduced capacitance in the function of the applied magnetic field is presented in Figure 8 for the pure 6CHBT host and for 6CHBT-based ferronematic samples. It is clearly seen that the Fréedericksz threshold field decreases with the increase of the concentration of the magnetic nanoparticles. The Burylov-Raikher expression in Eq. (15) applies also here.

Therefore we can conclude that (χ_a being positive) the threshold reduction proves the initial $\boldsymbol{m} \parallel \boldsymbol{n}$ condition. Using $\chi_a = 4.805 \cdot 10^{-7}$ for 6CHBT at 35 °C, Eq. (15) yields $W \sim 10^{-5}$ Nm^{-1} for the anchoring energy density. Using $K_1 = 6.71$ pN as the splay elastic constant one obtains $\omega \sim 10^{-1} < 1$. This again indicates soft anchoring at the liquid crystal - magnetic particle interface.

6.3. ZLI1695 ($\varepsilon_a > 0$, $\chi_a < 0$)

ZLI1695 is a commercial, room temperature nematic mixture from Merck Co. Its nematic - isotropic phase transition temperature is T_{NI} = 73.5 °C. The speciality of this mixture is that, in contrast to all nematics mentioned so far, it has a negative anisotropy of the diamagnetic susceptibility ($\chi_a = -2.55 \cdot 10^{-8}$ at T = 35 °C), while its dielectric anisotropy is positive.

ZLI1696-based ferronematics were prepared in three different volume concentrations of the magnetic nanoparticles: $\phi_1 = 2 \cdot 10^{-4}$, $\phi_2 = 5 \cdot 10^{-4}$ and $\phi_3 = 1 \cdot 10^{-3}$. Measurements were carried out at T = 35 °C in two different geometries [28].

In the first geometry the magnetic field was applied parallel to the director $\boldsymbol{n}$ (Figure 1a). Due to $\chi_a < 0$ this allows for a magnetic splay Fréedericksz transition. Indeed, as it is presented in Figure 9, the reduced capacitance increases when the magnetic field exceeds the threshold value H_C for the pure liquid crystal, or the threshold H_{CFN} for the ferronematics. Surprisingly, however, unlike any ferronematics discussed so far, doping with nanoparticles leads to an increase of the critical magnetic field. The bigger the concentration of the nanoparticles, the bigger the increment. These results imply that the admixture of magnetic particles to this kind of liquid crystal ($\chi_a < 0$) hinders the reorientation of the director by the external magnetic field.

This scenario can also be interpreted using the *Burylov-Raikher theory*. It has indeed been shown [24] that for this case the relation in Eq. (16) holds, i.e. ferronematics should have higher thresholds.

$$H_{CFN}^2 - H_C^2 = \frac{2W\phi}{\mu_0 \, |\, \chi_a \,|\, d} \tag{16}$$

Actually comparing this formula with Eq. (15) one can notice that the two equations are equivalent, in view of the negative diamagnetic susceptibility of the studied host. The agreement between the theoretical prediction and the experiment again implies that the assumption of parallel $\boldsymbol{n}$ and $\boldsymbol{m}$ holds for this compound too.

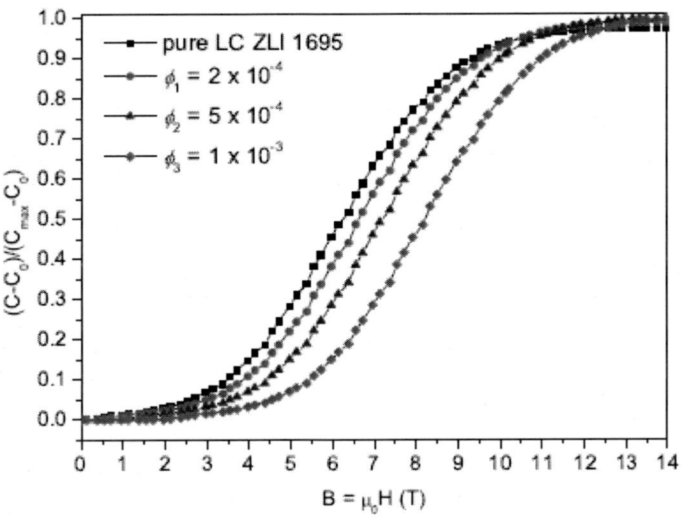

Figure 9. Dependence of the reduced capacitance on the external magnetic field for pure ZLI1695 and for ZLI1695-based ferronematics.

In the second experimental geometry the magnetic field was applied perpendicular to the initial alignment of the director and of the magnetic moments of the magnetic particles (Figure 1b). Due to $\chi_a < 0$ the planar orientation is now a stable state in magnetic field for the host nematic. In case of ferronematics, however, the situation becomes modified. In the initial state the direction of the magnetic moments is unfavoured by the magnetic field. Therefore, upon applying the magnetic field $\boldsymbol{m}$ will rotate toward $\boldsymbol{H}$. As the magnetic moments and the director prefer to be parallel, due to this coupling a torque appears which forces the director to follow the rotation of $\boldsymbol{m}$; as a result a distortion of the director field is induced around the magnetic particles.

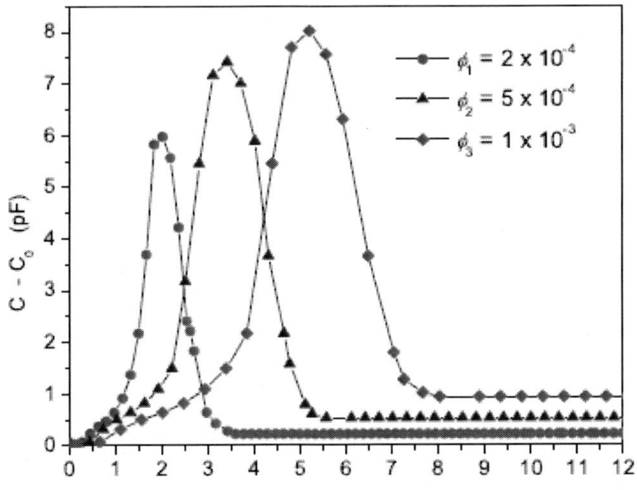

Figure 10. Dependence of the change of the capacitance of liquid crystal doped with different volume concentrations of spherical particles on magnetic field applied perpendicular to the surface of electrodes.

The distortion manifests itself in the increase of the capacitance, as it is shown in Figure 10. It is also seen, however, that the capacitance grows with H only up to some critical field intensity H_{max}, where it reaches a maximal value. On further increase of the magnetic field, i.e. $H > H_{max}$, the capacitance decreases again. This latter behaviour indicates that the diamagnetic torque arising due to the negative χ_a of the host liquid crystal overcomes the effect of the coupling between the director and the magnetic moment favouring parallel bonding between $\boldsymbol{m}$ and $\boldsymbol{n}$. Consequently at high magnetic fields the initial planar director alignment is restored in the cell resulting in $\boldsymbol{m} \perp \boldsymbol{n}$.

It is clearly seen from Figure 10 as well as from Table 1 that lists the measured values of B_{CFN} and B_{max} for ferronematics with different volume concentrations of the magnetic particles, that a higher doping concentration increases H_{max} substantially. We would like to emphasize that this latter effect is regarded as an example, definitely showing that the role of magnetic particles cannot be fully described by simply rescaling the diamagnetic anisotropy of the system.

The magnetic field corresponding to the maximum in capacitance can be estimated [21] using the free energy expression in Eq. (11), by equating the *magnetic energy density* of liquid crystal,

$$\frac{1}{2}\mu_0 |\chi_a| H_{max}^2, \tag{17}$$

with the *density of anchoring energy*, which is roughly

$$\frac{W}{d}\phi. \tag{18}$$

This comparison yields

$$H_{max}^2 \approx \frac{2W\phi}{\mu_0 |\chi_a| d}. \tag{19}$$

Comparing this result with Eq. (16) the following identity can be picked up:

$$H_{max}^2 \approx H_{CFN}^2 - H_C^2. \tag{20}$$

As an experimental test of this identity,

$$\frac{H_{CFN}^2 - H_C^2}{H_{max}^2} = \frac{B_{CFN}^2 - B_C^2}{B_{max}^2} \tag{21}$$

can be calculated from the data in Table 1. The calculation yields 0.97, 0.77 and 0.61 for the volume concentrations of magnetic particles $\phi_1 = 2 \cdot 10^{-4}$, $\phi_2 = 5 \cdot 10^{-4}$ and $\phi_3 = 1 \cdot 10^{-3}$, respectively. It shows that Eq. (20) fulfills nicely for the lowest concentration of the magnetic particles, but holds less for the higher concentrations.

Table 1. The critical magnetic inductions B_{CFN} and B_{max} and the surface density of the anchoring energy W calculated from B_{CFN} and B_{max} for different volume concentrations of magnetic particles

volume concentration of magnetic particles	B_{CFN} [T]	B_{max} [T]	$W_{B_{CFN}}$ [N/m]	$W_{B_{max}}$ [N/m]
$\phi_1 = 2 \cdot 10^{-4}$	3.9	2.0	$2.00 \cdot 10^{-6}$	$2.02 \cdot 10^{-6}$
$\phi_2 = 5 \cdot 10^{-4}$	4.6	3.4	$2.13 \cdot 10^{-6}$	$2.33 \cdot 10^{-6}$
$\phi_3 = 1 \cdot 10^{-3}$	5.6	5.2	$2.20 \cdot 10^{-6}$	$2.73 \cdot 10^{-6}$

Table 1 also lists the anchoring energy densities calculated according to Eqs. (16) and (19) [28]. The values obtained for different particle concentrations are in good agreement. Calculating the parameter ω with the splay elastic constant $K_1 = 7$ pN, $\omega \sim 10^{-4}$ is obtained, which again indicates soft anchoring at the particle's surface.

6. Influence of Size and Shape of Magnetic Particles on the Structural Phase Transition in 6CHBT-Based Ferronematics

This section is devoted to experimental studies aiming to explore the influence of the type, shape and size of the magnetic particles on the structural phase transitions in ferronematics [29, 30]. The used magnetic particles were spherical, rod-like and chain-like magnetite, as well as single walled carbon nanotubes (SWCNT) and SWCNT functionalized with Fe_3O_4 particles (SWCNT/Fe_3O_4); their synthesis was described in Sec. 11.4. At preparing the ferronematic suspensions these particles were mixed into the same host nematic liquid crystal, 6CHBT.

6.1. Spherical, Rod-Like and Chain-Like Magnetite Nanoparticles

Let us first focus our attention to the influence of the shape of the magnetic particles. Measurements were carried out using ferronematics made of spherical, rod-like and chain-like magnetite nanoparticles of volume concentration $\phi = 2 \cdot 10^{-4}$. The mean diameter of the spherical particles was 11.6 nm, rod-like particles had an average diameter of 25 nm and their mean length was 1200 nm, as calculated from the histogram of their length distribution. Chain-like particles were obtained from magnetotactic bacteria; they consisted of several spherical magnetic particles with a mean diameter of 34 nm which formed a chain with the mean length of 446 nm.

The various ferronematics were subjected to the combined action of electric and magnetic fields at the temperature of 35 °C in the experimental geometry of Figure 3. The initial alignment of the magnetic moments was achieved by applying a temporary strong magnetic field before starting the measurement.

Then the external magnetic field was switched off and the initially planar nematic layer was realigned by electric field applying a bias voltage U_B exceeding the purely electric splay Fréedericksz threshold voltage $U_c(0)$. Then a magnetic field was applied perpendicular to $\boldsymbol{E}$, along the initial surface alignment. In the presence of the magnetic field the threshold voltage $U_c(H)$ increases according to Eq. (9), reducing therefore the distortion angle in the cell. At a critical magnetic field H_c, where $U_c(H_c) = U_B$, the distortion of the director ceased and the initial planar texture was restored.

Figure 11a and b exhibit the dependence of the dimensionless reduced capacitance on the magnetic field at two different bias voltages for pure 6CHBT as well as for 6CHBT doped with magnetite particles of different shape. On the one hand, it is seen that a higher bias voltage increases the magnetic threshold field as expected. On the other hand, doping with magnetic nanoparticles evidently reduces the critical field, confirming that the coupling between the director and the magnetic moment favours $\boldsymbol{m} \parallel \boldsymbol{n}$. The reduction is the smallest for spherical particles and the largest for rod-like particles. Using the Burylov-Raikher expression in Eq. (15) the surface density of the anchoring energy could be calculated, yielding $W_s \sim 10^{-5}$ Nm^{-1} for spherical particles, $W_c \sim 10^{-7}$ Nm^{-1} for chain-like particles and $W_r \sim 10^{-2}$ Nm^{-1} for rod-like particles. These results indicate that the anchoring energy increases as the shape of the particles becomes more anisotropic.

Knowing the elastic modulus of the *host liquid crystal* ($K_1 = 6.71$ pN for pure 6CHBT), the obtained values of W were used for calculating the parameter ω. For spherical particles one obtains $\omega \sim 10^{-1} < 1$, which indicates soft anchoring of nematic molecules on the surfaces of magnetic particles. In contrast to this result, however, for chain-like particles we get $\omega \sim 10^2$ and for rod-like particles $\omega \sim 10^4$; in both these cases $\omega >> 1$, which corresponds to *rigid anchoring*. Thus one can conclude that the larger *shape anisotropy* of the particles changes the character of the anchoring at the liquid crystal - particle interface from soft to rigid.

The presented results show that doping with magnetic particles shaped similarly to the molecules of the host liquid crystal is more effective and thus offer better perspectives for ferronematics in applications where the magnetic field is necessary to control the orientation of the liquid crystal.

We have also studied the influence of the size of the particle on the magnetic response. Figure 12 shows the change of the capacitance during a classical magnetic *Fréedericksz transition* for two ferronematic samples doped with rod-like particles of different size. Sample A contained rod-like particles with a mean diameter of 18 nm and length of 300 nm, while sample B had also rod-like particles, however, with a mean diameter of 8 nm and length of 100 nm. Both ferronematics were based on the nematic 6CHBT and were doped in a volume concentration of $2 \cdot 10^{-4}$.

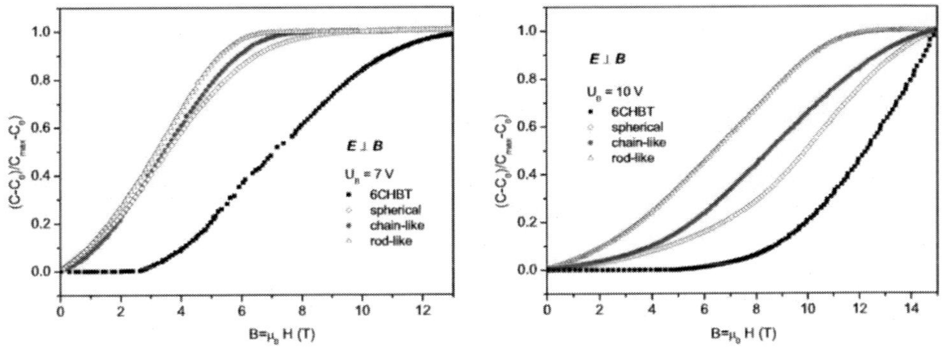

(a) (b)

Figure 11. Dependence of the reduced capacitance on the magnetic field (a) at $U_B = 7$ V, (b) at $U_B = 10$ V bias voltage for pure 6CHBT and for 6CHBT doped with magnetic particles of different shape.

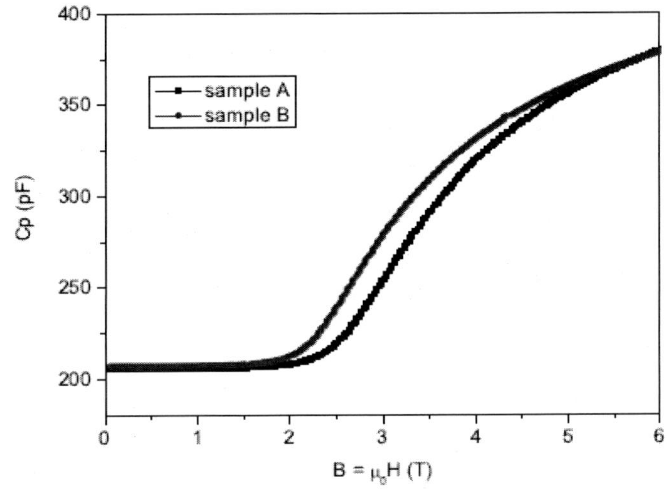

Figure 12. Dependence of the capacitance on the magnetic field for two 6CHBT-based ferronematics containing rod-like magnetic particles (Sample A: diameter 18 nm and length 300 nm and sample B: diameter 8 nm and length 100 nm.

Figure 12 shows that the ferronematic made of smaller particles has lower threshold field than that with larger particles. It has to be noted, however, that the two rod-like particles did not have the same shape anisotropy; the length to width ratio of the larger rods was about 30 % bigger than for the smaller rods.

Exploring the effect of the particle size is more unambiguous when using spherical magnetic particles [31]. The test was made using 6CHBT doped with spherical magnetic particles of different diameters: 4.2 nm, 11.4 nm, 31.4 nm and 112.7 nm. The volume concentration of magnetic particles was $2 \cdot 10^{-4}$ in all cases. Figure 13 shows again the capacitance variation during a classical magnetic Fréedericksz transition in these ferronematic samples. The size dependence is immediately noticeable from the figure. Interestingly, the critical magnetic field of the ferronematic doped with magnetic particles of 4.2 nm size are higher than that of the pure host 6CHBT.

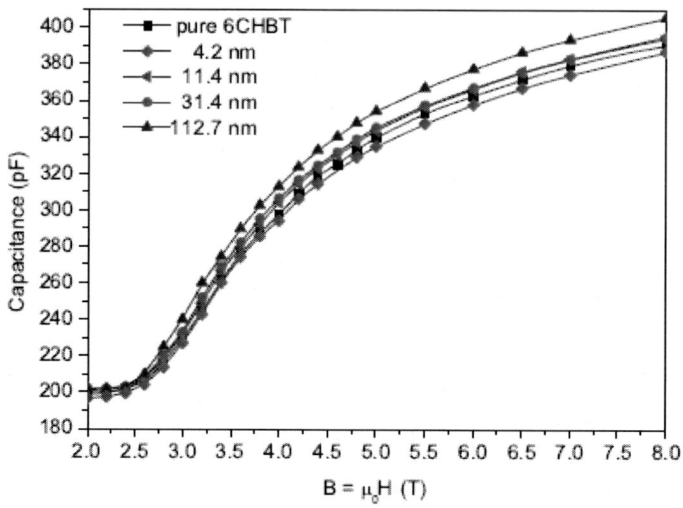

Figure 13. Dependence of the capacitance on the magnetic field for pure 6CHBT and for 6CHBT doped with spherical magnetic particles of different size.

In contrast to that, ferronematics with larger particle size have the critical field reduced compared to that of 6CHBT. The larger the particle's diameter, the bigger the reduction of the critical field in this latter case. The non-monotonic size dependence indicates a difference in the mechanisms, how the particles affect the magnetic response.

The motion of the magnetic moment relative to the particle is described by the *Landau-Lifshitz equation* [32]. When fine single domain magnetic particles in a magnetic fluid are subjected to an external magnetic field, two modes of rotation exists for their magnetic moments: Brownian and Néel rotation. When the particle rotates as a whole, it is called *Brownian rotation* and its relaxation time is given by

$$\tau_B = \frac{3V_h\eta}{k_BT}. \tag{22}$$

Here V_h denotes the *hydrodynamic volume* of the particles (including their surfactant layer); η is the *dynamic viscosity* of the liquid. On the other hand, if the magnetic moment is rotating inside the particle relative to its crystal structure, the rotation is called *Néel rotation* and its relaxation time is given by the expression

$$\tau_N = f_0^{-1} exp\left(\frac{\kappa V}{k_BT}\right). \tag{23}$$

Here κ is an *anisotropy constant* of the particles, V is their volume and f_0 is the *Larmour frequency* of the magnetization vector in the anisotropic field of the particle.

Both relaxation phenomena depend on the particle size. For smaller particles τ_N is smaller than τ_B and the relaxation will occur by the rotation of the magnetic moment inside the particle (Néel rotation), while for larger particles rotation of a whole particle will take place (Brownian rotation). There exists a critical size (called *Shliomis size*) that determines where the transition from Néel to Brownian relaxation occurs. The Shliomis size depends on the kind of the magnetic particles. For example for magnetite particles of nearly spherical shape, the Shliomis size is approximately 10 nm [33].

If an external magnetic field is applied, one can expect different behaviours for the two types of relaxation. In the case of particles with size smaller than the Shliomis size, there will be no significant help to change the director structure under application of the magnetic field via the $\boldsymbol{m}$ to $\boldsymbol{n}$ coupling which we referred to in the previous sections. As the magnetic moments will relax inside the particles without mechanical motion, they have no influence on the anchoring at the particle's surface. The director remains in its initial direction and rather hinders director rotation in the bulk then helps to it. This might be the reason for the increase of the critical field.

In the case of particles with their size larger than the Shliomis size, however, the magnetic particle will rotate as a whole in the host liquid and consequently, due to the anchoring of liquid crystal molecules at the particle surface, it leads to an additional distortion of the bulk director state, i.e. to a decrease of the magnetic Freedericksz threshold.

6.2. SWCNT

In a recent paper [34] it has been indicated that the axes of rod-like nanotubes are aligned parallel to the *liquid crystal director* $\boldsymbol{n}$ and liquid crystal molecules are strongly anchored on the wall of the tubes. The effect is associated with the formation of helical wrapping to enhance hexagon-hexagon π -overlapping and charge transfer from liquid crystal molecule to nanotube. The liquid crystal molecules anchored at the nanotubes surface should follow their movements. In order to explore the role of nanotubes in the structural transitions, we have performed experiments in combined electric and magnetic fields. Two kinds of 6CHBT-based ferronematics were prepared. One of them contained simple single walled carbon nanotubes (SWCNT), the other had SWCNT functionalized by magnetite particles as dopant (SWCNT/Fe_3O_4). The volume concentration of dopant was $\phi = 2 \cdot 10^{-4}$ in both cases.

The experiment was carried out in the experimental geometry of Figure 3; the magnetic field was held constant during the measurement and the electric field was increasing. As the magnetic field is stabilizing, the critical electric field should fulfil Eq. (9). Figure 14 shows the dependence of the critical electric field on the bias magnetic field for pure 6CHBT and 6CHBT doped with SWCNT or SWCNT/Fe_3O_4. It is clearly seen that the $U_C(B_B)$ curve of the ferronematic containing neat SWCNT almost coincides with that of the host liquid crystal; i.e. single wall carbon nanotubes do not affect the magnetic response significantly. On the contrary, the strong influence of SWCNT functionalized with magnetite nanoparticles is immediately noticeably. The anchoring energy densities estimated from the Burylov-Raikher theory [22, 23] are $W \sim 10^{-3}\,\mathrm{Nm}^{-1}$ for 6CHBT+SWCNT and $W \sim 10^{-2}$ Nm^{-1} for 6CHBT+SWCNT/Fe_3O_4; these values yield $\omega \sim 10$ and $\omega \sim 10^2$, respectively. In both cases $\omega > 1$, therefore there is a rigid boundary condition at the particle's surface which permits only parallel initial orientation of $\boldsymbol{m}$ and $\boldsymbol{n}$.

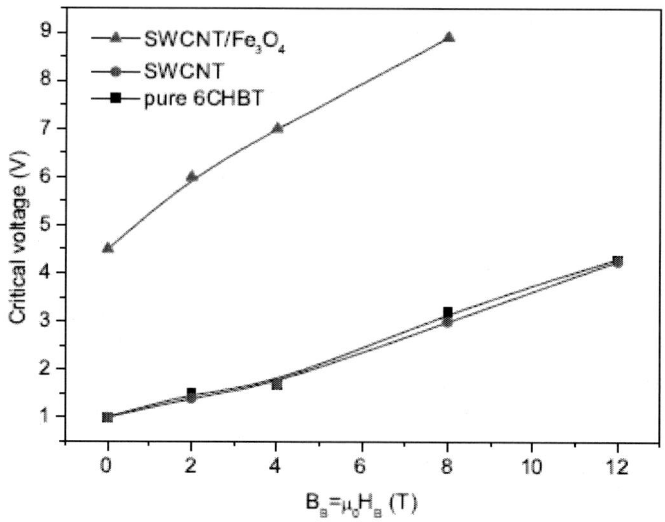

Figure 14. Dependence of critical electric field on bias magnetic field for pure 6CHBT and 6CHBT doped with SWCNT and SWCNT/Fe_3O_4.

From the obtained results we can conclude that admixture of SWCNT/Fe_3O_4 substantially increases the sensitivity of the host liquid crystal to the external magnetic field.

7. Magnetic Field Induced Isotropic to Nematic Phase Transition in Ferronematics and in PDLC

It has long been known that there exists a possibility in liquid crystals for an external field to substantially alter the *nematic-isotropic* transition temperature [35, 36] and the nematic-isotropic critical end point at a given temperature T and field strength [37]. However, this effect could not been induced by magnetic field in *thermotropic nematics* [38] until recently [39].

The theoretical description of the phenomenon is based on the basic mean-field theory of the nematic-isotropic transition: the *Landau-de Gennes model*. The model is a phenomenological Landau-type expansion of the free energy, including all symmetry-allowed scalar invariants which can be created from the second-rank nematic tensor order parameter $S_{\alpha\beta}$. It predicts the existence of a first-order phase transition, and can easily be expanded to include the effects of externally applied fields. The *Landau-de Gennes* free energy in the presence of a magnetic field H is

$$F = \frac{3}{4}a(T - T^*)S^2 - \frac{1}{4}bS^3 + \frac{9}{16}cS^4 - \frac{1}{2}\chi_a SH^2 \qquad (24)$$

where a, b and c are the *Landau coefficients*, S is the scalar order parameter of the uniaxial nematic, T^* is the temperature, below which the isotropic phase ($S = 0$) is unstable, and χ_a is the saturated *diamagnetic anisotropy*. The standard analytic approach is to find the value of S which minimizes the free energy F. The model predicts (at H = 0) a first order phase transition at the clearing point temperature

$$T_{cp} = T^* + \frac{b^2}{27ac}. \qquad (25)$$

When $H > 0$ the induction of the "paranematic" phase for $T > T_{cp}$, and a shift in the first order phase transition temperature $T_{NI}(H)$ are predicted according to

$$\frac{T_{NI}(H) - T^*}{T_{cp} - T^*} = \frac{1}{2}\left(\frac{H^2}{H_c^2} + 2\right) \quad (\text{for} \quad H \le H_c) \qquad (26)$$

where

$$H_c = \sqrt{\frac{b^3}{162\,\chi_a c^2}} \quad \text{and} \quad T_c = T^* + \frac{b^2}{18ac} \qquad (27)$$

are the critical magnetic field and the critical temperature at the critical end point [39]. Therefore, the first order nematic-isotropic phase transition temperature $T_{NI}(H)$ increases linearly with H^2.

The principal reason that the change of T_{NI} in magnetic field could not be proved experimentally is that the field strengths required are not accessible; the estimated critical magnetic induction necessary for a detectable temperature shift is well over 100 T for traditional liquid crystal materials. The first experimental observation of the predicted magnetic field dependence of the nematic-isotropic phase transition temperature thus required not only a powerful electromagnet ($B = \mu_0 H$ up to 30 T), but also a proper choice of a "non-traditional", bent-core, nematic liquid crystal material too [39]. The chosen bent-core nematic material has considerably different physical properties compared to those of "traditional" calamitic nematics; e.g. the first-order character of their nematic-isotropic transition is substantially weaker than for "traditional" nematics and the temperature difference between the clearing point and the supercooling limit $T_{cp} - T^*$ is more than twice as small as measured for any calamitic. These properties, combined with the high magnetic field have proved to be sufficient for the observation of a phase transition temperature shift of ~ 0.8 °C at B = 30 T.

In this section we demonstrate that the magnetic field induced shift of the isotropic-nematic phase transition is also detectable in a ferronematic based on a "traditional" calamitic liquid crystal 6CHBT doped with magnetic particles [40]. For the experiments, two types of ferronematic suspensions were prepared; the host nematic 6CHBT was doped either by spherical or by rod-like magnetite nanoparticles, both with a volume concentration of $2 \cdot 10^{-4}$. The influence of the magnetic particles on the isotropic-nematic phase transition of the calamitic liquid crystal 6CHBT was monitored by capacitance measurements. In this experiment the temperature was first increased above the nematic to isotropic phase transition temperature (T_{NI}) and then was slowly decreased, while a constant magnetic field, applied parallel to the capacitor electrodes (Figure 1a), was maintained. Figure 15a shows the dependence of the capacitance on the temperature for the neat 6CHBT liquid crystal and the ferronematic containing spherical particles, while Figure 15b depicts the same for 6CHBT doped with rod-like magnetic particles. On the one hand, it is seen from these figures that the presence of magnetic particles reduces the T_{NI} at $B = 0$. This is not surprising, as the magnetic particles are actually impurities in the host nematic, and impurities are known to shift T_{NI} of the liquid crystal to lower temperatures. Comparing Figure 15a and b one can deduce that this shift seems to be slightly larger in the case of doping with spherical magnetic particles.

On the other hand, as depicted in Figure 15b, in the case of doping with rod-like magnetic particles an additional, magnetic field dependent, shift was observed in T_{NI} by application of external magnetic fields up to $B = \mu_0 H = 12$ T. The largest shift, about 0.25 °C, was detected at the maximum of the applied magnetic field ($B = 12$ T) [40].

This value matches with the T_{NI} shift obtained in the bent-core nematic liquid crystal at the same magnetic field [39]. In contrast to these observations, a similar field induced shift in the phase transition temperatures was not observable in ferronematics doped with spherical particles (see Figure 15a).

Polymer dispersed liquid crystals (PDLCs) are another class of *optical composites* [41 - 45], which consist of a polymer and a liquid crystal material in an appropriate mixing ratio. In PDLCs nematic liquid crystals are present as uniformly dispersed, micron sized droplets in the transparent polymer matrix. The director within these nematic droplets is randomly oriented, such that the light incoming into the cell is strongly scattered, and the PDLC film appears opaque.

When an electric field is applied across the film, the liquid crystal aligns and the PDLC becomes transparent, pretending that the ordinary index of refraction of the liquid crystal is matched to the (isotropic) index of refraction of the polymer. The voltage and speed at which the PDLC switches from opaqueness to transparency, and the optical contrast between the opaque and transparent states is controlled by the size, shape, and anchoring energy of the liquid crystal droplets, as well as the by the *dielectric anisotropy* and viscosity of the liquid crystal.

While the latter properties depend on the selection of the liquid crystalline material, the former properties depend on the liquid crystal-monomer composition and the method of polymerization. In the previous sections we have seen several examples that doping by nanoparticles has a substantial effect on the behaviour of the system in a magnetic field. In order to explore the influence of magnetic nanoparticles on PDLCs, recently we have prepared a PDLC doped with magnetite labeled SWCNT (SWCNT/Fe_3O_4) [46].

The PDLC samples have been prepared by the following method. A small amount (0.05 ml) of liquid crystal (6CHBT) was added to 5 ml of the solution of 10% polyvinyl alcohol (PVA) in water. This mixture was stirred until a creamy white emulsion was obtained. The SWCNT/Fe_3O_4 nanotubes were added during the stirring.

The emulsion was let to degas, then a small drop of it was placed on a slide. After the evaporation of water we got a thin film. The sample thickness was 50 μm. The size of the 6CHBT microdroplets in the polymer matrix ranged from 5 to 8 μm. During the measurement the cell capacitance was recorded while cooling from the isotropic phase in a magnetic field parallel to the capacitor electrodes (Figure 1a), like above. Figure 16 shows the temperature dependence of the capacitance of the PDLC doped with SWCNT/Fe_3O_4 measured recently without magnetic field and at B = 12 T. As it is clearly seen from Figure 16, there is a significant magnetic field induced shift in the temperature of the isotropic to nematic phase transition due to doping [46], similarly to what has been found for ferronematics doped with rod-like magnetite particles. It proves that PDLCs can also be made more sensitive to magnetic fields by doping with magnetic nanoparticles.

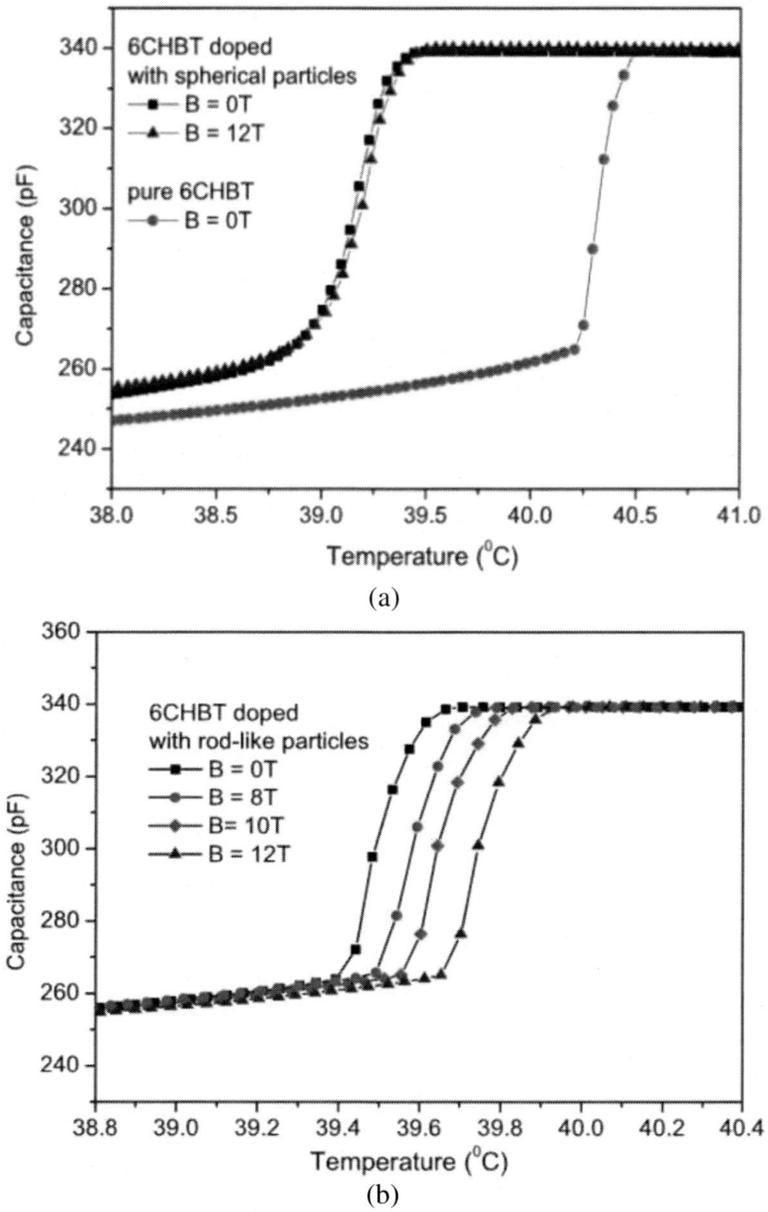

Figure 15. Capacitance vs. temperature (a) for pure 6CHBT and 6CHBT doped with spherical magnetic particles without magnetic field and at B = 12 T, (b) for 6CHBT doped with rod-like magnetic particles measured at different magnetic fields.

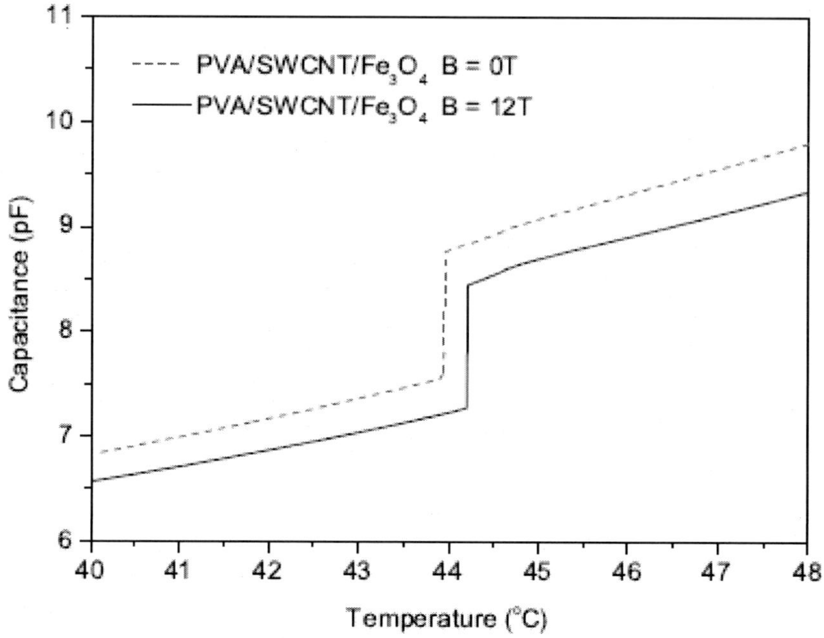

Figure 16. Temperature dependence of capacitance of mixture PVA with 6CHBT doped with SWCNT/Fe_3O_4 without magnetic field and in magnetic field 12 T.

CONCLUSION

In this chapter we reviewed our experimental work on ferronematics, i.e. liquid crystals doped with magnetic nanoparticles. It was demonstrated that addition of the nanoparticles has a substantial influence on their sensitivity to external magnetic field.

It was proven that the magnetic response depends on the diamagnetic susceptibility of the host nematic, as well as on the type, shape, size and concentration of the magnetic particles. An important ingredient of the interaction of the liquid crystal and the embedded particles is the type (strength) of the anchoring at the particle - liquid crystal interface which affects the mutual direction of the magnetic moments and the director. We could see that, accidentally, with all host materials tested $\boldsymbol{m} \parallel \boldsymbol{n}$ was favoured, independently whether the type of anchoring was soft (mostly for spherical particles) or rigid (e.g. for rod-like particles). It does not mean, however, that there are no counter examples with an $\boldsymbol{m} \perp \boldsymbol{n}$ preference [13, 47, 48].

The influence of the doping was mostly detected by monitoring field induced orientational (Fréedericksz) transitions in various geometries. For materials with positive diamagnetic anisotropy doping reduced the critical magnetic field, as if the effective diamagnetic anisotropy were increased. The role of the magnetic particles is, however, not so simple. We showed an example too for a geometry where such a rescaling of χ_a is not sufficient for the explanation of the findings. Nevertheless, the theoretical approach by Burylov and Raikher proved to be suitable for describing all experimental results.

There is still much work to be done in the field until all details of the behaviour of ferronematics are completely understood and/or until they find their way to practical applications. For example, the anchoring properties of nanoparticles with nontrivial shape, like chain-like magnetite or carbon nanotubes functionalized by magnetite particles need especially further clarification.

ACKNOWLEDGMENT

We would like to thank all those who, being co-authors of joint publications, contributed to the results presented in this chapter: Milan Timko, Martina Koneracká, Zuzana Mitróová, Vlasta Závišová, Anežka Hashim, Jan Jadzyn, Tibor Tóth-Katona, Anikó Vajda, Katalin Fodor-Csorba. We also would like to acknowledge the financial support provided for the research by the Slovak Academy of Sciences, in the framework of CEX-NANOFLUID, projects VEGA 0077, by the Slovak Research and Development Agency under the contract No. APVV-0171-10, by the Grenoble High Magnetic Field Laboratory (CRETA) and by the Hungarian Research Fund OTKA K-81250. The authors also acknowledge SAS-HAS bilateral projects for promoting their co-operations.

REFERENCES

[1] F. Reinitzer, *Monatshefte für Chemie* (Wien) 9, 421 (1888).
[2] G.H. Heilmeier, L.A. Zanoni, L.A. Barton, *Proc. IEEE*, 56, 1162 (1968).
[3] H. Kelker, B. Scheurle, *Angew. Chem. Int. Ed.* 8 (11), 884 (1969).
[4] M. Schadt, W. Helfrich, *Appl. Phys. Lett.* 18, 127 (1971).
[5] G.W. Gray, K.J. Harrison, J.A. Nash, *Electronics Lett.* 9(6), 130 (1973).
[6] P.G. de Gennes, *The Physics of Liquid Crystals*, (Clarendon Press, Oxford 1974).
[7] F. Brochard, P.G. de Gennes, *J. Phys. (Paris)* 31, 691 (1970).
[8] J. Rault, P.E. Cladis, J.P. Burger, *Phys.Lett. A* 32, 199 (1970)
[9] J. Liebert, A. Martinet, *J. Phys. Lett. (Paris)* 40, 363 (1979).
[10] A.M. Figueiredo Neto, L. Liebert, A.M. Levelut, *J.Phys. (Paris)* 45, 1505 (1984).
[11] S.H. Chen, N.M. Amer, *Phys. Rev. Lett.* 51, 2298 (1983).
[12] V. Fréedericksz, V. Zolina, *Trans. Faraday Soc.* 29, 919 (1933).
[13] P. Kopčanský, I. Potočová, M. Timko, M. Koneracká, A.M.G. Jansen, J. Jadzyn, G. Czechovski, *J. Magn. Magn. Mater.* 272-276, 2355 (2004).
[14] L.J. Martínez-Miranda, K. McCarthy, J.J. Harry, A. Noel, *Appl. Phys. Lett.* 89, 161917 (2006).
[15] N. Tomašovičová, P. Kopčanský, M. Koneracká, L. Tomčo, V. Závišová, M. Timko, N. Éber, K. Fodor-Csorba, T. Tóth-Katona, A. Vajda, J. Jadzyn, *J. Phys-Condens. Mat.* 20, 204123 (2008).
[16] L.J. Martinez-Miranda, L.K. Kurihara, *J. Appl. Phys.* 105, 084305 (2009).
[17] E. Peterscu, E.R. Bena, *J. Magn. Magn. Mater.* 321, 2757 (2009).

[18] N. Podoliak, O. Buchnev, O. Buluy, G. D'Alessandro, M. Kaczmarek, Y. Reznikov, T.J. Sluckin, *Soft Matter* 7, 4742 (2011).
[19] O. Buluy, S. Nepijko, V. Reshetnyak, E. Ouskova, V. Zadorozhnii, A. Leonhardt, M. Ritschel, G. Schonhense and Y. Reznikov, *Soft Matter* 7, 644 (2011).
[20] S.V. Burylov, Y.L. Raikher, *J. Phys. Lett. A* 149, 279 (1990).
[21] S.V. Burylov, Y.L. Raikher, *J. Magn. Magn. Mater.* 122, 62 (1993).
[22] S.V. Burylov, Y.L. Raikher, *Mol. Cryst. Liq. Cryst.* 258, 107 (1995).
[23] S.V. Burylov, Y.L. Raikher, *Mol. Cryst. Liq. Cryst.* 258, 123 (1995).
[24] S.V. Burylov, Y.L. Raikher, *J. Magn. Magn. Mater.* 85, 74 (1990).
[25] N. Tomašovičová, M. Koneracká, P. Kopčanský, M. Timko, V. Závišová, J. Jadzyn, *Phase Transitions* 79, 595 (2006).
[26] R. Dabrowski, J. Dziaduszek, T. Szczucinski, *Mol. Cryst. Liq. Cryst. Lett.* 102, 155 (1984).
[27] P. Kopčanský, N. Tomašovičová, M. Timko, M. Koneracká, V. Závišová, L. Tomčo, J. Jadzyn, *Journal of Physics: Conference Series* 200, 072055 (2010).
[28] P. Kopčanský, N. Tomašovičová, M. Koneracká, M. Timko, V. Závišová, N. Éber, K. Fodor-Csorba, T. Tóth-Katona, A. Vajda, J. Jadzyn, E. Beaugnon, X. Chaud, *J. Magn. Magn. Mater.* 322, 3696 (2010).
[29] P. Kopčanský, N. Tomašovičová, M. Koneracká, V. Závišová, M. Timko, A. Džarová, A. Šprincová, N. Éber, K. Fodor-Csorba, T. Tóth-Katona, A. Vajda, J. Jadzyn, *Phys Rev E* 78, 011702 (2008).
[30] Z. Mitróová, N. Tomašovičová, M. Timko, M. Koneracká, J. Kováč, J. Jadzyn, I. Vávra, N. Éber, T. Tóth-Katona, E. Beaugnon, X. Chaud, P. Kopčanský, *New J Chem.* 35, 1260 (2011).
[31] P. Kopčanský, N.Tomašovičová, M. Koneracká, V. Závišová, M. Timko, L. Tomčo, N. Éber, K. Fodor-Csorba, T. Tóth-Katona, A. Vajda, J.Jadzyn, E. Beaugnon, X. Chaud, *Physics Procedia* 9, 82 (2010).
[32] P.C. Fannin, C.N. Marin, I. Malaescu, N. Stefu, *Physica B* 388, 87 (2007).
[33] L.Vékás, M. Rasa, D. Bica, *J. Colloid and Interface Science* 231, 247 (2000).
[34] S.Y. Jeon, K.A. Park, I.S. Baik, S.J. Jeong, S.H. Jeong, K.H. An, S.H. Lee, Y.H. Lee, *Nano* 2, 41 (2007).
[35] W. Helfrich, *Phys. Rev. Lett.* 24, 201 (1970).
[36] I. Lelidis, G. Durand, *Phys. Rev. E* 48, 3822 (1993).
[37] H. Stanley, Oxford Science, New York, 1971.
[38] C. Rosenblatt, *Phys. Rev. A* 24, 2236 (1981).
[39] T. Ostapenko, D.B. Wiant, S.N. Sprunt, A. Jákli, J.T. Gleeson, *Phys. Rev. Lett.* 101, 247801 (2008).
[40] P. Kopčanský, N. Tomašovičová, M. Koneracká, V. Závišová, M. Timko, M. Hnatič, N. Éber, T. Tóth-Katona, J. Jadzyn, J. Honkonen, E. Beaugnon, X. Chaud, *IEEE Transactions on Magnetics* 47, 4409 (2011).
[41] J.L. Fergason, *SID Int. Symp. Dig. Tech. Pap.* 16, 68 (1985).
[42] P.S. Drzaic, *J. Appl. Phys.* 60, 2142 (1986).
[43] J.W. Doane, N.A. Vaz, B.G. Wu, S. Zumer, *Appl. Phys. Lett.* 48, 269 (1986).
[44] N.A. Vaz, G.W. Smith, G.P. Montgomery, *Mol. Cryst. Liq. Cryst.* 146, 1 (1987).

[45] P. Kumar, K.K. Raina, *Current Applied Physics* 7, 636 (2007).

[46] Z. Mitróová, N. Tomašovičová, M. Timko, M. Koneracká, L. Tomčo, J. Jadzyn, T. Palewski, P. Kopčanský, accepted for publication in *Acta Physica Polonica A*

[47] I. Potočová, M. Koneracká, P. Kopčanský, M. Timko, J. Jadzyn, G. Czechowski, *J. Magn. Magn. Mater.* 201, 163 (1999).

[48] P. Kopčanský, I. Potočová, M. Koneracká, M. Timko, A.G.M. Jansen, J. Jadzyn, G. Czechowski, *J. Magn. Magn. Mater.* 289, 101 (2005).

INDEX

A

B

C

D

E

F

G

H

I

N

O

P

Q

R

S

T

U

V

W

Y

Z